Kazhdan–Lusztig Cells with Unequal Parameters

Algebra and Applications

Volume 24

Algebra and Applications aims to publish well-written and carefully refereed monographs with up-to-date expositions of research in all fields of algebra, including its classical impact on commutative and noncommutative algebraic and differential geometry, K-theory and algebraic topology, and further applications in related domains, such as number theory, homotopy and (co)homology theory through to discrete mathematics and mathematical physics.

Particular emphasis will be put on state-of-the-art topics such as rings of differential operators, Lie algebras and super-algebras, group rings and algebras, Kac-Moody theory, arithmetic algebraic geometry, Hopf algebras and quantum groups, as well as their applications within mathematics and beyond. Books dedicated to computational aspects of these topics will also be welcome.

More information about this series at http://www.springer.com/series/6253

Cédric Bonnafé

Kazhdan–Lusztig Cells with Unequal Parameters

Springer

Cédric Bonnafé
Institut Montpelliérain Alexander
Grothendieck
CNRS–Université de Montpellier
France

ISSN 1572-5553 ISSN 2192-2950 (electronic)
Algebra and Applications
ISBN 978-3-030-09986-2 ISBN 978-3-319-70736-5 (eBook)
https://doi.org/10.1007/978-3-319-70736-5

Mathematics Subject Classification (2010): 20C08, 20F55

Printed on acid-free paper

This Springer imprint is published by the registered company Springer International Publishing AG part of Springer Nature
The registered company address is: Gewerbestrasse 11, 6330 Cham, Switzerland

Preface

In their seminal paper on representations of Coxeter groups and Hecke algebras [KaLu1], Kazhdan and Lusztig already had in mind lots of applications and connections between their theory (now called *Kazhdan–Lusztig theory*) and several problems in Lie theory: representations of Lie algebras and reductive groups, singularities of Schubert varieties, and also representations of Weyl groups in the cohomology of Springer fibers.[1] They were inspired by the works of Jantzen [Jan], Joseph [Jos1], [Jos2], and Vogan [Vog1], [Vog2], [Vog3]. The main objects introduced in their paper are the *Kazhdan–Lusztig basis* of a Hecke algebra (whose matrix in the standard basis is given by the *Kazhdan–Lusztig polynomials*) and the partitions of the Coxeter groups into *left, right, or two-sided cells*. They state the fundamental *Kazhdan–Lusztig conjecture*, which relates the multiplicities of simple modules of a semisimple complex Lie algebra in Verma modules to the Kazhdan–Lusztig polynomials, and they also characterize rationally smooth Schubert varieties in terms of Kazhdan–Lusztig polynomials.

In the decade following Kazhdan–Lusztig's paper, the theory developed incredibly quickly, with the proof of the Kazhdan–Lusztig conjecture by Brylinski–Kashiwara [BrKa1], [BrKa2] and Beilinson–Bernstein [BeBe]; the interpretation of Kazhdan–Lusztig polynomials in terms of the intersection cohomology of Schubert varieties by Kazhdan and Lusztig [KaLu2]; the work of Lusztig on the representation theory of finite and p-adic reductive groups [Lus2], [Lus3], [Lus4], [Lus5], [Lus8], [Lus11]; the works of several authors, such as Joseph [Jos3], Deodhar [Deo1], Deodhar–Gabber–Kac [DeGaKa], Douglass [Dou], Lusztig [Lus1], [Lus7], Shi [Shi1], Soergel [Soe1], Springer [Spr1], Xi [Xi1].

All these developments are concerned with the Hecke algebras with *equal parameters*. In 1983, Lusztig [Lus9] generalized the construction of the Kazhdan–Lusztig basis and the partition into cells in a framework that will be the one

[1]Even though this last point is not mentioned in the original paper [KaLu1], this is explained in [Lus22].

followed in this book: (W, S) is a Coxeter system, $\mathscr{A}$ is a totally ordered abelian group, and $\varphi : S \to \mathscr{A}_{>0}$ is a *weight function* (i.e., $\varphi(s) = \varphi(t)$ if s and t are conjugate in W). The equal parameter case corresponds to the case where φ is constant. In the general case, the important property of Kazhdan–Lusztig polynomials and structure constants to have nonnegative coefficients (which, in the equal parameter case, follows from the geometric interpretation) is false, and the development of the theory was more difficult and so much slower. Let us, however, note the works of Bremke [Bre1], [Bre2] and Xi [Xi2].

The interest in Kazhdan–Lusztig theory in the unequal parameter case was boosted by the publication of Lusztig's book on the subject [Lus20]. In this book, Lusztig

- proposed a series of conjectures (denoted by (P1) up to (P15)) involving Lusztig's **a**-function, whose proof would allow the extension of some constructions existing only in the equal parameter case (especially the *asymptotic algebra*);
- recalled from [Lus14] how these conjectures follow from the positivity properties in the equal parameter case;
- proved these conjectures in the *quasi-split case* (see Chapter 15 for the definition) for W finite[2];
- proposed a conjectural geometric interpretation of Kazhdan–Lusztig polynomials in terms of *character sheaves* on disconnected reductive groups (for finite Weyl groups and particular weight functions);
- proposed another conjecture linking this theory to *Harish-Chandra theory* for unipotent representations of finite reductive groups (again for finite Weyl groups and particular weight functions);
- computed the *constructible characters* and the *Lusztig families* of all finite Coxeter groups, for all choices of parameters.

This led to a new dynamics, where the central problems are to determine the partitions of W into cells and the interpretation of constructible characters: See, for instance Geck [Gec2], [Gec3], [Gec4], [Gec5], [Gec6], Geck–Iancu [GeIa1], Guilhot [Gui1], [Gui2], [Gui3], [Gui4], Leclerc–Miyachi [LeMi], Malle–Rouquier [MaRo], Shi [Shi5], Shi–Yang [ShYa1], Bonnafé–Iancu [BoIa], and Bonnafé [Bon1], [Bon2], [Bon3], [Bon4].

Last but not least, in their work on representations of *rational Cherednik algebras at* $t = 0$, Gordon and Martino stated the conjecture that Kazhdan–Lusztig families (defined using the partition into two-sided cells) coincide with *Calogero–Moser families* of finite Coxeter groups (defined as the blocks of the restricted Cherednik algebra), for *any* choices of parameters (see [Gor], [GoMa], [Mar], and the works of Bellamy and Thiel [Bell], [BeTh]). This conjecture was extended to

[2]G. Lusztig has posted in June 2014 an updated version of this book: See http://arxiv.org/abs/math/0208154v2. In this version, he uses the recent results of Elias–Williamson [ElWi] to extend the results on the quasi-split case to *bounded* Coxeter groups (it is conjectured that all Coxeter groups are bounded, and it is proved that affine Weyl groups are bounded).

constructible characters and the partition into cells by Rouquier and the author [BoRo1], [BoRo2], [BoRo3]: This would provide a completely different interpretation of cells. Even if it is not treated in this book, this possible interpretation is at the heart of our motivation for studying Kazhdan–Lusztig theory with unequal parameters (Chapter 18 about the action of the *cactus group* might be related to these conjectures).

The present book aims to introduce the reader to Kazhdan–Lusztig theory *with unequal parameters*, by using only algebraic/combinatorial methods. The goal is to go as far as possible using only these methods, by focusing on the general properties of the partition into cells ("general" in the sense that they are valid for all Coxeter systems, and not only for a class of Coxeter systems, such as finite or affine Weyl groups).[3]

We also tried as much as possible to provide proofs of the basic results of the theory which are different (sometimes, just slightly different) from those one can find in the literature. For instance, we try to avoid systematic induction arguments involving the Bruhat order, as most of them can be viewed as particular instances of a general lemma about existence and uniqueness of an invariant basis of a module under a semilinear involution (see Chapter 2): This is a classical result which is also used to construct *canonical bases* for representations of quantum groups.

Vailhauquès, France Cédric Bonnafé
October 2017

[3]There will be an exception to this: We will study the effect of multiplication by the longest element on cells whenever W is finite.

Acknowledgements

It would be stating the obvious to say that I would have never written this book if I had not been initiated to Kazhdan–Lusztig theory (especially the unequal parameters case) by Lacri Iancu and Meinolf Geck. Most of my knowledge in this field owes to the discussions I had with them, which led to several common articles. I would also like to thank warmly Jérémie Guilhot, not only for providing the type $\tilde{G}_2$ figures of Chapter 25, but also for explaining several applications of his induction process to me (I could not reproduce all of them here).

I have also included in Chapter 25 pictures that give some partitions into cells of some hyperbolic groups: They have been kindly provided by Paul Gunnells, and I am happy to thank him here for his invaluable help. I also thank Mikhail Belolipetsky for authorizing me to reproduce the picture of the hyperbolic pentagon group.

I thank Ben Elias and Geordie Williamson for clarifying subtle points of the theory of Soergel bimodules.

I also wish to thank Gunter Malle, Ulrich Thiel, and again Jérémie Guilhot for their very efficient and careful reading of a preliminary version of this book.

Contents

Part I Preliminaries

1 Preorders on Bases of Algebras 3
1.1 Definitions 3
1.1.A Preorders 3
1.1.B Cells 4
1.1.C Cell Modules 5
1.1.D Examples 7
1.2 Preorders and Symmetrizing Forms 8

2 Lusztig's Lemma 13
2.1 Invariant Bases 14
2.2 Base Change 16

Part II Coxeter Systems, Hecke Algebras

3 Coxeter Systems 19
3.1 Characterizations of Coxeter Systems 19
3.2 First Properties 21
3.3 Reflections, Roots 21
3.4 Braided Maps 23
3.5 Bruhat Order 27
3.6 Parabolic Subgroups 27
3.7 Longest Element, Left and Right Descent Set 30
3.8 More on Reflections 31
3.9 Conjugacy Classes of Involutions 32
3.10 Fixed Points under Automorphisms 34
3.11 Geometric Representation 36
3.11.A Coxeter Matrix, Coxeter Graph 36
3.11.B Reflection Representation 37
3.11.C Irreducibility 38

3.12 Some Classification Results 39
3.12.A Finiteness 39
3.12.B Tame Groups 41
3.12.C Hyperbolic Coxeter Systems 42
3.13 Tits Cone, Coxeter Complex 43
3.14 Drawing Coxeter Groups 44
3.14.A Rank 2 44
3.14.B Rank 3 44

4 Hecke Algebras 53
4.1 Definition, First Properties 53
4.1.A Basis 54
4.1.B Parabolic Subalgebras 55
4.1.C Anti-involution 55
4.1.D Central Form 56
4.2 The Case where A is a Group Algebra 56
4.2.A Index, Sign 57
4.2.B Central Form 57
4.2.C Invertibility 58
4.2.D Involutions 58
4.3 Functoriality, Parameters 59
4.3.A Changing the Group 59
4.3.B Generic Hecke Algebra 60
4.3.C Changing Signs 61
4.3.D Vanishing Parameters 61
4.4 r-Polynomials 62
4.5 Reflection Representation 66

Part III Kazhdan–Lusztig Cells

5 The Kazhdan–Lusztig Basis 73
5.1 Degree, Valuation, Properties of A 73
5.2 The Kazhdan–Lusztig Basis 74
5.3 Kazhdan–Lusztig Polynomials 75
5.4 Functoriality, Parameters 76
5.4.A Changing $\mathscr{A}$ 76
5.4.B Changing Signs 76
5.4.C Vanishing Parameters 77
5.5 Structure Constants 78
5.6 An Algorithm 81
5.7 Examples 85

6 Kazhdan–Lusztig Cells . . . 93
6.1 Preorders on W . . . 93
6.1.A Cells . . . 94
6.1.B Cell Modules . . . 94
6.2 Functoriality, Parameters . . . 96
6.2.A Changing the Group . . . 96
6.2.B Changing Signs . . . 97
6.2.C Vanishing Parameters . . . 97
6.3 Descent Sets . . . 99
6.4 About the Structure Constants $\lambda_{x,y}^s$ and $\rho_{x,y}^s$ . . . 100
6.5 Some Conjectures . . . 101
6.6 Drawing Kazhdan–Lusztig Cells . . . 102
6.7 Examples . . . 103

7 Semicontinuity . . . 111
7.1 Faces, Chambers . . . 112
7.2 Cells . . . 112
7.3 Examples . . . 113

Part IV General Properties of Cells

8 Cells and Parabolic Subgroups . . . 119
8.1 The Kazhdan–Lusztig–Geck Basis . . . 120
8.2 Induction Theorems . . . 122
8.3 The Restriction Theorem . . . 124
8.4 Induction of Isomorphic Cells . . . 126
8.4.A Definition, Examples . . . 126
8.4.B Induction . . . 128
8.4.C About Strongness . . . 130

9 Descent Sets, Knuth Relations and Vogan Classes . . . 133
9.1 Enhanced Descent Sets . . . 134
9.2 Knuth Relations . . . 134
9.3 Vogan Classes . . . 135

10 The Longest Element and Duality in Finite Coxeter Groups . . . 139
10.1 About ω_0 . . . 139
10.2 The Inversion Formula . . . 140
10.3 Cells, Structure Constants . . . 142
10.4 Cell Modules . . . 144
10.5 The Element $\mathbf{C}_{w_0}$ . . . 145

11 The Guilhot Induction Process . . . 147
11.1 Main Result . . . 147
11.2 Application to the Asymptotic Case . . . 150

12 Submaximal Cells (Split Case) . . . 155
12.1 About the Elements of Γ^{uni} . . . 156
12.1.A Elementary Properties . . . 156
12.1.B Description of Γ^{uni} . . . 156
12.1.C Further Properties . . . 157
12.1.D Finiteness . . . 158
12.2 Submaximal Cells . . . 160
12.3 Cell Modules, W-graphs . . . 162
12.3.A Description . . . 162
12.3.B Homomorphisms . . . 164
12.3.C Reflection Representation . . . 165
12.4 Unfolding Coxeter Graphs . . . 166

13 Submaximal Cells (General Case) . . . 169
13.1 Submaximal Two-Sided Cells . . . 169
13.2 Submaximal Left Cells . . . 171

Part V Lusztig's a-Function

14 Lusztig's Conjectures . . . 175
14.1 Bounds . . . 175
14.2 Lusztig's **a**-Function and Cells . . . 176
14.2.A Definition . . . 176
14.2.B Base Change . . . 179
14.2.C Conjectures . . . 179
14.3 Reductions . . . 181
14.3.A First Consequences of the Conjectures . . . 182
14.3.B Some Implications between Lusztig's Conjectures . . . 183
14.3.C About Conjecture (P15) . . . 185

15 Split and Quasi-Split Cases . . . 189
15.1 Recollection on Soergel Bimodules . . . 190
15.1.A Polynomial Functions . . . 190
15.1.B Graded Bimodules . . . 191
15.1.C Soergel Bimodules . . . 191
15.1.D Action of σ . . . 193
15.2 Lusztig's Conjectures: the Split Case . . . 195
15.3 Lusztig's Conjectures: the Quasi-Split Case . . . 197

Part VI Applications of Lusztig's Conjectures

16 Miscellanea . . . 205
16.1 About Cells . . . 205
16.2 About Duflo Involutions . . . 206
16.3 Strongness of Cellular Pairs . . . 207

17 Multiplication by T_{w_0} 209
17.1 The Mathas–Lusztig Theorem 209
17.2 Further Properties of the Pairs (λ, η) and (ρ, η) 214
17.3 Some Examples 216

18 Action of the Cactus Group 219
18.1 Cellularity 219
18.2 Action 222

19 Asymptotic Algebra 225
19.1 Definition 225
19.2 A Morphism from $\mathscr{H}$ to $A \otimes_{\mathbb{Z}} \mathscr{J}$ 227
19.3 Examples 230

20 Automorphisms 235
20.1 Statement 236
20.2 Brauer Quotient 236
20.2.A Definition 236
20.2.B Applications to Hecke Algebras 237
20.3 Proof of Theorem 20.1.1 239
20.4 Asymptotic Algebra 240

Part VII Examples

21 Finite Dihedral Groups 243
21.1 The Structure of W 244
21.1.A About Elements 244
21.1.B About Characters 245
21.2 The Kazhdan–Lusztig Basis 246
21.3 Cells, Cell Modules 250
21.4 Lusztig's **a**-Function and Lusztig's Conjectures 256
21.5 Action of T_{w_0} 259

22 The Symmetric Group 263
22.1 Some Combinatorics 263
22.1.A Tableaux 264
22.1.B The Robinson–Schensted–Knuth Correspondence 264
22.1.C Descent Set 265
22.1.D Dominance Order 266
22.2 Kazhdan–Lusztig Cells 268
22.2.A Knuth Relations 268
22.2.B Vogan Classes 269
22.2.C Cells 269
22.2.D Complements 270
22.2.E Example: Type $\boldsymbol{A}_3$ 270

23 Affine Weyl Groups of Type $\tilde{A}_2$ . . . 273
23.1 The Partition into Cells . . . 274
23.1.A Submaximal Cells . . . 274
23.1.B Parabolic Induction . . . 276
23.1.C Cellular Maps (*-Operation) . . . 276
23.1.D Lowest Two-sided Cell . . . 277
23.2 Complements . . . 278
23.2.A Duflo Involutions . . . 278
23.2.B Other Conjectures . . . 278

24 Free Coxeter Groups . . . 281
24.1 Group-Theoretic Properties . . . 282
24.1.A Reduced Expressions . . . 282
24.1.B Involutions . . . 282
24.1.C Some Partitions of W . . . 283
24.2 Boundedness . . . 284
24.3 About Kazhdan–Lusztig Basis Elements . . . 286
24.4 Duflo Involutions . . . 288
24.5 Cells . . . 289
24.6 Proof of the Conjectures . . . 292
24.7 The Infinite Dihedral Group . . . 294
24.7.A Preliminary: Representations of $\mathbf{SL}_2(\mathbb{C})$ and $\mathbf{PGL}_2(\mathbb{C})$ . . . 294
24.7.B The Equal Parameter Case . . . 295
24.7.C The Unequal Parameter Case . . . 297
24.8 Rank 3 . . . 298

25 Rank 3 . . . 303
25.1 Spherical Groups . . . 304
25.1.A Type B_3 . . . 304
25.1.B Type H_3 . . . 306
25.2 Affine Weyl Groups . . . 306
25.2.A Type $\tilde{B}_2$ . . . 307
25.2.B Type $\tilde{G}_2$ . . . 308
25.3 Hyperbolic Coxeter Systems . . . 309
25.4 Pictures . . . 311

26 Some Bibliographical Comments . . . 319
26.1 Finite Groups . . . 319
26.1.A Split Case . . . 319
26.1.B Unequal Parameter Case . . . 320
26.2 Affine Weyl Groups . . . 320
26.2.A Split Case . . . 320
26.2.B Unequal Parameter Case . . . 321

26.3 Other Groups . . . 321
26.3.A Split Case . . . 321
26.3.B Unequal Parameter Case . . . 321

Appendix A: Symmetric Algebras . . . 323

A.1 Symmetrizing Form . . . 323
A.2 Dual Bases . . . 325

Appendix B: Reflection Subgroups of Coxeter Groups . . . 327

B.1 Reflection Systems . . . 327
B.2 Reflection Subgroups of Coxeter Groups . . . 330
B.3 Semidirect Product Decomposition . . . 332

References . . . 335

Index . . . 343

General Notation

If $\mathscr{P}$ is a statement, we define

$$\delta_{\mathscr{P}} = \begin{cases} 1 & \text{if } \mathscr{P} \text{ is true,} \\ 0 & \text{otherwise.} \end{cases}$$

For instance, $\delta_{x=y}$ replaces the Kronecker symbol $\delta_{x,y}$ while $\delta_{a\in A}$ is 1 if $a \in A$ and 0 otherwise.

We set $\mathbb{N} = \{0, 1, 2, 3, \ldots\}$ and $\mathbb{N}^* = \{1, 2, 3, \ldots\} = \mathbb{N} \backslash \{0\}$. If $n \in \mathbb{N}^*$, the group of permutations of the set $\{1, 2, \ldots, n\}$ is denoted by $\mathfrak{S}_n$.

If $\mathscr{E}$ is a set, the cardinality of $\mathscr{E}$ (possibly infinite) is denoted by $|\mathscr{E}|$. The set of subsets of $\mathscr{E}$ is denoted by $\mathscr{P}(\mathscr{E})$. The *symmetric difference* is denoted by $\dot{+}$: $X \dot{+} Y = (X \cup Y) \backslash (X \cap Y)$. It defines a group law on $\mathscr{P}(\mathscr{E})$. We often write $X \dot{\cup} Y$ to denote the union of X and Y whenever $X \cap Y = \emptyset$. The group of permutations of $\mathscr{E}$ is denoted by $\mathfrak{S}_{\mathscr{E}}$. If $\sim$ is an equivalence relation on $\mathscr{E}$, we denote by $[\mathscr{E}/\sim]$ a set of representatives for the equivalence classes under $\sim$. When this notation is used, the reader will have no difficulty verifying that the statement does not depend on the choice of representatives. In an expression of the form $\sum_{x \in [\mathscr{E}/\sim]} f(x)$, the reader will also be able to verify that the element $f(x)$ depends only on the equivalence class of x.

If Γ is a group, we denote by $\mathrm{Z}(\Gamma)$ its center and by $\mathrm{D}(\Gamma)$ its derived subgroup. If $\mathscr{E}$ is a subset of Γ, we denote by $\mathrm{N}_\Gamma(\mathscr{E})$ (resp. $\mathrm{C}_\Gamma(\mathscr{E})$) its normalizer (resp. centralizer). We also set $\mathscr{E}^{-1} = \{\gamma^{-1} | \gamma \in \mathscr{E}\}$. If $\gamma \in \Gamma$, we set ${}^\gamma\mathscr{E} = \gamma\mathscr{E}\gamma^{-1}$. The order (possibly infinite) of γ is denoted by $o(\gamma)$. If $\gamma' \in \Gamma$, we denote by $[\gamma, \gamma']$ the commutator $\gamma\gamma'\gamma^{-1}\gamma'^{-1}$.

Unless otherwise specified, all the rings and algebras in this book are unital and associative. If A is a ring, we denote by A^+ its underlying additive group and by $A^\times$ the (multiplicative) group of invertible elements in A. The center of A is denoted by $\mathrm{Z}(A)$. We denote by A-mod the category of left A-modules of finite type. Given another ring B, we denote by (A, B) -bimod the category of (A, B)-bimodules of finite type, that is, the category whose objects are simultaneously a left A-module of finite type and a right B-module of finite type satisfying $(a \cdot m) \cdot b = a \cdot (m \cdot b)$ for all $a \in A$, $b \in B$ and $m \in M$. Given a commutative ring R, an R-algebra A, a commutative R-algebra R', and an A-module M we often simplify the notation and denote by $R'M$ the extension of scalars $R' \otimes_R M$: It is an $R'A = R' \otimes_R A$-module.

If $\mathscr{M}$ is a monoid and R is a commutative ring, the monoid algebra of $\mathscr{M}$ over R is denoted by $R\mathscr{M}$.

If R is a ring and M is an R-module, we denote by $\mathbf{GL}_R(M)$ the group of its R-linear automorphisms. If n is a positive integer, we denote by $\mathrm{Mat}_n(R)$ the ring of $n \times n$ square matrices with coefficients in R and set $\mathbf{GL}_n(R) = \mathrm{Mat}_n(R)^{\times}$. The identity matrix is denoted I_n. Given $M \in \mathrm{Mat}_n(R)$, its transpose is denoted ${}^tM$.

If R is commutative, the R-algebra $\mathrm{Mat}_n(R)$ as well as the group $\mathbf{GL}_n(R)$ are naturally identified with $\mathrm{End}_R(R^n)$ and $\mathbf{GL}_R(R^n)$, respectively, using the canonical identification of R^n with the R-module of $n \times 1$ column vectors. We denote by $\mathbf{SL}_n(R)$ the subgroup of $\mathbf{GL}_n(R)$ consisting of matrices of determinant 1.

List of Figures

Fig. 3.a Type A_2 . . . 44
Fig. 3.b Type B_2 . . . 44
Fig. 3.c Type A_3 . . . 45
Fig. 3.d Type $\tilde{A}_2$ . . . 46
Fig. 3.e Type $\mathbf{Hyp}(4,4,4)$ . . . 47
Fig. 6.a Cells in type A_2 . . . 104
Fig. 6.b Cells in type B_2 . . . 106
Fig. 22.a Cells in type A_3 . . . 271
Fig. 23.a Small length elements in type $\tilde{A}_2$ . . . 274
Fig. 23.b Cells in type $\tilde{A}_2$ . . . 275
Fig. 23.c Parabolic induction in type $\tilde{A}_2$ . . . 276
Fig. 23.d Action of $\bigstar_{s,u}$ on $X_{s,t}sts$ . . . 277
Fig. 24.a Cells in free Coxeter group of rank 3 . . . 299
Fig. 24.b Cells in free Coxeter group of rank 3 . . . 300
Fig. 25.a Essential hyperplanes in type B_3 . . . 304
Fig. 25.b Essential hyperplanes in type $\tilde{B}_2$ . . . 308
Fig. 25.c Essential hyperplanes in type $\tilde{G}_2$ . . . 309
Fig. 25.d Cells in type B_3 . . . 311
Fig. 25.e Cells in type H_3 . . . 312
Fig. 25.f Simple reflections in type $\tilde{G}_2$ . . . 313
Fig. 25.g Cells in type $\tilde{G}_2$ . . . 313
Fig. 25.h Cells in type $\tilde{G}_2$ . . . 314
Fig. 25.i Cells in type $\mathbf{Hyp}(3,3,4)$ . . . 315
Fig. 25.j Conjectural cells in type $\mathbf{Hyp}(2,3,\infty)$ . . . 315
Fig. 25.k Conjectural cells in type $\mathbf{Hyp}(2,3,7)$ . . . 316
Fig. 25.l Conjectural cells in type $\mathbf{Hyp}(4,4,4)$ . . . 316
Fig. 25.m Conjectural cells in type $\mathbf{Hyp}(2,2,2,3)$ . . . 317
Fig. 25.n Cells in type $\mathbf{Hyp}(2,2,2,2,2)$ . . . 317

Summary

Part I deals with general preliminary results. In Chapter 1, we study preorders built on algebras endowed with a basis, which will serve as a general model for the construction of left, right, and two-sided cells of Coxeter groups. Chapter 2 explains the construction of an invariant basis of modules endowed with particular semi-linear involutions. Despite their simplicity, the results of this chapter are not only key tools for constructing the Kazhdan–Lusztig basis of a Hecke algebra, but also provide proofs of lots of basic results of the theory without appealing to inductive arguments. Note that this is also used in the theory of canonical bases of modules for quantum groups.

Part II gathers basic results from the theory of Coxeter groups (Chapter 3) and Hecke algebras (Chapter 4). This part contains almost no proof (except for the results which are not contained in reference books): It just aims to fix notation that will be used throughout the book. We have decided to mainly follow [Bou] and [GePf].

Part III starts the Kazhdan–Lusztig theory. In Chapter 5, we show how the Kazhdan–Lusztig basis is constructed and investigate some first properties of the structure constants. Chapter 6 is concerned with the definition of the partitions into left, right, or two-sided cells and some easy properties (invariance of descent sets), as well as their associated cell modules. We also state some classical conjectures about this partition (connectedness, involutions). One of the other aims of this part is to study the functoriality properties of these constructions (Kazhdan–Lusztig basis, cells, cell modules) in terms of the parameters. This allows us to reduce the classical problems of the theory (computing the partition into cells, describing the structure of the cell modules) to the case where the parameters are strictly positive, which will be the context of the rest of the book.

Part IV studies the first properties of cells. In Chapter 8, we study the links between the cells of a Coxeter group and those of a parabolic subgroup (Geck's induction theorem, Lusztig's restriction theorem, cellular maps). We note that the result on the induction of cellular maps (see Theorems 8.4.4 and 8.4.8) generalizes the construction of the $*$-operation of Kazhdan–Lusztig in two directions: by allowing parabolic subgroups of any rank and by allowing unequal parameters. This is explained in the next Chapter 9, where Knuth relations are recalled and where an extension of Vogan's generalized τ-invariant is constructed. In Chapter 10, we prove that multiplication by the longest element in a finite Coxeter group permutes the cells and describes its effect on cell modules. In Chapter 11, we explain the Guilhot induction process and provide an application to the asymptotic case. Another application will be given in Chapter 24, where part of the proof of Theorem 24.5.1 (which describes the partition into cells of a free Coxeter group) is different from the original one of Shi and Yang [ShYa1]. In Chapters 12 and 13, we do not study a particular class of groups, but we study a particular class of two-sided cells, namely those which are maximal among the non-trivial two-sided cells. We also describe the left cells they contain in terms of the Coxeter graph.

Part V contains the construction of the most subtle invariant of cells that has been constructed so far, namely Lusztig's **a**-function. In the general case, its behavior seems to be governed by very precise statements (see Lusztig's Conjectures (Pk), $1 \leqslant k \leqslant 15$), which are known to hold only in a few cases: the equal parameter case and the quasi-split case (the proofs, using the fundamental results of Elias–Williamson [ElWi], are given in Chapter 15, except for (P15) in the quasi-split case);[4] dihedral groups (see Chapter 21); free Coxeter groups [ShYa1] (see Chapter 24)

Part VI studies several applications of Lusztig's Conjectures: So most of the results of this part are valid under the assumption that Lusztig's Conjectures (Pk) hold for $1 \leqslant k \leqslant 14$ (we do not use the difficult Conjecture (P15)). So, for instance, they hold in the equal parameter case and the quasi-split case. Chapter 16 is a collection of isolated results in this spirit. Chapter 17 explains how the multiplication by T_{w_0} (where w_0 is the longest element of a finite Coxeter group) acts on Kazhdan–Lusztig basis elements: In particular, we show that it induces a non-trivial left (or right) cellular map, even if w_0 is central. This is used in the next Chapter 18 where an action of two copies of the cactus group on the Coxeter group is constructed, which respects the partition into left or right cells. The important asymptotic algebra is constructed in Chapter 19: This is the only chapter where Lusztig's Conjecture (P15) is used. In Chapter 20, we show that taking the fixed points under a group of Coxeter system automorphisms is essentially compatible with the partition into cells (and we relate to the asymptotic algebras).

Part VII is devoted to the detailed study of some examples. Chapter 21 deals with finite dihedral groups: cells, cell modules, Lusztig's Conjectures, etc. The case of the symmetric group is treated in Chapter 22. In Chapter 23, we explain how to derive the partition into cells of the affine Weyl group of type $\tilde{A}_2$. Chapter 24 is mostly taken from an article of Shi and Yang [ShYa1]: It contains the full description of cells and the proof of Lusztig's Conjectures for free Coxeter groups (often called universal Coxeter groups). If, moreover, the rank is equal to 2 (infinite dihedral groups), we also describe the structure of the asymptotic algebra in terms of representation rings of the algebraic groups $\mathbf{SL}_2(\mathbb{C})$ and $\mathbf{PGL}_2(\mathbb{C})$. Chapter 25

[4](P15) is stated without proof in [Lus20, §16.24].

contains examples of the partition into cells for groups of rank 3 (finite, affine, or hyperbolic).

The last part contains two appendices: The first is about symmetric algebras, while the main result of the second is the description of the semidirect product decomposition of a Coxeter group. This last result is the key to reducing the general parameter case to the case where all parameters are nonzero.

Remark. Throughout this book, we illustrate the notions introduced with examples of groups of type A_2 and B_2, and we leave to the reader the case of groups of type G_2 through different exercises.

What is not contained in this book? The wide variety of subjects related to this theory forced us to make drastic choices. Because we wanted to use only algebraic methods and focus on results that are valid for general Coxeter systems, we do not cover:

- The different geometric interpretations (Schubert varieties, character sheaves, Springer fibers, ...).
- The particularly rich cases of finite and affine Weyl groups.
- Other links with representation theory (category $\mathcal{O}$, finite or p-adic reductive groups, rational Cherednik algebras).

Any of these subjects would deserve its own book, and the author is not competent to treat them completely.

Prerequisites. Together with the classical notions of algebras (groups, rings, modules), we assume that the reader is familiar with the theory of Coxeter groups and Hecke algebras. Two chapters, which contain almost no proofs, give an account of this theory, but they do not replace a real knowledge of the subject: Their main aim is to fix notation and gather the results that will be used throughout the book.

Apart from Chapter 15 (which uses deep results of Elias–Williamson [ElWi] on Soergel bimodules), the book is self-contained and relies on rather elementary methods: It can be used as a textbook for the algebraic and combinatorial aspects of Kazhdan–Lusztig theory (even for equal parameters).

Part I
Preliminaries

Kazhdan–Lusztig bases, polynomials or cells are objects associated to Hecke algebras. However, their definition relies on constructions which can be done in a very general context. For instance, the notion of cells or associated cell modules can be defined as soon as one encounters an algebra endowed with a basis. These notions can become totally uninteresting or very powerful according to the choice of the basis. Similarly, Kazhdan–Lusztig polynomials can be associated to any poset endowed with some extra datum.

The purpose of this part is to develop these notions in the most general context, in order to separate clearly what is general from what is specific to the Hecke algebra case.

Chapter 1
Preorders on Bases of Algebras

We fix in this chapter a commutative ring A as well as an A-algebra $\mathcal{H}$, which is assumed to be free as an A-module. We also fix an A-basis $\mathcal{C} = (c_x)_{x \in X}$ of $\mathcal{H}$. We write

$$c_x c_y = \sum_{z \in X} a_{x,y,z} c_z,$$

where the $a_{x,y,z}$ are in A.

1.1. Definitions

1.1.A. Preorders

We aim to define preorders $\leqslant_L$, $\leqslant_R$ and $\leqslant_{LR}$ on the set X by using the basis $\mathcal{C}$ of $\mathcal{H}$. First, if $x, y \in X$, we write $x \overset{L}{\twoheadleftarrow} y$ (resp. $x \overset{R}{\twoheadleftarrow} y$) if there exists an $h \in \mathcal{H}$ such that the coefficient of c_x in the expansion of hc_y (resp. $c_y h$) in the basis $\mathcal{C}$ is non-zero: in other words,

$$x \overset{L}{\twoheadleftarrow} y \Longleftrightarrow \exists\ u \in X,\ a_{u,y,x} \neq 0 \tag{1.1.1}$$

and

$$x \overset{R}{\twoheadleftarrow} y \Longleftrightarrow \exists\ u \in X,\ a_{y,u,x} \neq 0. \tag{1.1.2}$$

Note that

$$x \overset{L}{\twoheadleftarrow} x \quad \text{and} \quad x \overset{R}{\twoheadleftarrow} x \tag{1.1.3}$$

for all $x \in X$.

C. Bonnafé, *Kazhdan–Lusztig Cells with Unequal Parameters*,
Algebra and Applications 24, https://doi.org/10.1007/978-3-319-70736-5_1

Definition 1.1.4. *We denote by $\leqslant_L$ (resp. $\leqslant_R$) the transitive closure of the relation $\stackrel{L}{\twoheadleftarrow}$ (resp. $\stackrel{R}{\twoheadleftarrow}$). If x, $y \in X$, we write $x \leqslant_{LR} y$ if there exists a sequence $x = x_0, x_1, \dots, x_n = y$ of elements of X such that $x_{i-1} \leqslant_L x_i$ or $x_{i-1} \leqslant_R x_i$ for all $i \in \{1, 2, \dots, n\}$.*

The relations $\leqslant_L$, $\leqslant_R$ and $\leqslant_{LR}$ are transitive (by definition) and reflexive (by 1.1.3). So they are ***preorders***.

Remark 1.1.5. Assume that the A-algebra $\mathscr{H}$ is generated by a family of elements $\mathbf{h} = (h_i)_{i \in I}$. If x, $y \in X$, write $x \stackrel{\mathbf{h},L}{\twoheadleftarrow} y$ if there exists an $i \in I$ such that the coefficient of c_x in the expansion of $h_i c_y$ in the basis $\mathscr{C}$ is non-zero. Then $\leqslant_L$ is the reflexive and transitive closure of $\stackrel{\mathbf{h},L}{\twoheadleftarrow}$. To see this, let us denote by $\leqslant_L^{\mathbf{h}}$ this reflexive and transitive closure and fix two elements x and y of X. If $h \in \mathscr{H}$ and $u \in X$, write $hc_u = \sum_{v \in X} \lambda_h(u, v) c_v$, with $\lambda_h(u, v) \in A$.

First of all, it is clear that, if $x \stackrel{\mathbf{h},L}{\twoheadleftarrow} y$, then $x \stackrel{L}{\twoheadleftarrow} y$: consequently, if $x \leqslant_L^{\mathbf{h}} y$, then $x \leqslant_L y$. Conversely, assume that $x \leqslant_L y$ and let us prove that $x \leqslant_L^{\mathbf{h}} y$. For this, we may, and we will, assume that $x \stackrel{L}{\twoheadleftarrow} y$. Then there exists an $h \in \mathscr{H}$ such that $\lambda_h(y, x) \neq 0$. Since $\mathscr{H}$ is generated by $\mathbf{h}$, there exists a sequence $(i_1, \dots, i_r)$ of elements of I such that $\lambda_{h_{i_1} \cdots h_{i_r}}(y, x) \neq 0$. But,

$$\lambda_{h_{i_1} \cdots h_{i_r}}(y, x) = \sum_{x_1, \dots, x_{r-1} \in X} \lambda_{h_{i_r}}(y, x_1) \lambda_{h_{i_{r-1}}}(x_1, x_2) \cdots \lambda_{h_{i_2}}(x_{r-2}, x_{r-1}) \lambda_{h_{i_1}}(x_{r-1}, x).$$

So there exists a sequence $(x_1, \dots, x_{r-1})$ of elements of X such that

$$\lambda_{h_{i_r}}(y, x_1) \lambda_{h_{i_{r-1}}}(x_1, x_2) \cdots \lambda_{h_{i_2}}(x_{r-2}, x_{r-1}) \lambda_{h_{i_1}}(x_{r-1}, x) \neq 0.$$

In particular, $x \stackrel{\mathbf{h},L}{\twoheadleftarrow} x_{r-1} \stackrel{\mathbf{h},L}{\twoheadleftarrow} x_{r-2} \stackrel{\mathbf{h},L}{\twoheadleftarrow} \cdots \stackrel{\mathbf{h},L}{\twoheadleftarrow} x_1 \stackrel{\mathbf{h},L}{\twoheadleftarrow} y$, and so $x \leqslant_L^{\mathbf{h}} y$, as desired.

Notice that this remark also applies to the relations $\stackrel{R}{\twoheadleftarrow}$ and $\leqslant_R$, after obvious changes. □

1.1.B. Cells

In general, these preorders are not anti-symmetric, that is, they are not order relations on X (see the examples of subsection 1.1.D).

Definition 1.1.6. *If $? \in \{L, R, LR\}$, we denote by $\sim_?$ the relation on X defined by*

$$x \sim_? y \Longleftrightarrow x \leqslant_? y \text{ and } y \leqslant_? x.$$

From the construction, it follows that $\sim_L$, $\sim_R$ and $\sim_{LR}$ are equivalence relations.

Definition 1.1.7. *A **left cell** (resp. **right cell**, resp. **two-sided cell**) in X is an equivalence class for the relation $\sim_L$ (resp. $\sim_R$, resp. $\sim_{LR}$): the set of these cells is denoted by $\mathrm{Cell}_L(X)$ (resp. $\mathrm{Cell}_R(X)$, resp. $\mathrm{Cell}_{LR}(X)$).*

Remark 1.1.8. Even though the relation $\leqslant_{LR}$ is the preorder generated by $\leqslant_L$ and $\leqslant_R$, it might happen that $\sim_{LR}$ is not the equivalence relation generated by $\sim_L$ and $\sim_R$ (see Exercise 1.3). □

Let $? \in \{L, R, LR\}$. We also denote by $\leqslant_?$ the order on $\mathrm{Cell}_?(X)$ induced by the corresponding preorder: in other words, if C, $C' \in \mathrm{Cell}_?(X)$, we write $C \leqslant_? C'$ if $x \leqslant_? y$, where $x \in C$ and $y \in C'$ (it does not depend on the choice of x and y). Also, we write $x <_? y$ if $x \leqslant_? y$ and $x \not\sim_? y$ (and we define similarly the relation $C <_? C'$). If $z \in X$, one can also define $C \leqslant_? z$, $z \leqslant_? C$, $C <_? z$ and $z <_? C$.

Remark 1.1.9. If $\mathscr{H}$ is commutative, then the relations $\leqslant_L$, $\leqslant_R$ and $\leqslant_{LR}$ coincide. Hence, the relations $\sim_L$, $\sim_R$ and $\sim_{LR}$ coincide and $\mathrm{Cell}_L(X) = \mathrm{Cell}_R(X) = \mathrm{Cell}_{LR}(X)$. □

1.1.C. Cell Modules

The notions introduced in the previous subsection allow us to construct some representations of the algebra $\mathscr{H}$ (i.e. $\mathscr{H}$-modules). If $\mathfrak{I}$ is a subset of X, we denote by $\mathscr{H}(\mathfrak{I})$ the sub-A-module of $\mathscr{H}$ equal to $\bigoplus_{x \in \mathfrak{I}} A\, c_x$. Of course, if $(\mathfrak{I}_i)_{i \in \mathscr{I}}$ is a family of subsets of X, then

$$\mathscr{H}\big(\bigcap_{i \in \mathscr{I}} \mathfrak{I}_i\big) = \bigcap_{i \in \mathscr{I}} \mathscr{H}(\mathfrak{I}_i) \qquad \text{and} \qquad \mathscr{H}\big(\bigcup_{i \in \mathscr{I}} \mathfrak{I}_i\big) = \sum_{i \in \mathscr{I}} \mathscr{H}(\mathfrak{I}_i). \tag{1.1.10}$$

Definition 1.1.11. *A subset $\mathfrak{I}$ of X is called a **left ideal** if for all $y \in \mathfrak{I}$ and $x \in X$ such that $x \leqslant_L y$, we have $x \in \mathfrak{I}$. We similarly define **right** and **two-sided ideals**.*

Of course, a left (resp. right, resp. two-sided) ideal of X is a union of left (resp. right, resp. two-sided) cells of X. Also, if $(\mathfrak{I}_i)_{i \in \mathscr{I}}$ is a family of left (resp. right, resp. two-sided) ideals of W, then $\bigcap_{i \in \mathscr{I}} \mathfrak{I}_i$ and $\bigcup_{i \in \mathscr{I}} \mathfrak{I}_i$ are left (resp. right, resp. two-sided) ideals of W. Thanks to the next proposition, this fact is compatible with 1.1.10:

Proposition 1.1.12. *A subset $\mathfrak{I}$ of X is a left (resp. right, resp. two-sided) ideal of X if and only if $\mathscr{H}(\mathfrak{I})$ is a left (resp. right, resp. two-sided) ideal of $\mathscr{H}$.*

Proof. This is clear from the definition. ■

Definition 1.1.13. *A subset C of X is said to be **left convex** if, for all x, $y \in C$, and for all $z \in X$ such that $x \leqslant_L z \leqslant_L y$, we have $z \in C$. We similarly define **right convex** and **two-sided convex** subsets of X.*

Example 1.1.14. A left (resp. right, resp. two-sided) cell is left (resp. right, resp. two-sided) convex. Note also that a left (resp. right, resp. two-sided) convex subset of X is a union of left (resp. right, resp. two-sided) cells. □

If C is a left convex subset of X, we set

$$X^{\leqslant_L C} = \{x \in X \mid \exists y \in C \text{ such that } x \leqslant_L y\}$$

and

$$X^{<_L C} = X^{\leqslant_L C} \setminus C.$$

For simplification, we then define the following A-submodules of $\mathscr{H}$

$$\mathscr{H}^{\leqslant_L C} = \mathscr{H}(X^{\leqslant_L C}), \quad \mathscr{H}^{<_L C} = \mathscr{H}(X^{<_L C})$$

as well as the following A-module

$$\mathscr{H}^L[C] = \mathscr{H}^{\leqslant_L C} / \mathscr{H}^{<_L C}.$$

If C is a right (resp. two-sided) convex subset of X, we similarly define the sets $X^{\leqslant_R C}$ and $X^{<_R C}$ (resp. $X^{\leqslant_{LR} C}$ and $X^{<_{LR} C}$) as well as the A-modules $\mathscr{H}^{\leqslant_R C}$, $\mathscr{H}^{<_R C}$ and $\mathscr{H}^R[C]$ (resp. $\mathscr{H}^{\leqslant_{LR} C}$, $\mathscr{H}^{<_{LR} C}$ and $\mathscr{H}^{LR}[C]$).

Proposition 1.1.15. *Let C be a subset of X. Then:*

(a) *If C is left convex, then $\mathscr{H}^{\leqslant_L C}$ and $\mathscr{H}^{<_L C}$ are left ideals of $\mathscr{H}$ and $\mathscr{H}^L[C]$ is a left $\mathscr{H}$-module.*
(b) *If C is right convex, then $\mathscr{H}^{\leqslant_R C}$ and $\mathscr{H}^{<_R C}$ are right ideals of $\mathscr{H}$ and $\mathscr{H}^R[C]$ is a right $\mathscr{H}$-module.*
(c) *If C is two-sided convex, then $\mathscr{H}^{\leqslant_{LR} C}$ and $\mathscr{H}^{<_{LR} C}$ are two-sided ideals of $\mathscr{H}$ and $\mathscr{H}^{LR}[C]$ is an $(\mathscr{H}, \mathscr{H})$-bimodule.*

Proof. As the proofs of (a), (b) and (c) are entirely similar, we prove only (a). So assume that C is left convex. It is an immediate consequence of the Definitions 1.1.11 and 1.1.13 that $X^{\leqslant_L C}$ and $X^{<_L C}$ are left ideals of X, so (a) follows from Proposition 1.1.12. ■

Remark 1.1.16. If C is left convex then, as an A-module, $\mathscr{H}^L[C]$ is free with basis $(e_x)_{x \in C}$, where e_x denotes the image of c_x in the quotient $\mathscr{H}^L[C]$. This similarly applies to right or two-sided cells, with obvious changes. □

Remark 1.1.17. A consequence of the definitions is that a subset C of X is left (resp. right, resp. two-sided) convex if and only if there exist two left (resp. right, resp. two-sided) ideals $\mathfrak{I}$, $\mathfrak{I}'$ of X such that $C = \mathfrak{I}' \setminus \mathfrak{I}$. In this case, we necessarily have that $X^{\leqslant_L C} \subset \mathfrak{I}'$ (resp. $X^{\leqslant_R C} \subset \mathfrak{I}'$, resp. $X^{\leqslant_{LR} C} \subset \mathfrak{I}'$) and it is readily seen that the natural inclusion $\mathscr{H}^{\leqslant_L C} \hookrightarrow \mathscr{H}(\mathfrak{I}')$ (resp. $\mathscr{H}^{\leqslant_R C} \hookrightarrow \mathscr{H}(\mathfrak{I}')$, resp. $\mathscr{H}^{\leqslant_{LR} C} \hookrightarrow \mathscr{H}(\mathfrak{I}')$) induces an isomorphism of left $\mathscr{H}$-modules $\mathscr{H}^L[C] \xrightarrow{\sim} \mathscr{H}(\mathfrak{I}')/\mathscr{H}(\mathfrak{I})$ (resp. of right $\mathscr{H}$-modules $\mathscr{H}^R[C] \xrightarrow{\sim} \mathscr{H}(\mathfrak{I}')/\mathscr{H}(\mathfrak{I})$, resp. of $(\mathscr{H}, \mathscr{H})$-bimodules $\mathscr{H}^{LR}[C] \xrightarrow{\sim} \mathscr{H}(\mathfrak{I}')/\mathscr{H}(\mathfrak{I})$). □

Definition 1.1.18. *If C is a left cell, then the left $\mathscr{H}$-module $\mathscr{H}^L[C]$ is called the* ***left cell module*** *associated with C. We similarly define* ***right cell modules*** *and* ***two-sided cell bimodules***. *If there is no ambiguity, they are called* ***cell modules***.

1.1.D. Examples

We present here some standard examples, to illustrate the notions introduced in this section.

Example 1.1.19 (Group algebra). Let G be a group and assume here that $\mathscr{H} = AG$ and that $\mathscr{C} = (g)_{g \in G}$. Then, if $g, h \in G$, we have $(gh^{-1})h = g$, so $g \overset{L}{\twoheadleftarrow} h$. Similarly, $g \overset{R}{\twoheadleftarrow} h$. Therefore, all the elements of G are related through $\leqslant_L$ (resp. $\leqslant_R$, resp. $\leqslant_{LR}$).

In this case, G itself is the unique left, right and two-sided cell and $\mathscr{H}^{LR}[G] = \mathscr{H}$ is the natural $(\mathscr{H}, \mathscr{H})$-bimodule. □

Example 1.1.20 (Polynomial algebra). Let us fix an indeterminate T over A and assume here that $\mathscr{H} = A[T]$ and that $\mathscr{C} = (T^n)_{n \in \mathbb{N}}$. Then the preorders $\leqslant_L$, $\leqslant_R$ and $\leqslant_{LR}$ all coincide with the natural order $\geqslant$ on $\mathbb{N}$. So the cells are singletons and the left cell modules $\mathscr{H}^L[C]$ are all isomorphic to $A[T]/\langle T \rangle \simeq A$. □

Example 1.1.21 (Matrix algebra). Let n be a non-zero natural number and assume here that $\mathscr{H} = \mathrm{Mat}_n(A)$, that $\mathscr{C} = (E_{ij})_{1 \leqslant i,j \leqslant n}$, where E_{ij} is the elementary matrix whose (i,j)-entry is 1 and all others are zero. The multiplication rule

$$E_{ij} E_{kl} = \delta_{j=k} E_{il}$$

shows that

$$(i,j) \leqslant_L (i',j') \Longleftrightarrow j = j'$$

and

$$(i,j) \leqslant_R (i',j') \Longleftrightarrow i = i'.$$

In particular, they are already equivalence relations. Moreover,

$$\forall i, i', j, j' \in \{1, 2, \ldots, n\},\ (i,j) \leqslant_{LR} (i',j').$$

So there is only one two-sided cell, n left cells and n right cells. If C is a left (resp. right, resp. two-sided) cell, then $\mathscr{H}^L[C] \simeq A^n$ (resp. $\mathscr{H}^R[C] \simeq A^n$, resp. $\mathscr{H}^{LR}[C] = \mathrm{Mat}_n(A)$), endowed with its natural structure of left $\mathrm{Mat}_n(A)$-module (resp. right $\mathrm{Mat}_n(A)$-module, resp. $(\mathrm{Mat}_n(A), \mathrm{Mat}_n(A))$-bimodule). □

1.2. Preorders and Symmetrizing Forms

*From now on, and until the end of this section, we assume moreover that $\mathcal{H}$ is a **symmetric** algebra and we fix a symmetrizing form $\tau : \mathcal{H} \to A$ (see the Appendix A). Note that this implies that X is **finite**. If $x \in X$, we denote by $c_x^\vee$ the element of $\mathcal{H}$ such that $\tau(c_y c_x^\vee) = \delta_{x=y}$ for all $y \in X$.*

Under this hypothesis, the relation $\overset{L}{\twoheadleftarrow}$ can be redefined as follows: $x \overset{L}{\twoheadleftarrow} y$ if and only if there exists an $h \in \mathcal{H}$ such that $\tau(hc_y c_x^\vee) \neq 0$. But, since τ is a symmetrizing form, this is equivalent to saying that $c_y c_x^\vee \neq 0$. This is summarized in the next lemma.

Lemma 1.2.1. *Let x and y be two elements of X. Then $x \overset{L}{\twoheadleftarrow} y$ (resp. $x \overset{R}{\twoheadleftarrow} y$) if and only if $c_y c_x^\vee \neq 0$ (resp. $c_x^\vee c_y \neq 0$).*

Using the duality induced by τ, we define new modules associated with left convex subsets. If C is a left convex subset of X, we set

$$X^{\geqslant_L C} = \{x \in X \mid \exists y \in C \text{ such that } x \geqslant_L y\}$$

and
$$X^{>_L C} = X^{\geqslant_L C} \setminus C.$$

We then define

$$\mathcal{H}^{\vee,\geqslant_L C} = \bigoplus_{x \in X^{\geqslant_L C}} A c_x^\vee, \quad \mathcal{H}^{\vee,>_L C} = \bigoplus_{x \in X^{>_L C}} A c_x^\vee,$$

$$\mathcal{H}^{\vee,L}[C] = \mathcal{H}^{\vee,\geqslant_L C} / \mathcal{H}^{\vee,>_L C},$$

and
$$\mathfrak{h}^L[C] = \sum_{x,y \in C} A c_x c_y^\vee.$$

Similarly, if C is a right (resp. two-sided) convex subset of X, we define the subsets $X^{\geqslant_R C}$ and $X^{>_R C}$ (resp. $X^{\geqslant_{LR} C}$ and $X^{>_{LR} C}$) of X and the A-modules $\mathcal{H}^{\vee,\geqslant_R C}$, $\mathcal{H}^{\vee,>_R C}$, $\mathcal{H}^{\vee,R}[C]$ (resp. $\mathcal{H}^{\vee,\geqslant_{LR} C}$, $\mathcal{H}^{\vee,>_{LR} C}$, $\mathcal{H}^{\vee,LR}[C]$) and $\mathfrak{h}^R[C]$. There is no analogue of $\mathfrak{h}^?[C]$ for two-sided convex subsets. The next result can be seen as a dual to Proposition 1.1.15.

Proposition 1.2.2. *Let C be a subset of X.*

(a) *If C is left convex, then the A-modules $\mathcal{H}^{\vee,\geqslant_L C}$ and $\mathcal{H}^{\vee,>_L C}$ are **right** ideals of $\mathcal{H}$ and $\mathcal{H}^{\vee,L}[C]$ is a **right** $\mathcal{H}$-module, which is isomorphic to the dual*

$\mathrm{Hom}_A(\mathcal{H}^L[C],A)$ *of* $\mathcal{H}^L[C]$.

Moreover, $\mathfrak{h}^L[C]$ *is a two-sided ideal of* $\mathcal{H}$.

(b) *If* C *is right convex, then the A-modules* $\mathcal{H}^{\vee,\geqslant_R C}$ *and* $\mathcal{H}^{\vee,>_R C}$ *are* **left** *ideals of* $\mathcal{H}$ *and* $\mathcal{H}^{\vee,R}[C]$ *is a* **left** $\mathcal{H}$*-module, which is isomorphic to the dual* $\mathrm{Hom}_A(\mathcal{H}^R[C],A)$ *of* $\mathcal{H}^R[C]$.

Moreover, $\mathfrak{h}^R[C]$ *is a two-sided ideal of* $\mathcal{H}$.

(c) *If* C *is two-sided convex, then the A-modules* $\mathcal{H}^{\vee,\geqslant_{LR} C}$ *and* $\mathcal{H}^{\vee,>_{LR} C}$ *are two-sided ideal of* $\mathcal{H}$ *and* $\mathcal{H}^{\vee,LR}[C]$ *is an* $(\mathcal{H},\mathcal{H})$*-bimodule, which is isomorphic to the dual* $\mathrm{Hom}_A(\mathcal{H}^{LR}[C],A)$ *of* $\mathcal{H}^{LR}[C]$.

Proof. We prove only the statement (a), as the other statements can be proved similarly. So assume that C is left convex. Let $x \in X^{\geqslant_L C}$ and let $h \in \mathcal{H}$. Write

$$c_x^\vee h = \sum_{y \in X} a_{xy} c_y^\vee.$$

If $a_{xy} \neq 0$, then $\tau(c_y c_x^\vee h) \neq 0$ and so $c_y c_x^\vee \neq 0$. Therefore $x \overset{L}{\twoheadleftarrow} y$ by Lemma 1.2.1, and so $y \in X^{\geqslant_L C}$. This shows that $\mathcal{H}^{\vee,\geqslant_L C}$ is a right $\mathcal{H}$-submodule of $\mathcal{H}$. If, moreover, $x \notin C$, then $y \notin C$, which shows that $\mathcal{H}^{\vee,>_L C}$ is also a right $\mathcal{H}$-submodule of $\mathcal{H}$. In particular, $\mathcal{H}^{\vee,L}[C]$ is a right $\mathcal{H}$-module.

Now, let

$$\begin{aligned} \pi_C^L : \mathcal{H}^{\leqslant_L C} \otimes_A \mathcal{H}^{\vee,\geqslant_L C} &\longrightarrow \mathcal{H} \\ h \otimes_A k &\longmapsto hk. \end{aligned}$$

It is clear that π_C^L is a morphism of $(\mathcal{H},\mathcal{H})$-bimodules. On the other hand, by Lemma 1.2.1, we have:

(1.2.3) $\quad$ *if* $x \in X^{<_L C}$ *or if* $y \in X^{>_L C}$, *then* $\pi_C^L(c_x \otimes_A c_y^\vee) = 0$.

This shows that π_C^L factorizes through a morphism of $(\mathcal{H},\mathcal{H})$-bimodules

$$\bar{\pi}_C^L : \mathcal{H}^L[C] \otimes_A \mathcal{H}^{\vee,L}[C] \longrightarrow \mathcal{H},$$

whose image is exactly $\mathfrak{h}^L[C]$. So $\mathfrak{h}^L[C]$ is an $(\mathcal{H},\mathcal{H})$-sub-bimodule of $\mathcal{H}$.

It remains only to show that $\mathcal{H}^{\vee,L}[C] \simeq \mathrm{Hom}_A(\mathcal{H}^L[C],A)$. For this, note that the map

$$\tau \circ \bar{\pi}_C^L : \mathcal{H}^L[C] \otimes_A \mathcal{H}^{\vee,L}[C] \longrightarrow A$$

satisfies

$$\tau \circ \bar{\pi}_C^L(hv \otimes_A v^\vee) = \tau(v \otimes_A v^\vee h)$$

for all $h \in \mathcal{H}$, $v \in \mathcal{H}^L[V]$ and $v^\vee \in \mathcal{H}^{\vee,L}[C]$ (since τ is central). It is then easy to check that it induces a perfect pairing between $\mathcal{H}^L[C]$ and $\mathcal{H}^{\vee,L}[C]$ (since $\tau(c_x c_y^\vee) = \delta_{x=y}$ for all $x, y \in C$). ∎

Proposition 1.2.4. *If C is a left (resp. right) convex subset of X, then the left (resp. right) module $\mathscr{H}^L[C]$ (resp. $\mathscr{H}^R[C]$) is isomorphic to a submodule of $\bigoplus_{x\in C}\mathfrak{h}^L[C]$ (resp. $\bigoplus_{x\in C}\mathfrak{h}^R[C]$).*

Proof. Again, we may, and we will, assume that C is left convex. Let $y \in C$. We set

$$\begin{array}{rcl}\varphi_y : \mathscr{H}^{\leqslant_L C} & \longrightarrow & \mathscr{H} \\ h & \longmapsto & hc_y^\vee.\end{array}$$

Then φ_y is a morphism of left $\mathscr{H}$-modules, whose image is contained in $\mathfrak{h}^L[C]$ (see 1.2.3). Again by using 1.2.3, φ_y factorizes through a morphism

$$\bar{\varphi}_y : \mathscr{H}^L[C] \longrightarrow \mathfrak{h}^L[C].$$

Let

$$\begin{array}{rcl}\varphi : \mathscr{H}^L[C] & \longrightarrow & \bigoplus_{y\in C}\mathfrak{h}^L[C] \\ v & \longmapsto & (\bar{\varphi}_y(v))_{y\in C}.\end{array}$$

Then φ is a morphism of left $\mathscr{H}$-modules. If, moreover, $\varphi(v)=0$, and if we denote by $\sum_{x\in C} r_x c_x$ a representative of v in $\mathscr{H}^{\leqslant_L C}$ (here, $r_x \in A$), then we have $\sum_{x\in C} r_x c_x c_y^\vee = 0$ for all $y \in C$. By applying τ to all these equalities, we obtain that $r_y = 0$ for all $y \in C$, so $v = 0$, and thus φ is injective. ■

Exercises

Exercise 1.1. Let n be a non-zero natural number and let $X_1, \ldots, X_n$ be indeterminates over A. We assume in this exercise that $\mathscr{H} = A[X_1, \ldots, X_n]$ and that $\mathscr{C} = (X_1^{i_1} \cdots X_n^{i_n})_{(i_1,\ldots,i_n)\in\mathbb{N}^n}$.

Compute the preorder $\leqslant_L$ (which is equal to $\leqslant_R$ and $\leqslant_{LR}$) and the equivalence relation $\sim_L$ (which is equal to $\sim_R$ and $\sim_{LR}$) on $\mathbb{N}^n$. Describe the cell modules.

Exercise 1.2. Let $s = (1,2)$ and $t = (2,3)$ be two permutations in $\mathfrak{S}_3$ and assume here that $\mathscr{H} = A\mathfrak{S}_3$. Then $\mathfrak{S}_3 = \{1, s, t, st, ts, sts\}$ and let

$$c_1 = 1,\ c_s = 1+s,\ c_t = 1+t,\ c_{st} = c_s c_t,\ c_{ts} = c_t c_s \quad \text{and} \quad c_{sts} = \sum_{\sigma\in\mathfrak{S}_3} \sigma.$$

Show that $(c_\sigma)_{\sigma\in\mathfrak{S}_3}$ is an A-basis of $\mathscr{H}$ and compute the preorders $\leqslant_L$, $\leqslant_R$, $\leqslant_{LR}$, as well as the left, right and two-sided cells. Describe the cell modules.

Exercise 1.3. This exercise aims to justify Remark 1.1.8. Let $\mathbf{k}$ be a field, let v be an indeterminate and assume here that $A = \mathbf{k}[v]/\langle v^2 \rangle$. We denote by ε the image of v in A and we set

$$x = \begin{pmatrix} 0 & 0 & 0 \\ 1 & 0 & 0 \\ 0 & 0 & \varepsilon \end{pmatrix} \qquad \text{and} \qquad y = \begin{pmatrix} 0 & 0 & 0 \\ 0 & \varepsilon & 0 \\ 1 & 0 & 0 \end{pmatrix}.$$

(a) Show that $xy = \varepsilon y$, $yx = \varepsilon x$ and $x^2 = y^2 = 0$.
(b) Deduce that $\mathscr{H} = A \oplus Ax \oplus Ay \subset \mathrm{Mat}_3(A)$ is an A-subalgebra of $\mathrm{Mat}_3(A)$.
(c) We assume here that $\mathscr{C} = (1, x, y)$. Show that the left (resp. right) cells are singletons and show that $x \sim_{LR} y$.
(d) Show that $\sim_{LR}$ is not the equivalence relation generated by $\sim_L$ and $\sim_R$.

Chapter 2
Lusztig's Lemma

Lusztig's Lemma is an easy result which aims to construct invariant elements (and in general invariant bases) in modules endowed with a semilinear involution. We will state here the most general version we need, and its direct consequences. They will be used systematically in this book not only for constructing the Kazhdan–Lusztig basis, but also for unifying some proofs of basic results in the theory of Kazhdan–Lusztig cells.

Let R be a commutative ring. We fix a totally ordered abelian group $\mathscr{A}$, whose law is denoted by $+$ and whose set of positive (resp. negative) elements is denoted by $\mathscr{A}_{>0}$ (resp. $\mathscr{A}_{<0}$). We similarly define $\mathscr{A}_{\geqslant 0}$ and $\mathscr{A}_{\leqslant 0}$, and we denote by A the group algebra $R[\mathscr{A}]$. We use an exponential notation for the algebra A:

$$A = \bigoplus_{a \in \mathscr{A}} R\, v^a,$$

where $v^a \cdot v^{a'} = v^{a+a'}$. We set $A_{\leqslant 0} = \bigoplus_{a \in \mathscr{A}_{\leqslant 0}} R\, v^a$ and $A_{<0} = \bigoplus_{a \in \mathscr{A}_{<0}} R\, v^a$. We denote by $\overline{} : A \to A$ the unique (involutive) R-algebra automorphism such that $\overline{v^a} = v^{-a}$ for all $a \in \mathscr{A}$ and we set

$$A_{\text{anti}} = \{\alpha \in A \mid \alpha + \overline{\alpha} = 0\}.$$

We also fix an A-module $\mathscr{M}$ and we assume that it admits an A-basis $(m_x)_{x \in X}$, where X is a *poset*. We set

$$\mathscr{M}_{<0} = \bigoplus_{x \in X} A_{<0}\, m_x.$$

For $\mathscr{X} \subset X$, we set $\mathscr{M}^{\mathscr{X}} = \bigoplus_{x \in \mathscr{X}} A\, m_x$ and $\mathscr{M}^{\mathscr{X}}_{<0} = \bigoplus_{x \in \mathscr{X}} A_{<0}\, m_x$. We also assume that we are given a semilinear involution $\overline{} : \mathscr{M} \to \mathscr{M}$ and we set

$$\mathscr{M}_{\text{anti}} = \{m \in \mathscr{M} \mid m + \overline{m} = 0\}.$$

Here, saying that $\overline{} : \mathscr{M} \to \mathscr{M}$ is *semilinear* means that $\overline{a\,m} = \overline{a}\;\overline{m}$ for all $a \in A$ and $m \in \mathscr{M}$.

C. Bonnafé, *Kazhdan–Lusztig Cells with Unequal Parameters*,
Algebra and Applications 24, https://doi.org/10.1007/978-3-319-70736-5_2

Hypothesis. *From now on, and until the end of this chapter, we assume that for every $x \in X$, the following two properties hold:*
(I1) *The set $\{y \in X \mid y \leqslant x\}$ is finite.*
(I2) $\overline{m}_x \equiv m_x \mod \Big(\bigoplus_{y<x} A m_y\Big)$.

2.1. Invariant Bases

Our aim in this section is to construct an A-basis of $\mathcal{M}$ whose elements are invariant under the semilinear involution $\overline{}$. This will be a consequence of the following result, which compares the R-modules $\mathcal{M}_{<0}$ and $\mathcal{M}_{\mathrm{anti}}$:

Lusztig's Lemma 2.1.1. *The R-linear map*

$$\begin{aligned} \mathcal{M}_{<0} &\longrightarrow \mathcal{M}_{\mathrm{anti}} \\ m &\longmapsto m - \overline{m} \end{aligned}$$

is an isomorphism.

Proof. First, note that the corresponding result for the A-module A holds. In other words,

(2.1.2) *The map $A_{<0} \to A_{\mathrm{anti}}, a \mapsto a - \overline{a}$ is an isomorphism.*

Indeed, if $\alpha \in A_{\mathrm{anti}}$, write $\alpha = \sum_{a \in \mathcal{A}} \alpha_a v^a$, with $\alpha_a \in R$. So, if we set $\alpha_- = \sum_{a<0} \alpha_a v^a$, then $\alpha = \alpha_- - \overline{\alpha}_-$. This shows the surjectivity, while the injectivity is trivial.

Now, let $\Lambda : \mathcal{M}_{<0} \to \mathcal{M}_{\mathrm{anti}}$, $m \mapsto m - \overline{m}$. If $\mathcal{X}$ is an ideal of the poset X (this means that, if $y \leqslant x$ and $x \in \mathcal{X}$, then $y \in \mathcal{X}$) then Hypothesis (I2) implies that $\mathcal{M}^{\mathcal{X}}$ is stabilized by the involution $\overline{}$. Since X is the union of such finite $\mathcal{X}$ (by (I1)), it shows that we may, and we will, assume that X is finite. Let us write $X = \{x_0, x_1, \dots, x_n\}$ in such a way that, if $x_i \leqslant x_j$, then $i \leqslant j$ (this is always possible). To simplify the notation, we set $m_{x_i} = m_i$. Note that, by (I2),

$$(*) \qquad \overline{m}_i \in m_i + \Big(\bigoplus_{0 \leqslant j < i} A m_j\Big).$$

In particular, $\overline{m}_0 = m_0$.

Now, let $m \in \mathcal{M}_{<0}$ be such that $\overline{m} = m$ and assume that $m \neq 0$. Write $m = \sum_{i=0}^{r} a_i m_i$, with $r \leqslant n$, $a_i \in A_{<0}$ and $a_r \neq 0$. Then, by (I2),

$$\overline{m} \equiv \overline{a}_r m_r \mod \Big(\bigoplus_{0 \leqslant j < i} A m_j\Big).$$

Since $\overline{m} = m$, this forces $\overline{a}_r = a_r$, which is impossible (because $a_r \in A_{<0}$ and $a_r \neq 0$). So Λ is injective.

Let us now show that Λ is surjective. So, let $m \in \mathcal{M}_{\text{anti}}$, and assume that $m \neq 0$ (for otherwise there is nothing to prove). Write $m = \sum_{i=0}^{r} a_i m_i$, with $r \leqslant n$, $a_i \in A$ and $a_r \neq 0$. We will prove by induction on r that there exists a $\mu \in \mathcal{M}_{<0}$ such that $m = \mu - \overline{\mu}$. If $r = 0$, then the result follows from 2.1.2 and the fact that $\overline{m}_0 = m_0$. So assume that $r > 0$. By (I2),

$$m + \overline{m} \equiv (a_r + \overline{a}_r) m_r \mod \mathcal{M}^{\mathcal{Y}},$$

where $\mathcal{Y} = \{x_0, x_1, \ldots, x_{r-1}\}$. Since $m + \overline{m} = 0$, this forces $a_r \in A_{\text{anti}}$. So, by 2.1.2, there exists an $a \in A_{<0}$ such that $a - \overline{a} = a_r$. Now, let $m' = m - a m_r + \overline{a m_r}$. Then $m' + \overline{m}' = 0$ and $m' \in \bigoplus_{0 \leqslant j < r} A m_j$. So, by the induction hypothesis, there exists a $\mu' \in \mathcal{M}_{<0}$ such that $m' = \mu' - \overline{\mu}'$. Now, set $\mu = a m_r + \mu'$. Then $\mu \in \mathcal{M}_{<0}$ and $m = \mu - \overline{\mu} = \Lambda(\mu)$, as desired. ■

Corollary 2.1.3. *Let $m \in \mathcal{M}$. Then there exists a unique $M \in \mathcal{M}$ such that*

$$\begin{cases} \overline{M} = M, \\ M \equiv m \mod \mathcal{M}_{<0}. \end{cases}$$

Proof. Since $\overline{m} - m \in \mathcal{M}_{\text{anti}}$, it follows from Lusztig's Lemma that there exists a unique $\mu \in \mathcal{M}_{<0}$ such that $\mu - \overline{\mu} = \overline{m} - m$. Then $M = m + \mu$ is the unique solution to the above system. ■

Remark 2.1.4. Corollary 2.1.3 can be applied to the A-module A itself. However, in this case, its proof becomes obvious: if $\alpha = \sum_{a \in \mathcal{A}} \alpha_a v^a$, then $\beta = \sum_{a<0} \alpha_{-a} v^a + \sum_{a \geqslant 0} \alpha_a v^a$ is the unique element of A such that $\overline{\beta} = \beta$ and $\beta \equiv \alpha \mod A_{<0}$. □

Corollary 2.1.5. *Let $\mathcal{X}$ be an ideal of the poset X. Let $M \in \mathcal{M}$ be such that $\overline{M} = M$ and $M \in \mathcal{M}^{\mathcal{X}} + \mathcal{M}_{<0}$. Then $M \in \mathcal{M}^{\mathcal{X}}$.*

Proof. Let $M_0 \in \mathcal{M}^{\mathcal{X}}$ be such that $M \equiv M_0 \mod \mathcal{M}_{<0}$. From the existence statement of Corollary 2.1.3 applied to $\mathcal{M}^{\mathcal{X}}$, there exists an $M' \in \mathcal{M}^{\mathcal{X}}$ such that $\overline{M}' = M'$ and $M' \equiv M_0 \mod \mathcal{M}^{\mathcal{X}}_{<0}$. The fact that $M = M' \in \mathcal{M}^{\mathcal{X}}$ now follows from the uniqueness statement of Corollary 2.1.3. ■

Corollary 2.1.6. *Let $x \in X$. Then there exists a unique element $M_x \in \mathcal{M}$ such that*

$$\begin{cases} \overline{M}_x = M_x, \\ M_x \equiv m_x \mod \mathcal{M}_{<0}. \end{cases}$$

Moreover, $M_x \in m_x + \bigoplus_{y<x} A_{<0} m_y$ and $(M_x)_{x \in X}$ is an A-basis of $\mathcal{M}$.

Proof. The existence and uniqueness of M_x follow from Corollary 2.1.3. The statement about the base change follows by applying this existence and uniqueness to $\mathcal{M}^{X_x}$, where $X_x = \{y \in X \mid y \leqslant x\}$.

Finally, the fact that $(M_x)_{x\in X}$ is an A-basis of $\mathcal{M}$ follows from the fact that the base change from $(m_x)_{x\in X}$ to $(M_x)_{x\in X}$ is unitriangular. ■

2.2. Base Change

For $y \in X$, let us write

$$\overline{m}_y = \sum_{x \in X} \overline{\rho}_{x,y}\, m_x$$

and

$$M_y = \sum_{x \in X} \pi_{x,y}\, m_x,$$

with $\rho_{x,y}$, $\pi_{x,y} \in A$. Then, the fact that $\overline{}$ is an A-semilinear involution together with (I2) is equivalent to

$$\text{(2.2.1)} \qquad \begin{cases} \sum_{x \leqslant z \leqslant y} \overline{\rho}_{x,z}\rho_{z,y} = \delta_{x=y}, \\ \rho_{y,y} = 1, \\ \rho_{x,y} \neq 0 \Longrightarrow x \leqslant y \end{cases}$$

for all x, $y \in X$. On the other hand, the properties of the basis $(M_x)_{x\in X}$ are equivalent to the following:

$$\text{(2.2.2)} \qquad \begin{cases} \pi_{x,y} \in A_{<0}, & \text{if } x \neq y, \\ \pi_{x,y} = \sum_{x \leqslant z \leqslant y} \overline{\rho}_{x,z}\overline{\pi}_{z,y}, \\ \pi_{x,x} = 1, \\ \pi_{x,y} \neq 0 \Longrightarrow x \leqslant y \end{cases}$$

for all $x, y \in X$.

Notes

Corollary 2.1.6 gathers in a single general statement the argument given by Lusztig [Lus9] for the construction of the *Kazhdan–Lusztig basis* of a Hecke algebra (which is different from the argument contained in the original paper by Kazhdan and Lusztig [KaLu1]) and the construction, still due to Lusztig [Lus17, Theorem 3.2], of the *canonical basis* associated with modules over quantum groups.

Part II
Coxeter Systems, Hecke Algebras

Numerous excellent books deal with the theory of Coxeter groups, their underlying geometry and their associated Hecke algebras [Bah], [Bou], [BjBr], [Dav], [GePf], [Hil], [Hum],... They cover a wide scope of points of view and treat many interesting problems of different natures (structure, combinatorics, representations, geometry...).

The goal of our book is not to provide another full treatment, with proofs included, of the foundations of the theory: we only wish to continue this theory in one particular and precise direction, namely the Kazhdan–Lusztig theory with unequal parameters. Therefore, the present part aims essentially to fix notation and to gather, without proofs (and maybe sometimes in a way that is more convenient for our needs but which might not reflect the logical development of the theory), some of the results about Coxeter groups and Hecke algebras that will be needed throughout this book.

However, as we wish *all* the proofs in our book to rely only on results which appear in reference books[1] (and not in research papers), we include proofs of some results on reflection subgroups of Coxeter groups: this is, for instance, the purpose of Appendix B, whose main aim is to show how to decompose a Coxeter group as a semidirect product of two Coxeter groups (see Theorem B.3.1).

[1] There will be an important exception to this principle: in Chapter 15, we rely on deep results about Soergel bimodules proved by Elias and Williamson, as well as on some consequences obtained by Lusztig.

Chapter 3
Coxeter Systems

> *We fix in this **book** a group W together with a generating subset S of W consisting of elements of order 2. If s, $t \in S$, we denote by m_{st} the order (possibly infinite) of st in W. If $w \in W$, we set $w^{\infty} = 1$.*

It must be emphasized that S is not assumed to be finite.

3.1. Characterizations of Coxeter Systems

Since the elements of S have order 2, note that

(3.1.1) $$\begin{cases} m_{st} \in \mathbb{N}^* \mathbin{\dot{\cup}} \{\infty\} \\ m_{ss} = 1 \\ m_{st} = m_{ts} \\ s \neq t \Longrightarrow m_{st} \geqslant 2 \end{cases}$$

for all s, $t \in S$. We denote by $\ell : W \to \mathbb{N}$ (or ℓ_W, or $\ell_{W,S}$ if necessary) the *length* function on W (it depends on S) which is defined as follows: if $w \in W$, $\ell(w)$ is the minimal $r \in \mathbb{N}$ such that there exist $s_1, \ldots, s_r \in S$ satisfying $w = s_1 \cdots s_r$. As the elements of S are involutions,

(3.1.2) $$\ell(w^{-1}) = \ell(w)$$

for all $w \in W$. Also

(3.1.3) $$\ell(w) = 0 \textit{ if and only if } w = 1.$$

C. Bonnafé, *Kazhdan–Lusztig Cells with Unequal Parameters*,
Algebra and Applications 24, https://doi.org/10.1007/978-3-319-70736-5_3

Definition 3.1.4. *Let $w \in W$. A **reduced expression** of w is a sequence $(s_1, s_2, \ldots, s_r)$ of elements of S such that $w = s_1 s_2 \cdots s_r$ and $r = \ell(w)$.*

We follow [Bou, Chapter IV, §1, Definition 3]:

Definition 3.1.5. *We say that (W,S) is a **Coxeter system** if W admits the following presentation:*

$$\begin{cases} \textit{Generators:} & S, \\ \textit{Relations:} & \forall s,t \in S,\ (st)^{m_{st}} = 1. \end{cases}$$

Exchange Condition 3.1.6. *The following are equivalent:*

(1) *(W,S) is a Coxeter system.*
(2) *For all s, $s_1, \ldots, s_r$ in S such that $\ell(s_1 \cdots s_r) = r$ and $\ell(ss_1 \cdots s_r) \leqslant r$, there exists a $j \in \{1,2,\ldots,r\}$ such that*

$$ss_1 s_2 \cdots s_{j-1} = s_1 s_2 \cdots s_j.$$

Proof. See [Bou, Chapter IV, §1.6, Theorem 1]. ■

Corollary 3.1.7. *The following are equivalent:*

(1) *(W,S) is a Coxeter system.*
(2) *There exists a family $(\mathscr{E}_s)_{s\in S}$ of subsets of W such that:*
   (2.1) *If $s \in S$, then $1 \in \mathscr{E}_s$ and $\mathscr{E}_s \cap s\mathscr{E}_s = \varnothing$.*
   (2.2) *If s, $s' \in S$ and $w \in \mathscr{E}_s$ are such that $ws' \notin \mathscr{E}_s$, then $sw = ws'$.*

If both conditions hold, then

$$\mathscr{E}_s = \{w \in W \mid \ell(sw) > \ell(w)\}$$

for all $s \in S$.

Proof. See [Bou, Chapter IV, §1.7, Proposition 4]. ■

> *From now on, and until the end of this **book**, we assume that (W,S) is a Coxeter system.*

Definition 3.1.8. *The cardinality of S is called the **rank** of (W,S) (or of W).*

Commentary 3.1.9. Unless explicitly stated, we do not assume that S is finite. The reason is that, even if S is finite, there are interesting subgroups of W generated by reflections which are Coxeter systems with an infinite set of simple reflections (see Example B.3.6). Moreover, these subgroups appear in the reduction of the Kazhdan–Lusztig theory with arbitrary parameters to the case of positive parameters (see Commentary 6.2.12).

We will assume that S is finite only from Part V to the end of the book (except for the appendices). □

3.2. First Properties

Cancellation Law 3.2.1. *Let $w = s_1 s_2 \cdots s_n$ with $s_i \in S$. Then $\ell(w) \equiv n \mod 2$ and there exists a sequence $1 \leqslant i_1 < i_2 < \cdots < i_{\ell(w)} \leqslant n$ such that $w = s_{i_1} s_{i_2} \cdots s_{i_{\ell(w)}}$ (that is, $(s_{i_1}, s_{i_2}, \ldots, s_{i_{\ell(w)}})$ is a reduced expression of w).*

Proof. Using an induction argument on n, this follows immediately from the Exchange Condition 3.1.6. ∎

From the presentation given in Definition 3.1.5, it follows that the next definition is valid:

Definition 3.2.2. *We define the* **sign character** *of W to be the unique morphism of groups $\varepsilon : W \to \{1, -1\}$ such that $\varepsilon(s) = -1$ for all $s \in S$. If necessary, ε is denoted by ε_W.*

Note that

(3.2.3) $$\varepsilon(w) = (-1)^{\ell(w)}$$

for all $w \in W$. Consequently,

(3.2.4) $$\ell(ww') \equiv \ell(w) + \ell(w') \mod 2 \qquad \text{and} \qquad \ell({}^{w}w') \equiv \ell(w') \mod 2.$$

Note also that

(3.2.5) $$|\ell(w) - \ell(w')| \leqslant \ell(ww') \leqslant \ell(w) + \ell(w').$$

It then follows from 3.2.4 and 3.2.5 that, if $s \in S$ and $w \in W$, then

(3.2.6) $$\ell(sw), \ell(ws) \in \{\ell(w) - 1, \ell(w) + 1\}.$$

Recall also the following result [Bou, Chapter IV, §1, Proposition 3]:

Lemma 3.2.7. *Let s, $t \in S$. Then s and t are conjugate in W if and only if there exists a sequence $s_0, s_1, \ldots, s_n$ of elements of S such that $s = s_0$, $t = s_n$ and $m_{s_{i-1} s_i}$ is finite and odd for all $i \in \{1, 2, \ldots, n\}$.*

3.3. Reflections, Roots

Let

$$T = \bigcup_{w \in W} wSw^{-1}$$

and

$$\Phi = \{1, -1\} \times T.$$

The elements of T are called the *reflections* of W (while the elements of S are called the *simple reflections* of W) and the elements of Φ are called the *roots* of W. If $s \in S$, we set, following [Bou, Chapter IV, §1.4, Lemma 1],

$$U_s(\mu, t) = (\mu \cdot (-1)^{\delta_{s=t}}, sts^{-1}) \in \Phi$$

for all $(\mu, t) \in \Phi$. It is easily checked that $U_s^2 = \mathrm{Id}_\Phi$ and it follows from [Bou, Chapter IV, §1.4, Proof of Lemma 1] that $(U_s U_t)^{m_{st}} = \mathrm{Id}_\Phi$ for all s, $t \in S$. So $U_s \in \mathfrak{S}_\Phi$ and the map $s \mapsto U_s$ extends uniquely to a morphism of groups

$$\begin{aligned} U : W &\longrightarrow \mathfrak{S}_\Phi \\ w &\longmapsto U_w. \end{aligned}$$

In other words, we have defined an action of W on the set of roots Φ. Again by [Bou, Chapter IV, §1.4, Lemma 1], there exists a unique map

$$\eta : W \times T \longrightarrow \{1, -1\}$$

such that

$$U_w(\mu, t) = (\mu \cdot \eta(w^{-1}, t), wtw^{-1}).$$

Finally, we set

$$N(w) = \{t \in T \mid \eta(w, t) = -1\}.$$

Of course, $N(1) = \varnothing$ and it follows from the definition of U_s that

(3.3.1) $$N(s) = \{s\}.$$

Note also that, since $U : W \to \mathfrak{S}_\Phi$ is a morphism of groups, we have

(3.3.2) $$\eta(xy, t) = \eta(x, t)\eta(y, x^{-1}tx).$$

Taking $x = y = 1$ gives

(3.3.3) $$\eta(1, t) = 1$$

for all $t \in T$. Taking now $x = w^{-1} = y^{-1}$ in 3.3.2 implies that

(3.3.4) $$\eta(w^{-1}, t) = \eta(w, wtw^{-1}).$$

Now, endow the set $\mathcal{P}(T)$ of subsets of T with the group law given by the symmetric difference $\dotplus$. Then, $\mathcal{P}(T)$ becomes a $\mathbb{Z}W$-module (for the action of W by conjugation on T) and:

Corollary 3.3.5. *The map $N : W \to \mathcal{P}(T)$ is a 1-cocycle: in other words,*

$$N(xy) = N(x) \dotplus {}^x N(y)$$

for all x, $y \in W$. Consequently,

$$N(w^{-1}) = {}^{w^{-1}}N(w).$$

Corollary 3.3.6. *Let $t \in T$. Then $t \in N(t)$ (in other words, $\eta(t,t) = -1$).*

Proof. Let $x \in W$ and $s \in S$ such that $t = xsx^{-1}$. Then, by 3.3.2,

$$\begin{aligned}\eta(t,t) = \eta(xsx^{-1}, xsx^{-1}) &= \eta(xs, xsx^{-1})\eta(x^{-1}, s) \\ &= \eta(x, xsx^{-1})\eta(s,s)\eta(x^{-1}, s).\end{aligned}$$

As $\eta(s,s) = -1$ by definition, the result now follows from 3.3.4. ∎

Corollary 3.3.7. *Let $t \in T$ and $w \in W$. Then $\eta(tw, t) = -\eta(w, t)$.*

Proof. This follows from 3.3.2 and Corollary 3.3.6. ∎

Then [Bou, Chapter IV, §1.4, Lemma 2]:

Lemma 3.3.8. *Let $w \in W$ and let $(s_1, s_2, \ldots, s_r)$ be a reduced expression of w. For $1 \leqslant k \leqslant r$, let*

$$t_k = (s_1 s_2 \cdots s_{k-1}) s_k (s_1 s_2 \cdots s_{k-1})^{-1},$$

so that $s_1 s_2 \cdots s_k = t_k t_{k-1} \cdots t_1$. Then

$$|N(w)| = r \quad \text{and} \quad N(w) = \{t_1, t_2, \ldots, t_r\}.$$

Corollary 3.3.9. *The action of W on Φ is faithful.*

Corollary 3.3.10. *If $t \in T$ and $w \in W$, then $\ell(tw) > \ell(w)$ if and only if $\eta(w,t)$=1.*

Proof. Let $(s_1, s_2, \ldots, s_r)$ be a reduced expression of w. If $\eta(w,t) = -1$, then, by Lemma 3.3.8, there exists a $k \in \{1, 2, \ldots, r\}$ such that

$$t = (s_1 s_2 \cdots s_{k-1}) s_k (s_1 s_2 \cdots s_{k-1})^{-1}.$$

Therefore $tw = s_1 s_2 \cdots s_{k-1} s_{k+1} s_{k+2} \cdots s_r$, so $\ell(tw) \leqslant r - 1 < \ell(w)$.

Conversely, if $\eta(w,t) = 1$, then $\eta(tw, t) = -1$ by Corollary 3.3.7 and so, by the above result, $\ell(w) = \ell(ttw) < \ell(tw)$, as desired. ∎

It follows from Corollary 3.3.10 that

(3.3.11) $$N(w) = \{t \in T \mid \ell(tw) < \ell(w)\}.$$

3.4. Braided Maps

We fix in this section, and only in this section, a monoid M, whose law is denoted multiplicatively, unless otherwise specified.

Definition 3.4.1. *A* ***braided map*** *on W (with values in M) is a map $\varphi : W \to M$ such that*

$$\forall\, w, w' \in W,\ \ell(ww') = \ell(w) + \ell(w') \Longrightarrow \varphi(ww') = \varphi(w)\varphi(w').$$

The set of braided maps on W with values in M is denoted by $\mathrm{Braid}(W, M)$.

Equivalently, a braided map $\varphi : W \to M$ is a map such that, whenever $(s_1, s_2, \ldots, s_r)$ and $(s'_1, s'_2, \ldots, s'_r)$ are two reduced expressions of the same element $w \in W$, then

$$\varphi(s_1)\varphi(s_2)\cdots\varphi(s_r) = \varphi(s'_1)\varphi(s'_2)\cdots\varphi(s'_r) \quad (= \varphi(w)). \tag{3.4.2}$$

In particular,

(3.4.3) *if M is* ***commutative*** *and $\varphi \in \mathrm{Braid}(W, M)$, then $\varphi(w) = \varphi(w^{-1})$*

for all $w \in W$.

Example 3.4.4. The map $\ell : W \to (\mathbb{N}, +)$ is a braided map. □

Following [GePf, Page 7], if a, b are two elements of M and $m \geqslant 0$, we define an element $\mathrm{Prod}(a, b; m)$ of M inductively by

$$\begin{cases} \mathrm{Prod}(a, b; 0) = 1, \\ \mathrm{Prod}(a, b; m+1) = a\,\mathrm{Prod}(b, a; m). \end{cases}$$

In other words,

$$\mathrm{Prod}(a, b; m) = \underbrace{aba\cdots}_{m \text{ factors}} \in M.$$

We add the convention that $\mathrm{Prod}(a, b; \infty) = 1$. With this notation, note that the presentation of W given in Definition 3.1.5 can be rewritten as follows:

$$\begin{cases} \textit{Generators:} & S, & \\ \textit{Relations:} & (Q) & \forall s \in S,\ s^2 = 1, \\ & (B) & \forall s, t \in S,\ \mathrm{Prod}(s, t; m_{st}) = \mathrm{Prod}(t, s; m_{st}). \end{cases} \tag{3.4.5}$$

The relations (Q) are called the *quadratic* relations while the relations (B) are called the *braid* relations. Note also that

$$\forall\, s, t \in S,\ \forall\, m \leqslant m_{st},\ \ell(\mathrm{Prod}(s, t; m)) = m. \tag{3.4.6}$$

So, if $\varphi : W \to M$ is a braided map, then

$$\mathrm{Prod}(\varphi(s), \varphi(t); m_{st}) = \mathrm{Prod}(\varphi(t), \varphi(s); m_{st})$$

for all $s, t \in S$. Matsumoto's Theorem is a converse to this property (for a proof, see for instance [GePf, Theorem 1.2.2]):

Matsumoto's Theorem 3.4.7. *Let $\varphi : S \to M$ be a map such that*

$$(\wp) \qquad \forall s,t \in S, \quad \mathrm{Prod}(\varphi(s),\varphi(t);m_{st}) = \mathrm{Prod}(\varphi(t),\varphi(s);m_{st}).$$

Then there exists a unique braided map $\varphi^\flat : W \to M$ extending φ.

If $w \in W$ and $(s_1, s_2, \ldots, s_r)$ is a reduced expression of w, then the function $\varphi^\flat$ defined in Matsumoto's Theorem is given by

$$\varphi^\flat(w) = \varphi(s_1)\varphi(s_2)\cdots\varphi(s_r).$$

Matsumoto's Theorem says that, under the hypothesis $(\wp)$, the above product $\varphi^\flat(w)$ does not depend on the choice of the reduced expression.

> **Notation.** *The braided map $\varphi^\flat$ of Matsumoto's Theorem will generally simply be denoted by φ.*

Example 3.4.8 (Braid monoid and reduced expressions). Let Mon(S) denote the set of finite sequences of elements of S, viewed as a monoid under the concatenation law. In other words, Mon(S) is the free monoid on the alphabet S. The length of an element γ of Mon(S) is denoted by $\ell^+(\gamma)$. This defines a morphism of monoids $\ell^+ : \mathrm{Mon}(S) \to \mathbb{N}$.

We denote by $\mathrm{B}^+(W,S)$ the monoid obtained as the quotient of Mon(S) by the monoid congruence $\equiv_+$ generated by the following congruences

$$\forall s,t \in S, \quad \mathrm{Prod}((s),(t);m_{st}) \equiv_+ \mathrm{Prod}((t),(s);m_{st}).$$

It is called the *braid monoid*. Here, (s) denotes the element of Mon(S) of length 1 whose unique term is s. If $(s_1,s_2,\ldots,s_r) \in \mathrm{Mon}(S)$, we denote by $[s_1,s_2,\ldots,s_r]$ its image in $\mathrm{B}^+(W,S)$. It follows from this presentation of $\mathrm{B}^+(W,S)$ that the length function ℓ^+ factorizes through a length function on $\mathrm{B}^+(W,S)$ that is still denoted by ℓ^+.

Now, let us consider the map $\beta : S \to \mathrm{B}^+(W,S)$, $s \mapsto [s]$. Then β satisfies the relation $(\wp)$ of Matsumoto's Theorem, so it extends to a unique braided map $\beta^\flat : W \to \mathrm{B}^+(W,S)$. Since ℓ^+ factorizes through $\mathrm{B}^+(W,S)$, this means precisely that, if $w \in W$ and $\gamma = (s_1,s_2,\ldots,s_r)$ and $\gamma' = (s'_1,s'_2,\ldots,s'_r)$ are two reduced expressions of w, then there exist reduced expressions $\gamma_0 = \gamma, \gamma_1, \ldots, \gamma_n = \gamma'$ of w such that γ_{i+1} is obtained from γ_i by applying only one braid relation.

This example is universal in the following sense. If $\varphi : S \to M$ is a map to a monoid M, then it extends naturally to a morphism of monoids $\varphi_1 : \mathrm{Mon}(S) \to M$ and this last map factors through a morphism of monoids $\varphi_2 : \mathrm{B}^+(W,S) \to M$ if and only if φ satisfies $(\wp)$. In this case, the braided map $\varphi^\flat$ of Matsumoto's Theorem is just $\varphi_2 \circ \beta^\flat$. □

Definition 3.4.9. *If $\mathscr{E}$ is a set, a* ***weight function*** *from S to $\mathscr{E}$ is a map $\varphi : S \to \mathscr{E}$ such that $\varphi(s) = \varphi(t)$ whenever s, $t \in S$ are conjugate in W. The set of weight functions from S to $\mathscr{E}$ is denoted by* $\mathrm{Weight}(S,\mathscr{E})$*, or* $\mathrm{Weight}_W(S,\mathscr{E})$ *if necessary.*

Example 3.4.10 (Commutative monoid). We assume in this example that M is *commutative* and change the law notation to an additive notation. If $\varphi \in \mathrm{Weight}(S,M)$, then, according to Lemma 3.2.7, φ satisfies the identity $(\wp)$ of Matsumoto's Theorem 3.4.7. So there is a well-defined injective map

$$\mathrm{Weight}(S,M) \hookrightarrow \mathrm{Braid}(W,M).$$

If, moreover, M is *regular* (i.e., for all $m, x, y \in M$, $m+x = m+y$ if and only if $x = y$), then the above map is surjective: indeed, if φ is a map $S \to M$, then φ satisfies the property $(\wp)$ of Matsumoto's Theorem 3.4.7 *if and only if* $\varphi \in \mathrm{Weight}(S,M)$. As an example, if M is a group, then M is regular. □

Example 3.4.11 (Support). In this example, we view the set $\mathscr{P}(S)$ of subsets of S as a monoid under the law $\cup$. Let $\mathrm{supp} : S \to \mathscr{P}(S)$, $s \mapsto \{s\}$. Then, in the monoid $(\mathscr{P}(S),\cup)$,

$$\mathrm{Prod}(\mathrm{supp}(s),\mathrm{supp}(t); m_{st}) = \{s,t\} = \mathrm{Prod}(\mathrm{supp}(t),\mathrm{supp}(s); m_{st}).$$

So, by Matsumoto's Theorem 3.4.7, the map supp can be extended to a braided map $\mathrm{supp} : W \to \mathscr{P}(S)$, which is defined as follows: if $(s_1, s_2, \ldots, s_r)$ is a reduced expression of $w \in W$, then the set

$$\mathrm{supp}(w) = \{s_1, s_2, \ldots, s_r\}$$

does not depend on the choice of the reduced expression. The subset $\mathrm{supp}(w)$ of S is called the *support* of w. Of course, $\mathrm{supp}(w) = \mathrm{supp}(w^{-1})$.

Note that the *commutative* monoid $(\mathscr{P}(S),\cup)$ is *not regular* and this example shows that the map $\mathrm{Weight}(S,\mathscr{P}(S)) \hookrightarrow \mathrm{Braid}(W,\mathscr{P}(S))$ defined in Example 3.4.10 is not surjective in general. Indeed, in general, the restriction of supp to S is not a weight function. □

We conclude this section with an application of Example 3.4.10.

Definition 3.4.12. *A subset I of S is called W*-**closed** *if, for all $s \in I$ and $t \in S$ such that s and t are conjugate in W, we have $t \in I$.*

Of course,

(3.4.13) *if I is W-closed, then $S \backslash I$ is also W-closed.*

If I is W-closed, let $\ell_I : S \to \mathbb{N}$ be the map defined by

$$\ell_I(s) = \begin{cases} 1 & \text{if } s \in I, \\ 0 & \text{if } s \in S \backslash I. \end{cases}$$

Then ℓ_I is a weight function so, by Example 3.4.10, ℓ_I extends uniquely to a braided map, still denoted by $\ell_I : W \to \mathbb{N}$. If $(s_1, s_2, \ldots, s_r)$ is a reduced expression of an element $w \in W$, then $\ell_I(w)$ is just the number of occurrences of elements of I in the sequence $(s_1, s_2, \ldots, s_r)$. For instance, $\ell_S = \ell$.

3.5. Bruhat Order

We follow the treatment in [GePf, §1.2]. Let $\mathscr{P}(W)$ denote the set of subsets of W and endow $\mathscr{P}(W)$ with the monoid structure given by the multiplication law in W: if $X, Y \subset W$, then

$$X \cdot Y = \{xy \mid x \in X \text{ and } y \in Y\}.$$

Let $\mathscr{B}r : S \to \mathscr{P}(W)$, $s \mapsto \{1, s\}$. Then, by [GePf, 1.2.4], for all $s \neq t \in S$,

$$\operatorname{Prod}(\mathscr{B}r(s), \mathscr{B}r(t); m_{st}) = \operatorname{Prod}(\mathscr{B}r(t), \mathscr{B}r(s); m_{st})$$

in the monoid $(\mathscr{P}(W), \cdot)$ and so, by Matsumoto's Theorem 3.4.7, $\mathscr{B}r$ extends uniquely to a braided map still denoted by

$$\mathscr{B}r : W \longrightarrow \mathscr{P}(W).$$

Definition 3.5.1. *If $x, y \in W$, we write $x \leqslant y$ if $x \in \mathscr{B}r(y)$. The relation $\leqslant$ is an order on W, called the* ***Bruhat order****.*

In other words, if $(s_1, s_2, \ldots, s_r)$ is a reduced expression of y, then $x \leqslant y$ if and only if there exists a sequence $1 \leqslant i_1 < \cdots < i_d \leqslant r$ such that $x = s_{i_1} \cdots s_{i_d}$.

The Bruhat order is compatible with taking the inverse:

(3.5.2) $$x \leqslant y \text{ if and only if } x^{-1} \leqslant y^{-1}.$$

Also,

(3.5.3) $$\text{If } x \leqslant y, \text{ then } \operatorname{supp}(x) \subset \operatorname{supp}(y), \ell(x) \leqslant \ell(y) \text{ and } \ell_I(x) \leqslant \ell_I(y)$$

for all W-closed subsets I of S. Finally, if $s \in S$ and $w \in W$, then

(3.5.4) $$\ell(sw) > \ell(w) \text{ if and only if } sw > w.$$

3.6. Parabolic Subgroups

If I is a subset of S, we denote by W_I the subgroup of W generated by I. Then [Bou, Chapter IV, §1, Theorem 2(i)]

(3.6.1) (W_I, I) *is a Coxeter system.*

If $s_1, \ldots, s_n$ belong to S, we simply write $W_{s_1,s_2,\ldots,s_n}$ for $W_{\{s_1,s_2,\ldots,s_n\}}$.

Definition 3.6.2. *A subgroup $\mathscr{W}$ of W is called* **parabolic** *if there exists a subset I of S such that $\mathscr{W}$ is conjugate to W_I. The subgroups of the form W_I for $I \subset S$ are called the* **standard parabolic** *subgroups.*

The Cancellation Law 3.2.1 (together with Example 3.4.11) shows that

(3.6.3) $w \in W_I$ *if and only if* $\mathrm{supp}(w) \subset I$.

Consequently,

(3.6.4) $$W_I \cap W_J = W_{I \cap J}$$

and

(3.6.5) *the length function in the Coxeter system (W_I, I) is the restriction of ℓ.*

Definition 3.6.6. *If I, J are two subsets of S, and if $x \in W$, we say that x is (I,J)-* **reduced** *if x has minimal length in $W_I x W_J$. The set of (I,J)-reduced elements of W is denoted by X_{IJ}. If K is another subset of S containing both I and J, then we set $X_{IJ}^K = X_{IJ} \cap W_K$: it is the set of (I,J)-reduced elements of W_K.*

For simplification, we set $X_I = X_{\varnothing I}$ and $X_I^K = X_{\varnothing I}^K$. Note that $X_{IJ} = X_{JI}^{-1}$ by 3.1.2.

Theorem 3.6.7. *Let I and J be two subsets of S. Then:*

(a) *If $w \in W$, there exist $u \in W_I$, $v \in W_J$ and $d \in X_{IJ}$ such that $w = udv$ and $\ell(w) = \ell(u) + \ell(d) + \ell(v)$.*
(b) *Every double coset $W_I w W_J$ contains a unique (I,J)-reduced element.*
(c) *Let $d \in W$. Then the following are equivalent:*

  (1) $d \in X_{IJ}$;
  (2) $d \in X_I^{-1} \cap X_J$;
  (3) *for all $s \in I$ and $t \in J$, $sd > d$ and $dt > d$.*

Proof. See, [GePf, Proposition 2.1.7] or [Bou, Chapter IV, §1, Exercise 3]. ■

By applying Theorem 3.6.7 to the case where $I = \varnothing$, one gets:

Corollary 3.6.8. *Let J be a subset of S. If $w \in W$, then there exists a unique pair $(x, \omega) \in X_J \times W_J$ such that $w = x\omega$. Moreover, $\ell(w) = \ell(x) + \ell(\omega)$.*

Corollary 3.6.9. *Let I and J be two subsets of S and let $w \in W$. Then there exists a unique triple (x, d, u) such that $u \in W_J$, $d \in X_{IJ}$, $x \in X_{I \cap {}^d J}^I$ and $w = xdu$. Moreover, $\ell(w) = \ell(x) + \ell(d) + \ell(u)$ and $W_I \cap {}^d W_J = W_{I \cap {}^d J}$.*

Proof. See, for instance, [GePf, Lemma 2.1.9] and [GePf, Theorem 2.1.12]. ■

Example 3.6.10. Let I be a subset of S and let $s \in S \setminus I$. Then

$$(\clubsuit) \qquad s \in X_I.$$

Indeed, by 3.4.6, $\ell(st) = 2 > 1 = \ell(s)$ for all $t \in I$. Therefore, by Corollary 3.6.8,

$$(\diamondsuit) \qquad \textit{if } s \in S \setminus I \textit{ and } w \in W_I, \textit{ then } \ell(sw) = \ell(ws) = 1 + \ell(w).$$

Also, if I and J are subsets of S such that $I \cap J = \varnothing$ and if $w \in W_I$ and $w' \in W_J$, then

$$(\heartsuit) \qquad \ell(ww') = \ell(w) + \ell(w').$$

Indeed, $ws > w$ for all $s \in J$ (by ($\diamondsuit$)). So $w \in X_J$ and the result follows from Corollary 3.6.8.

As a consequence, if $s_1, s_2, \dots, s_r$ are elements of S such that $s_i \neq s_j$ whenever $i \neq j$, then

$$(\spadesuit) \qquad \ell(s_1 s_2 \cdots s_r) = r.$$

This follows from ($\heartsuit$) by an obvious induction argument. □

Remark 3.6.11. If I is a subset of S such that $W_I = W$, then $I = S$. This follows from the property ($\clubsuit$) of Example 3.6.10. □

The next lemma is very useful (for its proof, see for instance [GePf, Lemma 2.1.2]):

Deodhar's Lemma 3.6.12. *Let $x \in X_I$ and $s \in S$. If $sx \notin X_I$, then there exists a $t \in I$ such that $sx = xt$. In particular, if $sx < x$ then $sx \in X_I$.*

Corollary 3.6.13. *Let s, $t \in S$ and $w \in W$ be such that $\ell(sw) = \ell(wt) = \ell(w) - 1$ and $\ell(swt) = \ell(w)$. Then $swt = w$.*

Proof. By hypothesis, $wt \in X_{\{t\}}$ and $swt \notin X_{\{t\}}$. So, by Deodhar's Lemma, there exists a $u \in \{t\}$ such that $swt = wtu$. But then $u = t$, so $swt = w$. ■

By 3.6.3, the Bruhat order on W_I is the restriction of the Bruhat order of W and

$$(3.6.14) \qquad \textit{if } y \in W_I \textit{ and } x \leqslant y, \textit{ then } x \in W_I.$$

The compatibility between the Bruhat order and the notion of standard parabolic subgroup is enhanced by the next proposition:

Proposition 3.6.15. *Let $I \subset S$, let $x, x' \in X_I$ and let $w, w' \in W_I$. Then:*

(a) *If $x'w' \leqslant xw$, then $x' \leqslant x$.*
(b) *$xw \leqslant xw'$ if and only if $w \leqslant w'$.*

Proof. (a) If $x'w' \leqslant xw$, then there exist $\xi \leqslant x$ and $\omega \leqslant w$ such that $x'w' = \xi\omega$. By Corollary 3.6.8, there exist $\xi' \in X_I$ and $\omega' \in W_I$ such that $\xi = \xi'\omega'$: since $\ell(\xi) = \ell(\xi') + \ell(\omega')$, we have $\xi' \leqslant \xi$. But, by 3.6.14, $\omega \in W_I$ and so $x'w' = \xi'(\omega'\omega)$ with $\xi' \in X_I$ and $\omega'\omega \in W_I$. Therefore, $x' = \xi' \leqslant \xi \leqslant x$.

(b) If $w \leqslant w'$, then $xw \leqslant xw'$ because $\ell(xw) = \ell(x) + \ell(w)$ and $\ell(xw') = \ell(x) + \ell(w')$. Conversely, if $xw \leqslant xw'$, then the proof of (a) (from which we take the notation) shows that $x \leqslant \xi \leqslant x$, so $\xi = x$ and so $\omega' = 1$ and $w' = \omega \leqslant w$. ∎

We conclude this section with a result about the normalizer of W_I in W. Let

$$W(I) = \{w \in X_I \mid {}^w I = I\}.$$

Then [GePf, Proposition 2.1.15(b)]

(3.6.16) *$W(I)$ is a subgroup of W and $N_W(W_I) = W(I) \ltimes W_I$.*

3.7. Longest Element, Left and Right Descent Set

A proof of the next proposition may be found in [GePf, Proposition 1.5.1 and Lemma 1.5.3]:

Proposition 3.7.1. *The following are equivalent:*

(1) *W is finite.*
(2) *ℓ is bounded.*
(3) *There exists an element $w_0 \in W$ such that $sw_0 < w_0$ for all $s \in S$.*

If these conditions hold, then w_0 is unique, satisfies $w_0^2 = 1$ and

$$\ell(w_0 w) = \ell(w w_0) = \ell(w_0) - \ell(w)$$

for all $w \in W$.

Corollary 3.7.2. *Assume that W is finite and let $x, y \in W$. Then $x \leqslant y$ if and only if $w_0 y \leqslant w_0 x$ (resp. $y w_0 \leqslant x w_0$).*

Proof. See, for instance, [Lus20, Lemma 11.3]. ∎

Definition 3.7.3. *A subset I of S is called* ***spherical*** *if W_I is finite. If S itself is spherical, we say that (W,S) is a* ***spherical Coxeter system****. The set of spherical subsets of S is denoted by $\mathscr{P}_{\mathrm{sph}}(S)$.*

If I is a spherical subset of S, then the unique element of maximal length of W_I (according to Proposition 3.7.1) is called the ***longest element*** *of W_I: it is denoted by w_I. If W itself is finite, we follow tradition and denote by w_0 the element w_S.*

If $w \in W$, we set

$$\mathscr{R}(w) = \{s \in S \mid ws < w\}$$

and

$$\mathscr{L}(w) = \{s \in S \mid sw < w\}.$$

The subset $\mathscr{R}(w)$ (resp. $\mathscr{L}(w)$) of S is called the *right descent set* (resp. *left descent set*) of w. Note the following immediate properties:

$$\mathscr{L}(w) = \mathscr{R}(w^{-1}) \quad \text{and} \quad \mathscr{R}(w) \subset \mathrm{supp}(w). \tag{3.7.4}$$

Consequently,

$$\mathscr{R}(w) \text{ and } \mathscr{L}(w) \text{ are finite subsets of } S. \tag{3.7.5}$$

Lemma 3.7.6. *Let $w \in W$. Then $\mathscr{R}(w)$ and $\mathscr{L}(w)$ are spherical subsets of S. Moreover, if $z \in W_{\mathscr{R}(w)}$ (resp. $z \in W_{\mathscr{L}(w)}$), then $\ell(wz) = \ell(w) - \ell(z)$ (resp. $\ell(zw) = \ell(w) - \ell(z)$).*

Proof. Let $I = \mathscr{R}(w)$ and write $w = xy$, with $x \in X_I$ and $y \in W_I$. Then, for all $s \in I$,

$$\ell(x) + \ell(y) - 1 = \ell(w) - 1 = \ell(ws) = \ell(xys) = \ell(x) + \ell(ys).$$

So $\ell(ys) = \ell(y) - 1$ for all $s \in I$, which means that I is spherical and $y = w_I$ by Proposition 3.7.1. Now the result on the length function follows again from Proposition 3.7.1. ■

3.8. More on Reflections

The next result is proved in [Spr2, Proposition 1] and [Dye, Lemma 2.7].

Proposition 3.8.1. *Let $t \in T$ and let $(s_1, s_2, \ldots, s_{2r+1})$ be a reduced expression of t (recall that $\ell(t)$ is odd by 3.2.4). Then $t = (s_1 s_2 \cdots s_r) s_{r+1} (s_1 s_2 \cdots s_r)^{-1}$ (in other words, $(s_1, s_2, \ldots, s_r, s_{r+1}, s_r, \ldots, s_2, s_1)$ is also a reduced expression of t).*

Proof. By Corollary 3.3.6, $t \in N(t)$ so, by Lemma 3.3.8, there exists a j in $\{1,2,\dots,2r+1\}$ such that

$$t=(s_1s_2\cdots s_{j-1})s_j(s_1s_2\cdots s_{j-1})^{-1}.$$

But,

$$1=t^2=(s_1s_2\cdots s_{j-1})s_j(s_1s_2\cdots s_{j-1})^{-1}\cdot s_1s_2\cdots s_{2r+1}=s_1s_2\cdots s_{j-1}s_{j+1}s_{j+2}\cdots s_{2r+1}.$$

In other words,

$$s_1s_2\cdots s_{j-1}=s_{2r+1}s_{2r}\cdots s_{j+1}.$$

By comparing the length of both sides, we obtain that $j-1=2r+1-j$, that is, $j=r+1$, as desired. ∎

We conclude with an easy result on (partial) centralizers of reflections. If $s\in S$, we set

$$s^{\perp}=\{t\in S \mid st=ts\}.$$

Proposition 3.8.2. *If $s\in S$ and I is a subset of S not containing s, then*

$$C_{W_I}(s)=W_{I\cap s^{\perp}}.$$

Proof. It is clear that $W_{I\cap s^{\perp}}\subset C_{W_I}(s)$. Conversely, let $w\in C_{W_I}(s)$. We will prove by induction on $\ell(w)$ that $w\in W_{I\cap s^{\perp}}$ (note that this is obvious whenever $\ell(w)=0$). So let $(s_1,s_2,\dots,s_r)$ be a reduced expression of w. It follows from Example 3.6.10 that $s_1s_2\cdots s_{r-1}s > s_1s_2\cdots s_{r-1}$ (since $s\notin I$) and, by hypothesis, $s_1s_2\cdots s_{r-1}s_r > s_1s_2\cdots s_{r-1}$. So $s_1s_2\cdots s_{r-1}\in X_{\{s_r,s\}}$ and so $\ell(s_1s_2\cdots s_{r-1}x)=r-1+\ell(x)$ for all $x\in W_{s_r,s}$. As $s_1s_2\cdots s_{r-1}(s_rss_r)=ss_1s_2\cdots s_{r-1}$, this forces $\ell(s_rss_r)=1$, which can only occur if $s_rs=ss_r$. So $s_r\in I\cap s^{\perp}$ and we conclude easily by using the induction hypothesis. ∎

3.9. Conjugacy Classes of Involutions

We aim to describe in this section the conjugacy classes of involutions in W. Let $\mathscr{P}_{\mathrm{sph}}^{\flat}(S)$ denote the set of $I\in\mathscr{P}_{\mathrm{sph}}(S)$ such that w_I is central in W_I. The following result is due to Richardson [Ric].

Theorem 3.9.1 (Richardson). *We have:*

(a) *If w is an involution which has minimal length in its conjugacy class, then there exists an $I\in\mathscr{P}_{\mathrm{sph}}^{\flat}(S)$ such that $w=w_I$.*

(b) *Let I and J in $\mathscr{P}_{\mathrm{sph}}^{\flat}(S)$. Then w_I and w_J are conjugate in W if and only if I and J are conjugate in W.*

Proof. (a) Let w be an involution which has minimal length in its conjugacy class. Let $I = \mathscr{R}(w)$. Since w is an involution, $I = \mathscr{L}(w)$. So, if $s \in I$, then $ws, sw < w$. But sws is conjugate to w, so $sws > ws$ by the minimality of the length of w in its conjugacy class. So $ws \in X_s$ and $sws \notin X_s$. It follows from Deodhar's Lemma 3.6.12 that $sws = wss = w$.

In particular, $ww_I = w_I w$. Now, let $x = ww_I = w_I w$, so that $x \in X_I$ by Lemma 3.7.6 and so $\ell(w) = \ell(x) + \ell(w_I)$. Moreover, x is an involution, so $\mathscr{L}(x) = \mathscr{R}(x)$. Assume that $x \neq 1$. Then $\mathscr{L}(x) \neq \varnothing$ so we can pick $s \in \mathscr{L}(x)$. By Deodhar's Lemma 3.6.12, $sx \in X_I$, so

$$\ell(sw) = \ell(sxw_I) = \ell(sx) + \ell(w_I) = \ell(x) - 1 + \ell(w_I) = \ell(w) - 1.$$

This implies that $s \in \mathscr{L}(w) = I$, which is impossible because $s \in \mathscr{L}(x) = \mathscr{R}(x)$ and $x \in X_I$. So $\mathscr{L}(x) = \varnothing$, which means that $x = 1$ and $w = w_I$.

It remains to prove that w_I is central in W_I, but we have proved that $w = w_I$ commutes with all the elements of I, so (a) follows.

(b) Let I and J in $\mathscr{P}^{\flat}_{\text{sph}}(S)$. Let us first assume that w_I and w_J are conjugate in W. Let $x \in W$ be such that $w_I = xw_J x^{-1}$. Let d be the element of minimal length in $W_I x W_J$. Since w_I is central in W_I and w_J is central in W_J, we get $w_I = dw_J d^{-1}$. This means that we may assume that $x = d$ or, in other words, that $x \in X_{IJ}$.

Now let $s \in I$. We want to prove that $x^{-1}sx \in J$. But

$$\ell(sw_I x) = \ell(sw_I) + \ell(x) = \ell(w_I) - 1 + \ell(x) = \ell(w_I x) - 1,$$

so $\ell(sxw_J) = \ell(xw_J) - 1$. But $sx > x$ because $x \in X_{IJ}$, so this means that $sx \notin X_J$. By Deodhar's Lemma 3.6.12, we get that $x^{-1}sx \in J$, as desired. So we have proved that $x^{-1}Ix \subset J$, but we can prove similarly that $xJx^{-1} \subset I$. In other words, $I = xJx^{-1}$.

Conversely, if I and J are conjugate in W, then there exists an $x \in W$ such that $I = xJx^{-1}$. So the length function on W_J is transported through x to the length function on W_I, and so conjugacy by x sends the longest element of W_J to the longest element of W_I. ■

Remark 3.9.2. One can also provide an algorithm to determine when two subsets I and J of S are conjugate in W: together with Theorem 3.9.1, this provides a classification of conjugacy classes of involutions in W, which we will not need in this book (we will only need this result in "small" examples, where this algorithm becomes trivial). □

3.10. Fixed Points under Automorphisms

Hypothesis and notation. *We fix a group G acting on W by group automorphisms such that $\sigma(S)=S$ for all $\sigma\in G$. We denote by $(S/G)_{\mathrm{sph}}$ the set of orbits $I\in S/G$ such that W_I is finite. Let*

$$S_G=\{w_I \mid I\in(S/G)_{\mathrm{sph}}\},$$

and let W^G denote the fixed point subgroup under the action of G.

Our aim is to prove that (W^G,S_G) is a Coxeter system. First of all, note that the length function is constant on the G-orbits. Therefore, if $I\in(S/G)_{\mathrm{sph}}$, then W_I is G-stable and the characterization of w_I as *the* longest element of W_I shows that $w_I\in W_I^G\subset W^G$. In other words,

$$S_G\subset W^G. \tag{3.10.1}$$

Also, it follows from Proposition 3.7.1 that

$$\forall\ s\in S_G,\ s^2=1. \tag{3.10.2}$$

We start with an easy lemma:

Lemma 3.10.3. *Let $w\in W^G$ and let $I\in(S/G)_{\mathrm{sph}}$ be non-empty. Then the following are equivalent:*

(1) $\ell(w_I w)=\ell(w)+\ell(w_I)$.
(2) $\ell(w_I w)\geqslant\ell(w)$.
(3) *There exists an $s\in I$ such that $\ell(sw)>\ell(w)$.*
(4) *For all $s\in I$, $\ell(sw)>\ell(w)$ (in other words, $w\in X_I^{-1}$).*

Also, the following are equivalent:

(1′) $\ell(w_I w)=\ell(w)-\ell(w_I)$.
(2′) $\ell(w_I w)\leqslant\ell(w)$.
(3′) *There exists an $s\in I$ such that $\ell(sw)<\ell(w)$.*
(4′) *For all $s\in I$, $\ell(sw)<\ell(w)$.*

Also, the statement "(2) or (2′)" always holds, so $\ell(w_I w)=\ell(w)\pm\ell(w_I)$.

Remark 3.10.4. Analogous statements hold by replacing left multiplication by w_I or $s\in I$ by right multiplication. □

Proof. The equivalence between (3) and (4) (resp. between (3′) and (4′)) follows from the fact that ℓ is constant on G-orbits and G acts transitively on I. The equivalence between (3) and (1) (resp. between (3′) and (1′)) follows from Theorem 3.6.7 (resp. from Lemma 3.7.6). Since the statement "(3) or (3′)″ is automatically true, this shows the last statement of the lemma, and this also shows that (1) and (2) (resp. (1′) and (2′)) are equivalent. ■

Corollary 3.10.5. *Let $w \in W^G$. Then there exist $s_1, s_2, \dots, s_r$ in S_G such that $w = s_1 s_2 \cdots s_r$ and $\ell(w) = \ell(s_1) + \ell(s_2) + \cdots + \ell(s_r)$. In particular,*

$$W^G = \langle S_G \rangle.$$

Proof. We prove the result by induction on $\ell(w)$. The result is obvious if $\ell(w) = 0$. So assume that $\ell(w) > 0$. Then $\mathscr{L}(w)$ is non-empty and G-stable. Let I be a G-orbit in $\mathscr{L}(w)$: it follows from Lemma 3.7.6 that W_I is finite and that $\ell(w_I w) = \ell(w) - \ell(w_I)$. Since $w_I \in S_G$, the result follows easily from the induction hypothesis. ■

Since W^G is generated by S_G, we can define a map $\ell_G : W^G \to \mathbb{N}$ as the length function on W^G associated with S_G.

Theorem 3.10.6. *The pair (W^G, S_G) is a Coxeter system. Moreover, if $w, w' \in W^G$, then $\ell(ww') = \ell(w) + \ell(w')$ if and only if $\ell_G(ww') = \ell_G(w) + \ell_G(w')$.*

Proof. If $s \in S_G$, we set

$$\mathscr{E}_s = \{w \in W^G \mid \ell(sw) > \ell(w)\}.$$

We will show that the family $(\mathscr{E}_s)_{s \in S_G}$ satisfies the conditions of Corollary 3.1.7.

First of all, it is clear that $1 \in \mathscr{E}_s$. Also, it follows from Lemma 3.10.3 that $\mathscr{E}_s \cap s\mathscr{E}_s = \varnothing$. This shows the property (2.1) of Corollary 3.1.7.

Now, let $s' \in S_G$ and let $w \in \mathscr{E}_s$ be such that $ws' \notin \mathscr{E}_s$. Then, again by Lemma 3.10.3, we get

$$\ell(sw) = \ell(s) + \ell(w) \qquad \text{and} \qquad \ell(sws') = \ell(ws') - \ell(s).$$

But $\ell(sws') \geqslant \ell(sw) - \ell(s')$, so this implies that $\ell(ws') \geqslant \ell(w) + 2\ell(s) - \ell(s') > \ell(w) - \ell(s')$. So it follows from Lemma 3.10.3 that $\ell(ws') = \ell(w) + \ell(s')$. Let I and I' be the subsets of S such that $s = w_I$ and $s' = w_{I'}$. Let $t \in I$. Then it follows from Lemma 3.10.3 that $w \in X_{I'}$ and $tw \notin X_{I'}$ (indeed, if $tw \in X_{I'}$, then $\ell(tww_{I'}) = \ell(tw) + \ell(w_{I'}) = \ell(t) + \ell(w) + \ell(w_{I'}) > \ell(ww_{I'})$). So, by Deodhar's Lemma 3.6.12, $w^{-1}tw \in I'$. But G acts transitively on I and I' and $w \in W^G$, so $w^{-1}Iw = I'$. This implies that $w^{-1}w_I w = w_{I'}$, that is, $sw = ws'$. This shows the property (2.2) of Corollary 3.1.7.

Therefore, it follows from Corollary 3.1.7 that (W^G, S_G) is a Coxeter system and that

$$\mathscr{E}_s = \{w \in W^G \mid \ell_G(sw) > \ell_G(w)\}$$

for all $s \in S_G$. Using Lemma 3.10.3, we see that $\ell_G(sw) = \ell_G(w) + 1$ is equivalent to $\ell(sw) = \ell(s) + \ell(w)$. An easy induction argument now proves the last statement of the theorem. ■

Corollary 3.10.7. *Let M be a monoid and let $\varphi : W \to M$ be a G-invariant braided map. Then the restriction of φ to W^G is a braided map (with respect to the Coxeter system (W^G, S_G)).*

This corollary applies, for instance, to the length function $\ell : W \to \mathbb{N}$.

3.11. Geometric Representation

3.11.A. Coxeter Matrix, Coxeter Graph

We follow [Bou, Chapter 4, §1.9, Definition 4]:

Definition 3.11.1. *Let X be a set. A **Coxeter matrix (of type X)** is a matrix $(\mu_{xy})_{x,y\in X}$ such that*

$$\begin{cases} \mu_{xy} \in \mathbb{N}^* \,\dot{\cup}\, \{\infty\}, \\ \mu_{xx} = 1, \\ \mu_{xy} = \mu_{yx}, \\ x \neq y \Longrightarrow \mu_{xy} \geqslant 2, \end{cases}$$

for all $x, y \in X$.

By 3.1.1, the matrix $(m_{st})_{s,t\in S}$ is a Coxeter matrix of type S.

Definition 3.11.2. *Let X be a set. A **Coxeter graph (of type X)** is a pair (Γ, f), where Γ is a graph whose vertex set is X and f is a map from the set of edges of Γ to $\{3,4,5,\ldots\} \,\dot{\cup}\, \{\infty\}$. The number $f(e)$ is called the **label** of the edge e.*

In other words, a Coxeter graph is an edge-labelled graph, the set of possible labels being equal to $\{3,4,5,\ldots\} \,\dot{\cup}\, \{\infty\}$. If (Γ, f) is a Coxeter graph of type X, we associate a Coxeter matrix $(\mu_{xy})_{x,y\in X}$ of type X as follows: we define $\mu_{xx} = 1$, $\mu_{xy} = 2$ if x and y are not connected by an edge and $\mu_{xy} = f(e)$ if x and y are related by the edge e. This clearly defines a bijection between the set of Coxeter graphs of type X and the set of Coxeter matrices of type X.

Definition 3.11.3. *We define the **Coxeter graph** of (W,S), and we denote by $\mathscr{C}ox(W,S)$ the Coxeter graph associated with the Coxeter matrix $(m_{st})_{s,t\in S}$. It is a Coxeter graph of type S.*

Following tradition, a Coxeter graph is drawn using the following convention: the number $f(e)$ is put on the edge e except if $f(e)=3$ (in which case, nothing is written). Also, if $f(e)=4$, the edge e may be replaced by a double edge while, if $f(e)=6$, the edge e may be replaced by a triple edge: in each case, the number $f(e)$ is omitted.

Example 3.11.4. Assume that $|X|=4$ and $X=\{x_1,x_2,x_3,x_4\}$ and let (Γ,f) be the Coxeter graph associated with the Coxeter matrix

$$M=\begin{pmatrix}1&3&4&2\\3&1&5&6\\4&5&1&\infty\\2&6&\infty&1\end{pmatrix},$$

where the (i,j)-entry is $\mu_{x_ix_j}$. Then the graphs

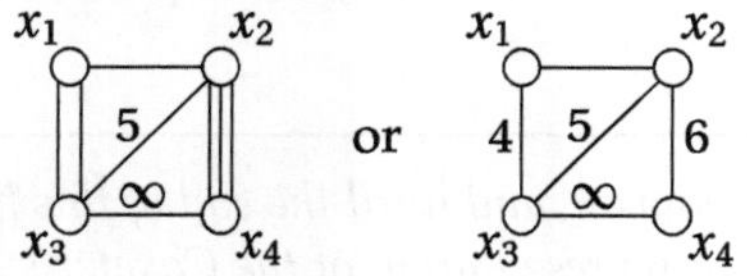

both represent (Γ,f). □

3.11.B. Reflection Representation

If $M=(\mu_{xy})_{x,y\in X}$ is a Coxeter matrix, then we define a symbol s_x associated to each $x\in X$ and we denote by $\mathscr{S}(M)$ the set $\{s_x \mid x\in X\}$, which is in bijection with X. We then denote by $\mathscr{W}(M)$ the group whose presentation is given by:

$$\begin{cases}\text{Generators:} & \mathscr{S}(M),\\ \text{Relations:} & \forall x,y\in X,\ (s_xs_y)^{\mu_{xy}}=1.\end{cases}$$

Let E_X be the $\mathbb{R}$-vector space endowed with a basis $(e_x)_{x\in X}$ and we define on E_X a symmetric bilinear form B_M by:

$$\forall x,y\in X,\ B_M(e_x,e_y)=-\cos\Big(\frac{\pi}{\mu_{xy}}\Big),$$

with $\pi/\infty=0$. As $B_M(e_x,e_x)=1$, we can define $\sigma_x\in\mathbf{O}(E_X,B_M)$ to be the B_M-orthogonal reflection such that $\sigma_x(e_x)=-e_x$.

Proposition 3.11.5. *The map $\mathscr{S}(M)\to\mathbf{O}(E_X,B_M)$, $s_x\mapsto\sigma_x$, is injective and, if x, $y\in X$, then the order of $\sigma_x\sigma_y$ (and of s_xs_y) is μ_{xy}. In particular, $(\mathscr{W}(M),\mathscr{S}(M))$ is a Coxeter system.*

Proof. See [Bou, Chapter V, §4.3, Proposition 4]. ■

This shows that every Coxeter matrix corresponds to a Coxeter system. Also, the map $\mathscr{S}(M) \to \mathbf{O}(E_X, B_M)$, $s_x \mapsto \sigma_x$, can be extended uniquely to a morphism of groups $\mathscr{W}(M) \to \mathbf{O}(E_X, B_M)$, $w \mapsto \sigma(w)$.

Theorem 3.11.6 (Tits). *The morphism $\sigma : \mathscr{W}(M) \to \mathbf{O}(E_X, B_M)$ is injective.*

Proof. See [Bou, Chapter V, §4.4, Theorem 1]. ∎

Definition 3.11.7. *The vector space E_X, endowed with the symmetric bilinear form B_M and with the (faithful) action of $\mathscr{W}(M)$ through σ, is called the* ***reflection representation*** *of the Coxeter system $(\mathscr{W}(M), \mathscr{S}(M))$.*

Remark 3.11.8 (Classification). Two Coxeter matrices $M = (\mu_{xy})_{x,y\in X}$ and $M' = (\mu'_{x'y'})_{x',y'\in X'}$ are called *equivalent* if there exists a bijection $\varphi : X \xrightarrow{\sim} X'$ such that $\mu'_{\varphi(x)\varphi(y)} = \mu_{xy}$ for all x, $y \in X$. The results in this section show that the Coxeter systems are classified by the equivalence classes of Coxeter matrices. □

Notation. *From now on, and until the end of this book, we denote by E the reflection representation of the Coxeter system (W, S). It admits $(e_s)_{s\in S}$ as a basis and the symmetric bilinear form is denoted by B_W: it satisfies*

$$B_W(e_s, e_t) = -\cos\left(\frac{\pi}{m_{st}}\right)$$

for all $s, t \in S$.

3.11.C. Irreducibility

Definition 3.11.9. *The Coxeter system (W, S) is said to be* ***irreducible*** *if its associated Coxeter graph $\mathscr{C}ox(W, S)$ is connected.*

In other words, (W, S) is irreducible if and only if there does not exist a partition $S = S_1 \dot\cup S_2$ where S_1, S_2 are non-empty and $m_{s_1 s_2} = 2$ for all $s_1 \in S_1$ and $s_2 \in S_2$. We will see in Exercise 3.2 that a decomposition of the Coxeter graph of (W, S) is equivalent to a decomposition of the group W as a direct product of Coxeter groups: it allows us to reduce most of the problems about Coxeter systems to the case where the Coxeter system is irreducible.

Remark 3.11.10. Assume that (W, S) is irreducible. Then it might happen that its reflection representation E constructed in §3.11.B is non-irreducible (this can occur only if W is infinite). It is, however, always indecomposable [Bou, Chapter V, §4.7, Proposition 7]. □

3.12. Some Classification Results

3.12.A. *Finiteness*

Recall the following result [Bou, Chapter V, §4.8, Theorem 2 and Chapter VI, §4.1, Theorem 1]:

Theorem 3.12.1. *The following are equivalent:*

(1) *The group W is finite.*
(2) *S is finite and the symmetric bilinear form B_W is positive definite.*
(3) *S is finite and all the irreducible components of $\mathscr{C}ox(W,S)$ are of type A_n $(n \geqslant 1)$, B_n $(n \geqslant 2)$, D_n $(n \geqslant 4)$, E_6, E_7, E_8, F_4, H_3, H_4 or $I_2(m)$ $(5 \leqslant m < \infty)$, as given in Table 3.12.3, where the index in the name gives the number of nodes.*

***Commentary* 3.12.2.** To simplify some statements or some arguments, it is usual to set $B_1 = A_1$, $D_2 = A_1 \times A_1$, $D_3 = A_3$, $I_2(2) = A_1 \times A_1$, $I_2(3) = A_3$, $I_2(4) = B_2$, $H_2 = I_2(5)$ and $G_2 = I_2(6)$. If Δ is a Coxeter graph, then we denote by $W(\Delta)$ the Coxeter system associated with Δ.

In Chapter 21, we will study the example of groups of type $I_2(m)$.

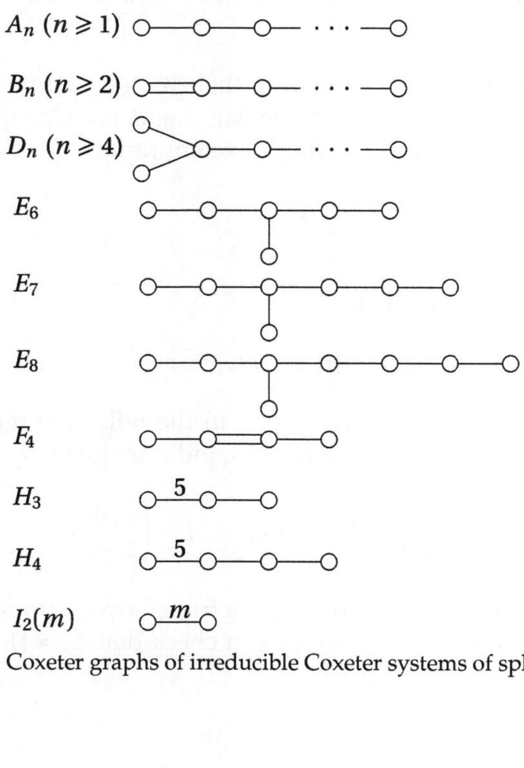

Table 3.12.3. Coxeter graphs of irreducible Coxeter systems of spherical type

Throughout this book, we will illustrate most of the notions involved with the easy examples given by type A_2 (see Examples 3.12.7, 4.4.13, 5.7.9, 6.7.6 and 8.4.2) and type B_2 (see Examples 3.12.8, 4.4.14, 5.7.10, 6.7.7 and 8.4.3), while the type $G_2 = I_2(6)$ will always be left as an exercise for the reader (see Exercises 3.1, 4.1, 5.1, 6.1 and 8.3). □

Definition 3.12.4. *The Coxeter graph of (W,S) is called a* ***forest*** *if it does not contain a loop, that is, if there do not exist n distinct elements $s_1, s_2, \ldots, s_n$ in S with $n \geqslant 3$ such that $m_{s_i s_{i+1}} \geqslant 3$ for $1 \leqslant i \leqslant n-1$ and $m_{s_n s_1} \geqslant 3$. It is called a* ***tree*** *if it is a connected forest.*

It is immediately checked using the classification of spherical Coxeter systems given in Theorem 3.12.1 that the following holds (it may also be checked without using the classification: see [Bou, Chapter V, §4.8, Corollary of Proposition 8]):

Proposition 3.12.5. *If W is finite, then $\mathscr{C}ox(W,S)$ is a forest.*

Finally, let us recall the following result about forests [Bou, Chapter IV, Annex, §3, Proposition 2]:

Proposition 3.12.6. *Assume that $|S| \geqslant 2$ and that $\mathscr{C}ox(W,S)$ is a forest. Then there exists a partition $S = S' \dot\cup S''$ with S' and S'' non-empty such that, if $s', t' \in S'$, then $m_{s't'} = 2$ and, if $s'', t'' \in S''$, then $m_{s''t''} = 2$.*

Example 3.12.7 (Type A_2). Assume here that (W,S) is of type A_2. This means that $S = \{s,t\}$ with $m_{st} = 3$ (in particular, s and t are conjugate in W, as $t = ststs$). Here is the corresponding Coxeter graph:

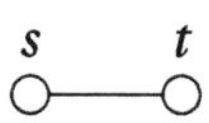

Then it is readily checked that

$$W = \{1, s, t, st, ts, sts\}$$

so that W is finite and $w_0 = sts = tst$. In the reflection representation of (W,S), with basis (e_s, e_t), the matrices of s and t are given by

$$s = \begin{pmatrix} -1 & 1 \\ 0 & 1 \end{pmatrix} \quad \text{and} \quad t = \begin{pmatrix} 1 & 0 \\ 1 & -1 \end{pmatrix}.$$

Of course, W is isomorphic to the symmetric group $\mathfrak{S}_3$ (by sending s and t to two different transpositions). One can check that $X_s = \{1, t, st\}$ and $X_t = \{1, s, ts\}$. Finally, note that $Z(W) = 1$. □

Example 3.12.8 (Type B_2). Assume here that (W,S) is of type B_2. This means that $S=\{s,t\}$ with $m_{st}=4$ (in particular, s and t are not conjugate in W by Lemma 3.2.7). Here is the corresponding Coxeter graph:

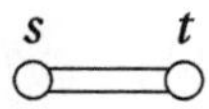

Then it is readily checked that

$$W=\{1,s,t,st,ts,sts,tst,stst\}$$

so that W is finite and $w_0=stst=tsts$. In the reflection representation of (W,S), with basis (e_s,e_t), the matrices of s and t are given by

$$s=\begin{pmatrix}-1 & \sqrt{2}\\ 0 & 1\end{pmatrix} \qquad \text{and} \qquad t=\begin{pmatrix}1 & 0\\ \sqrt{2} & -1\end{pmatrix},$$

where $\sqrt{2}$ is the positive square root of 2 in $\mathbb{R}$. Then W is isomorphic to the dihedral group of order 8. One can check that $X_s=\{1,t,st,tst\}$ and $X_t=\{1,s,ts,sts\}$. Finally, note that $\mathrm{Z}(W)=\langle w_0\rangle\simeq\mathbb{Z}/2\mathbb{Z}$.

Also, one can check that $C_W(s)=\langle s,tst\rangle$ so that $C_W(s)\neq W_{s^\perp}$ (compare with Proposition 3.8.2). □

3.12.B. Tame Groups

Definition 3.12.9. *We say that the Coxeter system (W,S) is* ***tame*** *if $B_W(e,e)\geqslant 0$ for all $e\in E$. If (W,S) is tame and all its irreducible components are infinite, we say that W is an* ***affine Weyl group****.*

In particular, if (W,S) is spherical, then (W,S) is tame. As for the classification of spherical Coxeter systems, the classification of tame Coxeter systems can easily be reduced to the irreducible case, and it is then given by the following theorem [Bou, Chapter V, §4.9, Proposition 10 and Chapter VI, §2.5, Proposition 8].

Theorem 3.12.10. *If (W,S) is irreducible and infinite, then (W,S) is tame (hence, affine) if and only if its Coxeter graph belongs to the list given in Table 3.12.11, where the number of nodes is one more than the index in the name.*

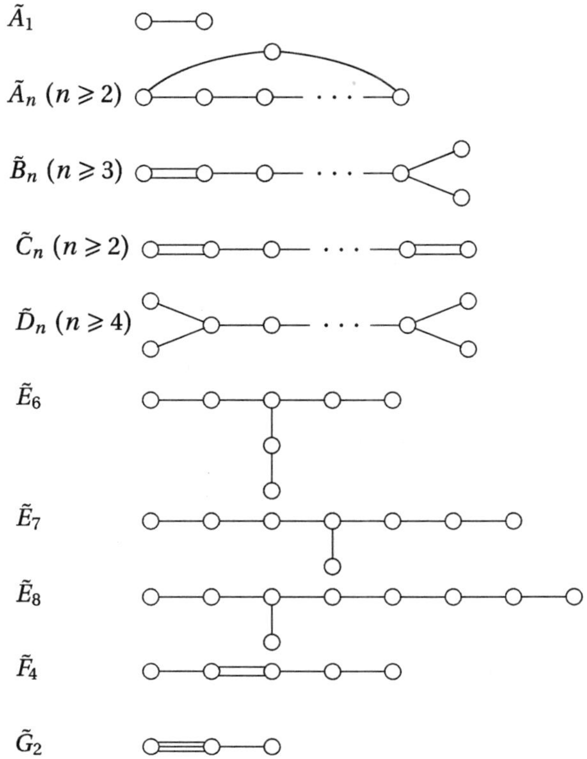

Table 3.12.11. Coxeter graphs of tame irreducible infinite Coxeter systems

3.12.C. *Hyperbolic Coxeter Systems*

Definition 3.12.12. *If S is finite of cardinality n, we say that (W,S) is a* ***hyperbolic Coxeter system*** *if the bilinear form B_W has signature $(n-1,1)$.*

There is a classification of hyperbolic Coxeter systems, but we do not recall it here. Just notice that there is no hyperbolic Coxeter system of rank 2. We will almost only encounter some hyperbolic groups of rank 3, coming from the following example (see [Bou, Chapter V, §4, Exercise 4]).

Example 3.12.13. Assume that $|S|=3$ and that (W,S) is irreducible. We write $S=\{s,t,u\}$ and we define

$$\xi = \frac{1}{m_{st}} + \frac{1}{m_{su}} + \frac{1}{m_{tu}}.$$

Then [Bou, Chapter IV, §4, Exercise 4]:

- If $\xi > 1$, then (W,S) is spherical.
- If $\xi = 1$, then W is an affine Weyl group.
- If $\xi < 1$, then (W,S) is hyperbolic.

This gives infinitely many examples of hyperbolic Coxeter systems. □

3.13. Tits Cone, Coxeter Complex

Hypothesis. *We assume in this section that S is finite.*

Recall that E denotes the $\mathbb{R}$-vector space affording the reflection representation of W as a reflection group (see §3.11.B). If X is a subset of the dual E^* of E, we denote by $\overline{X}$ its (topological) closure in E. Recall that W acts on E^* by the contragredient representation. We set $\mathfrak{C}_1 = \{x^* \in E^* \mid \forall\ s \in S,\ x^*(e_s) > 0\}$. Then $\mathfrak{C}_1$ is called the ***fundamental chamber***. Let

$$\mathscr{T}\!its(W,S) = \bigcup_{w \in W} w(\overline{\mathfrak{C}}_1).$$

Then $\mathscr{T}\!its(W,S)$ is called the *Tits cone* of (W,S). It follows from [Bou, Chapter V, §4.4, Theorem 1] that $\overline{\mathfrak{C}}_1$ is a fundamental domain for the action of W on $\mathscr{T}\!its(W,S)$. We then define

$$\Sigma(W,S) = (\mathscr{T}\!its(W,S) \setminus \{0\})/\mathbb{R}^{\times}_{>0}.$$

Then $\Sigma(W,S)$ is called the *Coxeter complex* of (W,S): it is acted on by W. We denote by $\mathfrak{C}_1^{\Sigma}$ the image of $\mathfrak{C}_1$ in $\Sigma(W,S)$: then $\overline{\mathfrak{C}}_1^{\Sigma}$ is a fundamental domain for the action of W on the Coxeter complex $\Sigma(W,S)$.

Theorem 3.13.1. *Assume that (W,S) is irreducible. Then:*

(a) *If (W,S) is spherical (i.e. if W is finite), then $\Sigma(W,S)$ is homeomorphic to a sphere of dimension $|S|-1$.*
(b) *If (W,S) is an affine Weyl group, then $\Sigma(W,S)$ is homeomorphic to an affine space of dimension $|S|-1$.*
(c) *If (W,S) is hyperbolic of rank 3, then $\Sigma(W,S)$ is homeomorphic to the Poincaré disk.*
(d) *If W is infinite, then $\Sigma(W,S)$ is contractible.*
(e) *W is finite if and only if $\mathscr{T}\!its(W,S) = E^*$.*

Proof. See, for instance, [Dav, Theorems 6.8.12, C.1.2, C.1.3, C.1.4]. ■

3.14. Drawing Coxeter Groups

3.14.A. *Rank* 2

We assume here, and only here, that $|S|=2$ and W is finite. We then identify E with its dual E^*, through the non-degenerate symmetric bilinear form B_W. Since $\overline{\mathfrak{C}}_1$ is a fundamental domain for the action of W on $\mathscr{T}its(W,S)=E$, this means that we can identify the set of elements of W and the family of chambers $(\mathfrak{C}_w)_{w\in W}$, where $\mathfrak{C}_w = w^{-1}(\mathfrak{C}_1)$. For instance, we will draw in Figures 3.a and 3.b the set of elements of Coxeter groups of type A_2 and B_2 as follows (following the notation of Examples 3.12.7 and 3.12.8).

3.14.B. *Rank* 3

We assume here, and only here, that $|S|=3$. In this case, we will identify the elements of W with the family of chambers $(\mathfrak{C}_w^\Sigma)_{w\in W}$ of the Coxeter complex $\Sigma(W,S)$, where $\mathfrak{C}_w^\Sigma = w^{-1}(\mathfrak{C}_1^\Sigma)$. In case W is infinite, we will draw only the elements of small length.

According to Example 3.12.13, three cases may occur (we use Theorem 3.13.1):

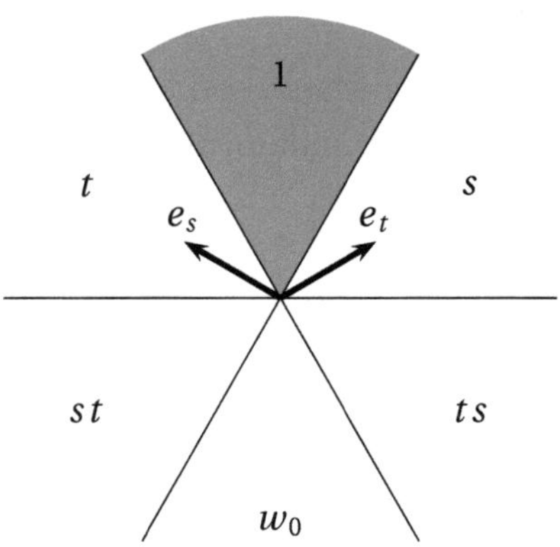

Fig. 3.a. Type A_2

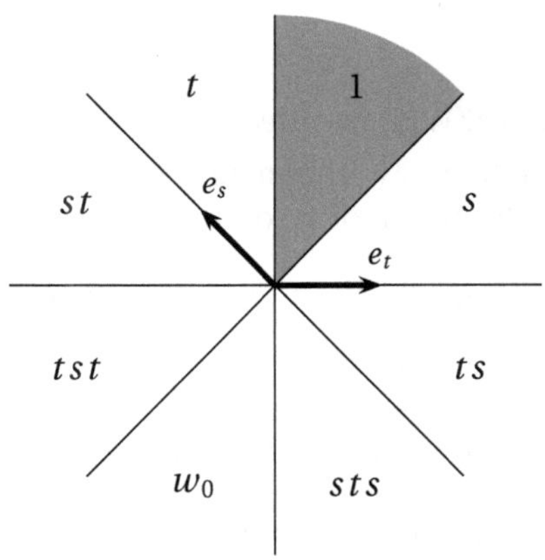

Fig. 3.b. Type B_2

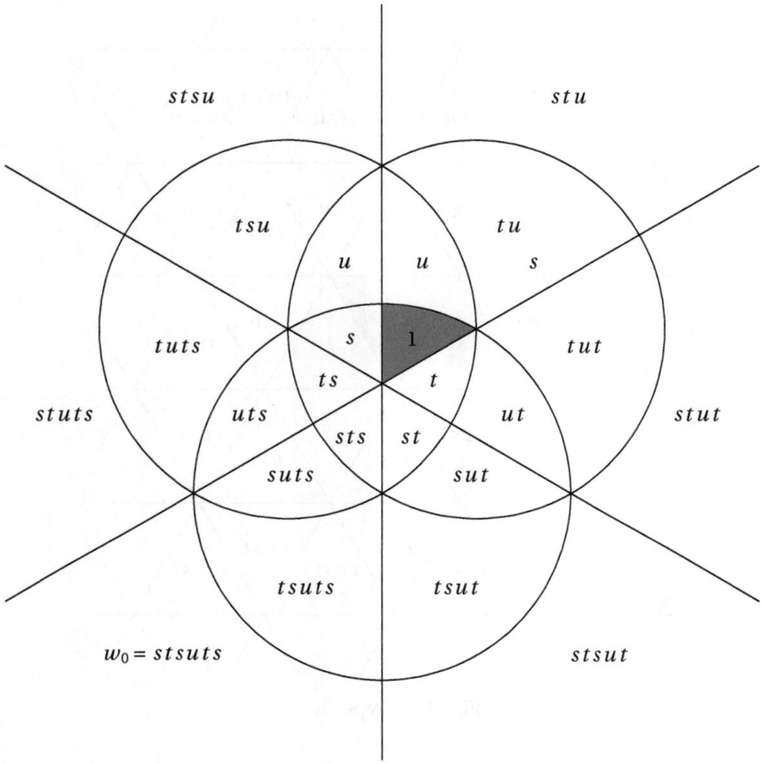

Fig. 3.c. Type A_3

Spherical Case. If W is finite, then $\Sigma(W,S)$ is a sphere of dimension 2, and we will represent $\Sigma(W,S)$ through a stereographic projection. For instance, if (W,S) is of type A_3, if $S=\{s,t,u\}$ and if the Coxeter graph is given by

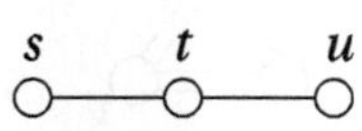

then the elements of W are given in Figure 3.c below.

Tame Case. If W is tame, then $\Sigma(W,S)$ is an affine space of dimension 2, and we will represent it in a natural way. For instance, if (W,S) is of type $\tilde{A}_2$, if $S=\{s,t,u\}$ and if the Coxeter graph is given by

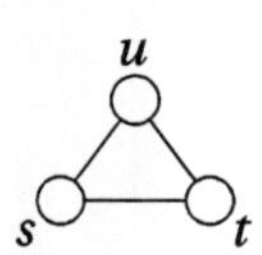

then the first elements of W are given in Figure 3.d below.

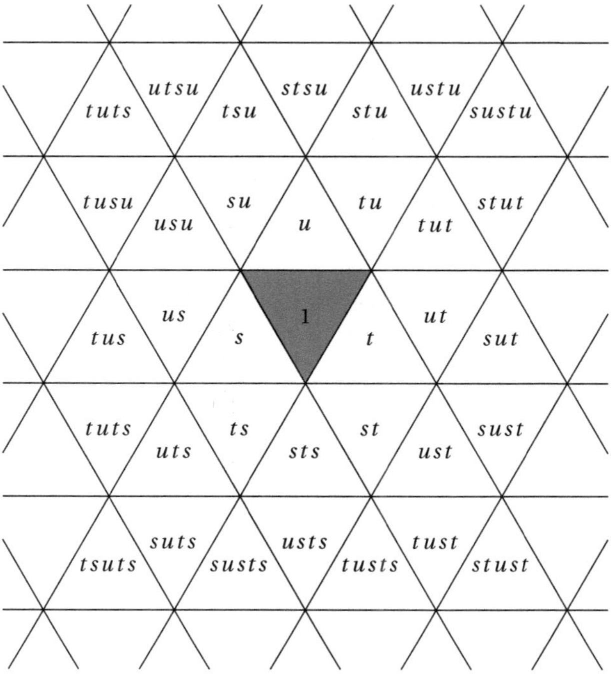

Fig. 3.d. Type $\tilde{A}_2$

Hyperbolic Case. If W is hyperbolic, then $\Sigma(W,S)$ is homeomorphic to the Poincaré disk, and we will represent it in a natural way. For instance, if $S = \{s,t,u\}$ and if the Coxeter graph is given by

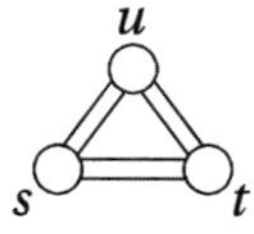

(as $m_{st} = m_{tu} = m_{su} = 4$, we say that (W,S) is of type **Hyp**$(4,4,4)$) then the first elements of W are given in Figure 3.e below.

Remark.— In all cases, we have coloured in gray the chamber of the neutral element: it then follows that the adjacent chambers to the gray chamber correspond to elements of S. □

Notes

As the reader might have noticed, we have mainly followed the treatment of [Bou] or [GePf, Chapters 1 and 2]. An exception is Section 3.10, where

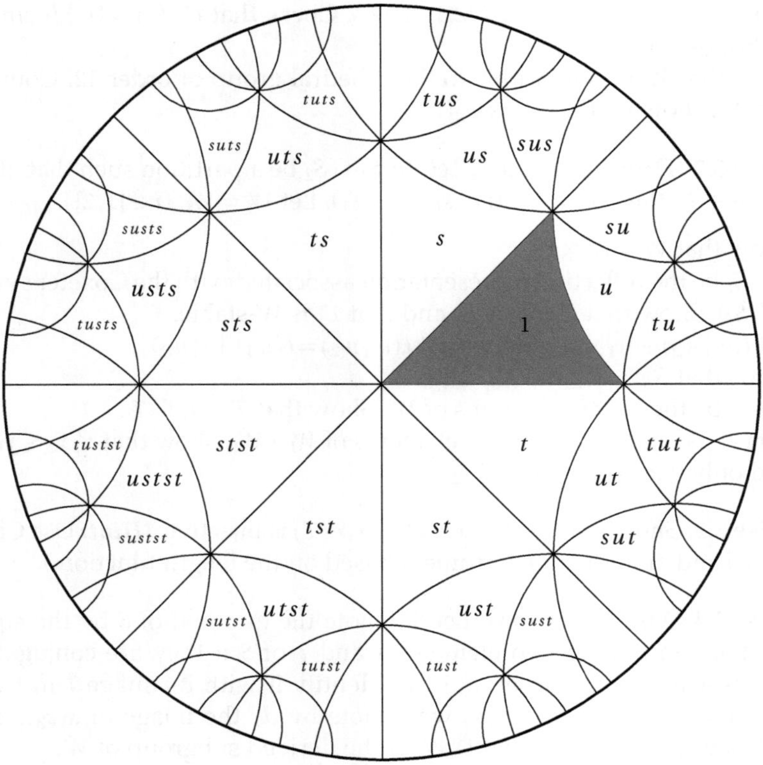

Fig. 3.e. Type **Hyp**(4, 4, 4)

fixed points subgroups under Coxeter system automorphisms are investigated: there are several proofs of Theorem 3.10.6 (see [Ste, §11] for W finite and [Hée, §3], [Lus20, Appendix] or [GeIa2, Theorem 1] for the general case). Our argument, which is based on the study of reduced expressions, mostly follows [GeIa2], but is slightly different.

Exercises

Exercises (or questions) marked with a star are more difficult than the others.

Exercise 3.1 (Type G_2). Assume here that $S = \{s, t\}$ and $m_{st} = 6$, so that (W, S) is of type $G_2 = I_2(6)$. Here is the corresponding Coxeter graph:

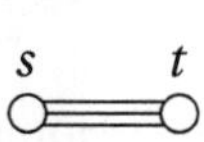

Show that s and t are not conjugate in W, that

$$W = \{1, s, t, st, ts, sts, tst, stst, tsts, ststs, tstst, ststst = tststs = w_0\}$$

and that $C_W(s) = \langle s, tstst \rangle \simeq \mathbb{Z}/2\mathbb{Z} \times \mathbb{Z}/2\mathbb{Z}$. Check that $C_W(s) \neq W_{s^\perp}$ (compare with Proposition 3.8.2).

Show that W is isomorphic to the dihedral group of order 12. Compute X_s and X_t; show that $\mathrm{Z}(W) = \{w_0\}$.

Exercise 3.2 (Direct products). Let $S = S_1 \dot{\cup} S_2$ be a partition such that, if $s_1 \in S_1$ and $s_2 \in S_2$, then $m_{s_1 s_2} = 2$ (i.e. $s_1 s_2 = s_2 s_1$). Let $W_i = W_{S_i}$ $(i \in \{1,2\})$.

(a) Show that $W = W_1 \times W_2$.
(b) Let E_i be the reflection representation associated with the Coxeter system (W_i, S_i). Show that $E \simeq E_1 \oplus E_2$ and that E_i is W-stable.
(c) Let $(w_1, w_2) \in W_1 \times W_2$. Show that $\ell(w_1 w_2) = \ell(w_1) + \ell(w_2)$.
(d) Show that $X_{S_2} = W_1$ and $X_{S_1} = W_2$.
(e) Let T_i be the set of reflections of W_i. Show that $T = T_1 \dot{\cup} T_2$.
(f) Let (x_1, x_2) and (y_1, y_2) be two elements of $W_1 \times W_2$. Show that $x_1 x_2 \leqslant y_1 y_2$ if and only if $x_1 \leqslant y_1$ and $x_2 \leqslant y_2$.

Exercise 3.3. Show that the map $N : W \to \mathscr{P}(T)$ is injective. (***Hint:*** Use Corollary 3.3.5 and an induction argument based on the length function.)

Exercise 3.4 (Abelianization). Let $\bar{S}$ denote the quotient of S by the equivalence relation linking two elements s and t of S if they are conjugate in W. If I is a W-closed subset of S, we identify it with its image $\bar{I}$ in $\bar{S}$ and, if $\bar{I} = \{\bar{s}_1, \bar{s}_2, \ldots, \bar{s}_r\}$ with $r = |\bar{I}|$, we denote by d_I the image of $s_1 s_2 \ldots s_r$ in $W/\mathrm{D}(W)$, where we recall that $\mathrm{D}(W)$ is the derived subgroup of W.

(a) Show that d_I depends only on I and not on the choice of the representatives $s_1, \ldots, s_r$.
(b) We identify $\mathscr{P}(\bar{S})$ with the set of W-closed subsets of S as before. Show that the map

$$\begin{array}{rcl} d : (\mathscr{P}(\bar{S}), \dot{+}) & \longrightarrow & W/\mathrm{D}(W) \\ I & \longmapsto & d_I \end{array}$$

is an isomorphism of groups.

Exercise 3.5 (Linear characters). We keep the notation of Exercise 3.4. Let I be a W-closed subset of S. We define $\varepsilon_I(w) = (-1)^{\ell_I(w)}$. Note that $\varepsilon_\varnothing = \mathbf{1}_W$ (where $\mathbf{1}_W(w) = 1$) and $\varepsilon_S = \varepsilon$.

(a) Show that $\varepsilon_I : W \to \{1, -1\}$ is a morphism of groups.
(b) Show that $\varepsilon_I \varepsilon_J = \varepsilon_{I \dot{+} J}$.
(c) We identify $\mathscr{P}(\bar{S})$ with the set of W-closed subsets of S as in Exercise 3.4. Show that the map

$$\begin{array}{rcl} (\mathscr{P}(\bar{S}), \dot{+}) & \longrightarrow & \mathrm{Hom}(W, \{1, -1\}) \\ I & \longmapsto & \varepsilon_I \end{array}$$

is an isomorphism of groups.

Exercise 3.6 (Bourbaki). Let $W^+ = \operatorname{Ker}(\varepsilon)$. Fix $s \in S$ and, for $t \in S \setminus \{s\}$, set $g_t = ts$. Show that W^+ admits the following presentation:

$$\begin{cases} \text{Generators:} & (g_t)_{t \in S\setminus\{s\}} \\ \text{Relations:} & \forall t \in S \setminus \{s\},\ g_t^{m_{ts}} = 1 \\ & \forall t \neq t' \in S \setminus \{s\},\ (g_{t'} g_t^{-1})^{m_{tt'}} = 1. \end{cases}$$

Exercise 3.7. Let $s, t \in S$ and $x \in W$ be such that $sx < x < tx < stx$. Show that $\mathscr{R}(x) = \mathscr{R}(tx)$. (***Hint:*** Use Deodhar's Lemma 3.6.12 or Corollary 3.6.13.)

Exercise 3.8*. If $\begin{pmatrix} a & b \\ c & d \end{pmatrix} \in \mathbf{GL}_2(\mathbb{Z})$, we denote by $\begin{bmatrix} a & b \\ c & d \end{bmatrix}$ its image in $\mathbf{PGL}_2(\mathbb{Z})$. Note that the determinant $\det : \mathbf{GL}_2(\mathbb{Z}) \to \{1, -1\}$ factorizes through $\det_0 : \mathbf{PGL}_2(\mathbb{Z}) \to \{1, -1\}$. For instance, $\mathbf{PSL}_2(\mathbb{Z}) = \operatorname{Ker}(\det_0)$. Now, let

$$s = \begin{bmatrix} 1 & 0 \\ 0 & -1 \end{bmatrix}, \quad t = \begin{bmatrix} 1 & 0 \\ -1 & -1 \end{bmatrix} \quad \text{and} \quad u = \begin{bmatrix} 0 & 1 \\ 1 & 0 \end{bmatrix}.$$

We denote by I_2 the 2×2 identity matrix.

The aim of this exercise is to show that $(\mathbf{PGL}_2(\mathbb{Z}), \{s, t, u\})$ is a Coxeter system.

(a) Show that $s^2 = t^2 = u^2 = 1$ and that $(su)^2 = (tu)^3 = 1$.
(b) Show that st has infinite order.
(c*) Show that $\mathbf{PGL}_2(\mathbb{Z})$ is generated by s, t and u.

Let (W, S) be the Coxeter system such that $S = \{\mathbf{s}, \mathbf{t}, \mathbf{u}\}$ and $m_{\mathbf{st}} = \infty$, $m_{\mathbf{su}} = 2$ and $m_{\mathbf{tu}} = 3$. In other words, the Coxeter graph $\mathscr{C}ox(W, S)$ is given by

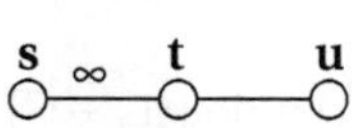

By (a) and (b), there exists a unique morphism of groups $p : W \longrightarrow \mathbf{PGL}_2(\mathbb{Z})$ such that $p(\mathbf{s}) = s$, $p(\mathbf{t}) = t$ and $p(\mathbf{u}) = u$. By (c), it is surjective. It remains to show that p is injective.

For this, let (x, y) be the canonical basis of $\mathbb{R}^2$, let $\mathscr{E} = \mathrm{S}^2(\mathbb{R}^2)$ be the second symmetric power (note that (x^2, xy, y^2) is an $\mathbb{R}$-basis of $\mathscr{E}$), let $\operatorname{disc} : \mathscr{E} \to \mathbb{R}$ be the quadratic form defined by $\operatorname{disc}(ax^2 + 2bxy + cy^2) = b^2 - ac$ (the *discriminant*) and let $\delta : \mathscr{E} \times \mathscr{E} \to \mathbb{R}$ be the associated symmetric bilinear form. Let

$$e_s = 2xy, \quad e_t = -2xy + y^2 \quad \text{and} \quad e_u = x^2 - y^2.$$

Finally, we denote by $\mathrm{S}^2 : \mathbf{GL}_2(\mathbb{R}) \to \mathbf{GL}_{\mathbb{R}}(\mathscr{E})$ the canonical morphism of groups.

(d) Show that $\operatorname{disc}(\mathrm{S}^2(\gamma) f) = \det(\gamma)^2 \operatorname{disc}(f)$ for all $\gamma \in \mathbf{GL}_2(\mathbb{R})$ and $f \in \mathscr{E}$.
(e) Show that $\operatorname{Ker}(\mathrm{S}^2) = \{I_2, -I_2\}$. We denote by $\bar{\mathrm{S}} : \mathbf{PGL}_2(\mathbb{Z}) \to \mathbf{GL}_{\mathbb{R}}(\mathscr{E})$ the factorization of the restriction of S^2 to $\mathbf{GL}_2(\mathbb{Z})$.

(f) Show that the representation $\bar{S} \circ p : W \longrightarrow \mathbf{GL}_{\mathbb{R}}(\mathscr{E})$ is isomorphic to the reflection representation of W.
(g) Deduce that p is injective and that $(\mathbf{PGL}_2(\mathbb{Z}), \{s, t, u\})$ is a Coxeter system.
(h) Show that $\det \circ p = \varepsilon_W$.
(i) Let $a = su$ and $b = tu$. Show that $\mathbf{PSL}_2(\mathbb{Z})$ has the following presentation:

$$\begin{cases} \text{Generators:} & a, b \\ \text{Relations:} & a^2 = b^3 = 1. \end{cases}$$

(***Hint:*** Use Exercise 3.6.)

Exercise 3.9* (Coxeter element). Assume here that S is finite, with cardinality n, and write $S = \{s_1, s_2, \ldots, s_n\}$. Let $c = s_1 s_2 \cdots s_n \in W$ (c is called a *standard Coxeter element*). One of the aims of this exercise is to prove that c has finite order if and only if W is finite‡ (a result which is due to Howlett [Howl, Theorem 4.1]). We also study the eigenvalues of c in the reflection representation.

For simplicity, we set $e_i = e_{s_i} \in E$. Let $A = (a_{ij})_{1 \leqslant i,j \leqslant n}$ be the matrix defined by:

$$a_{ij} = \begin{cases} 1 & \text{if } i = j, \\ 0 & \text{if } 1 \leqslant j < i \leqslant n, \\ 2B_W(e_i, e_j) & \text{if } 1 \leqslant i < j \leqslant n. \end{cases}$$

Let B be the symmetric matrix $\big(2B_W(e_i, e_j)\big)_{1 \leqslant i,j \leqslant n}$. Note that $\det(A) = 1$, that A is invertible and that

$$A + {}^t A = B.$$

Finally, let C denote the matrix of c (for its action on E) in the basis $\mathscr{E} = (e_1, e_2, \ldots, e_n)$ of E and let

$$\chi(\mathbf{t}) = \det(\mathbf{t}\mathrm{I}_n - C)$$

denote its characteristic polynomial.

(a) Show that A^{-1} is the matrix of the linear map $\psi : E \to E$ such that $\psi(e_i) = s_1 s_2 \cdots s_{i-1}(e_i)$ for $1 \leqslant i \leqslant n$.
(b) Show that $-{}^t A$ is the matrix of the linear map $\psi' : E \to E$ defined by $\psi' = c\psi^{-1}$.
(c) Deduce that $C = -{}^t A\, A^{-1}$.
(d) Show that ${}^t C^{-1} = A^{-1} C A$ and that $\mathbf{t}^n \chi(\mathbf{t}^{-1}) = \chi(\mathbf{t})$.
(e) Show that $\chi(\mathbf{t}) = \det(\mathbf{t}A + {}^t A)$.
(f) Deduce that 1 is an eigenvalue of C if and only if B_W is degenerate. If so, show that the multiplicity of 1 as an eigenvalue of C is then even (***Hint:*** use (d)).

‡ This result was first proved by A'Campo, provided that $\mathscr{C}ox(W, S)$ is a forest [A'Ca]. This was then generalized to all Coxeter systems by Howlett [Howl], and improved by Speyer [Spe], who proved that if W is infinite, then $\ell(c^r) = r\ell(c)$ for all $r \geqslant 0$. Note that this exercise is also inspired by [Bou, Chapter V, §6, Exercises 3 and 4].

From now on, we set, for $\lambda \in \mathbb{R}_{\geqslant 0}$,

$$M(\lambda) = (1+\lambda)\mathrm{I}_n - \lambda A - {}^tA.$$

We denote by $\rho(\lambda)$ the *spectral radius* of $M(\lambda)$ (i.e., if $\mu_1, \mu_2, \ldots, \mu_r$ denote the complex eigenvalues of $M(\lambda)$, then $\rho(\lambda) = \max(|\mu_1|, |\mu_2|, \ldots, |\mu_r|)$). Recall that $\rho(\lambda)$ is a continuous function of λ.

(g) Show that the coefficients of $M(\lambda)$ are non-negative.
(h) Show that $\rho(0) = 0$.

Note that, by the *Perron–Frobenius Theorem*, this implies that $\rho(\lambda)$ is an eigenvalue of $M(\lambda)$; moreover, if $M(\lambda)$ is irreducible,[‡] then the multiplicity of $\rho(\lambda)$ as an eigenvalue of $M(\lambda)$ is equal to one.

(h) Assume here that W is not finite (by Theorem 3.12.1, this implies that B_W is not positive definite or, in other words, that $B = A + {}^tA$ admits a non-positive eigenvalue).

 (h1) Show that $\rho(1) \geqslant 2$.
 (h2) Deduce that there exists a $\lambda_0 \in \mathbb{R}$, $0 < \lambda_0 \leqslant 1$, such that $\rho(\lambda_0) = 1 + \lambda_0$.
 (h3) Deduce from (e) that λ_0 is an eigenvalue of C.
 (h4) Prove that c has infinite order. (***Hint:*** if $\lambda_0 = 1$, then, after reduction to the case where (W,S) is irreducible, use (f) and the fact that $\rho(1) = 2$ is an eigenvalue of multiplicity one of $M(1)$ to show that C is not diagonalizable over $\mathbb{C}$.)

From now on, and until the end of this exercise, we assume that $\mathscr{C}ox(W,S)$ is a forest. By Proposition 3.12.6, this implies that there exists a partition $\{1,2,\ldots,n\} = I \mathbin{\dot\cup} J$ such that $m_{s_i s_{i'}} = m_{s_j s_{j'}} = 2$ for all i, $i' \in I$ and j, $j' \in J$. We set $B' = B - 2\mathrm{I}_n$ and

$$P(\mathbf{t}) = \det(\mathbf{t}\mathrm{I}_n + B').$$

For $\alpha \in \mathbb{C}$, let $d(\alpha)$ denote the diagonal matrix whose (i,i)-entry is equal to α if $i \in I$ and to 1 if $i \in J$.

(i) Show that $d(\alpha)(\alpha^2 A + {}^tA)d(\alpha)^{-1} = \alpha\big((\alpha-1)\mathrm{I}_n + {}^tA + A\big)$.
(j) Deduce that B' is equivalent to $-B'$ and that, for $\alpha \neq 0$,

$$\det(\alpha^2\mathrm{I}_n - C) = \alpha^n P(\alpha + \alpha^{-1}).$$

(k) Deduce that, if β is a complex eigenvalue of c which is not real, then $|\beta| = 1$.

[‡] A matrix $M = (m_{i,j})_{1 \leqslant i,j \leqslant n}$ is said to be *irreducible* if there does not exist a non-trivial partition $\{1,2,\ldots,n\} = X \mathbin{\dot\cup} Y$ such that $m_{i,j} = 0$ for all $(i,j) \in X \times Y$.

Chapter 4
Hecke Algebras

The aim of this chapter is to recall the basic properties of Hecke algebras (definition, parabolic subalgebras, involutions, symmetrizing form) in a suitable way with respect to our needs. In particular, the normalization of the parameters used here is tailor-made for applications to Kazhdan–Lusztig theory. As in Chapter 3 on Coxeter groups, most of the results are taken from [Bou, GePf], and we do not recall their proof. Of course, we keep our fixed Coxeter system (W, S).

4.1. Definition, First Properties

We fix in this section a commutative ring A as well as a weight function $\boldsymbol{\alpha} \in \mathrm{Weight}(S, A^2)$. If $s \in S$, we write $\boldsymbol{\alpha}(s) = (\alpha_s, \beta_s)$.

Definition 4.1.1. *The* **Hecke algebra** *(with parameter $\boldsymbol{\alpha}$), which is denoted by $\mathcal{H}_A(W, S, \boldsymbol{\alpha})$, is the A-algebra whose presentation is given as follows:*

$$\begin{cases} \textit{Generators}: & (T_s)_{s\in S} \\ \textit{Relations}: & (Q) \quad \forall s \in S, \quad T_s^2 = \alpha_s T_s + \beta_s, \\ & (B) \quad \forall s, t \in S, \quad \mathrm{Prod}(T_s, T_t; m_{st}) = \mathrm{Prod}(T_t, T_s; m_{st}), \end{cases}$$

where $\mathrm{Prod}(.,.;.)$ *is understood in the multiplicative monoid $(\mathcal{H}_A(W, S, \boldsymbol{\alpha}), \times)$.*

The relations of the form (Q) *(resp.* (B)*) are called the* **quadratic** *(resp.* **braid***) relations.*

C. Bonnafé, *Kazhdan–Lusztig Cells with Unequal Parameters*,
Algebra and Applications 24, https://doi.org/10.1007/978-3-319-70736-5_4

4.1.A. Basis

The braid relations together with Matsumoto's Theorem 3.4.7 show that the map $S \to (\mathscr{H},\times)$, $s \mapsto T_s$, can be extended to a braided map $W \to (\mathscr{H},\times)$, $w \mapsto T_w$. In other words, if $(s_1, s_2, \dots, s_r)$ is a reduced expression of w, then

$$T_w = T_{s_1} T_{s_2} \cdots T_{s_r}.$$

By definition of a braided map,

$$\text{(4.1.2)} \qquad \textit{if } \ell(ww') = \ell(w) + \ell(w'), \textit{ then } T_{ww'} = T_w T_{w'}.$$

Theorem 4.1.3. *The family* $(T_w)_{w \in W}$ *is an* A*-basis of* $\mathscr{H}_A(W,S,\boldsymbol{\alpha})$.

Proof. See [GePf, Theorem 4.4.6] (or [Bou, Chapter IV, §2, Exercise 23]). ■

Definition 4.1.4. *The* A*-basis* $(T_w)_{w \in W}$ *is called the* ***standard*** *basis of* $\mathscr{H}_A(W,S,\boldsymbol{\alpha})$.

Remark 4.1.5. If it is necessary to emphasize in which algebra we are working, we denote by $T_w^{\boldsymbol{\alpha}}$ the element T_w. For instance, if $\theta : A \to B$ is a morphism of rings, then there exists a unique morphism of B-algebras

$$\theta_{\mathscr{H}} : B \otimes_A \mathscr{H}_A(W,S,\boldsymbol{\alpha}) \longrightarrow \mathscr{H}_B(W,S,(\theta \times \theta) \circ \boldsymbol{\alpha})$$

such that $\theta_{\mathscr{H}}(1 \otimes_A T_w^{\boldsymbol{\alpha}}) = T_w^{(\theta \times \theta) \circ \boldsymbol{\alpha}}$. It is an isomorphism of B-algebras (by Theorem 4.1.3). □

The left multiplication by T_s in the A-basis $(T_w)_{w \in W}$ is given by

$$\text{(4.1.6)} \qquad T_s T_w = \begin{cases} T_{sw} & \text{if } sw > w, \\ \alpha_s T_w + \beta_s T_{sw} & \text{if } sw < w. \end{cases}$$

With the notation $\delta_{\mathscr{P}}$ defined in the *General Notation*, this formula can be rewritten in a compact form as follows:

$$\text{(4.1.7)} \qquad T_s T_w = T_{sw} + \delta_{sw<w}(\alpha_s T_w + (\beta_s - 1) T_{sw}).$$

Example 4.1.8. Assume here that $S = \{s\}$ and let $\alpha_s = \alpha$ and $\beta_s = \beta$. Then, if T is an indeterminate over A, the unique morphism of algebras $A[T] \to \mathscr{H}_A(W,S,\boldsymbol{\alpha})$, $T \mapsto T_s$, is surjective and its kernel is, by definition, $T^2 - \alpha T - \beta$. So

$$\mathscr{H}_A(W,S,\boldsymbol{\alpha}) \simeq A[T]/\langle T^2 - \alpha T - \beta \rangle.$$

If the polynomial $T^2 - \alpha T - \beta$ can be factorized in $A[T]$ as $T^2 - \alpha T - \beta = (T - \lambda)(T - \mu)$, with λ, $\mu \in A$, then

$$\mathcal{H}_A(W,S,\boldsymbol{\alpha}) \simeq \begin{cases} A[T]/T^2 & \text{if } \lambda = \mu, \\ \{(a,b) \in A^2 \mid a \equiv b \mod (\lambda - \mu)\} & \text{if } \lambda \neq \mu \text{ and } A \text{ is a domain.} \end{cases}$$

The proof of this result is left as an exercise (see Exercise 4.4). □

Example 4.1.9. Assume here, and only here, that $\boldsymbol{\alpha}(s) = (0,1)$ for all $s \in S$. The quadratic relations (Q) then become $T_s^2 = 1$, which implies that $\mathcal{H}_A(W,S,\boldsymbol{\alpha})$ is just the group algebra AW. □

Finally, note that it is possible to obtain an isomorphic Hecke algebra by multiplying the parameters by a suitable family of invertible elements. Indeed, if $\lambda = (\lambda_s)_{s \in S} \in \mathrm{Weight}(S, A^\times)$ then, by Example 3.4.10, λ can be extended to a braided map $W \to A^\times$, $w \mapsto \lambda_w$ and we obtain an isomorphism of A-algebras

$$\begin{array}{ccc} \mathcal{H}_A(W,S,\boldsymbol{\alpha}) & \xrightarrow{\sim} & \mathcal{H}_A(W,S,\lambda{\bullet}\boldsymbol{\alpha}) \\ T_w^{\boldsymbol{\alpha}} & \longmapsto & \lambda_w^{-1} T_w^{\lambda\bullet\boldsymbol{\alpha}}, \end{array} \tag{4.1.10}$$

where $\lambda{\bullet}\boldsymbol{\alpha}(s) = (\lambda_s \alpha_s, \lambda_s^2 \beta_s)$.

> *From now on, if we do not need to make precise the base ring, the Coxeter group, or the parameters, we denote by $\mathcal{H}$ the A-algebra $\mathcal{H}_A(W,S,\boldsymbol{\alpha})$.*

4.1.B. *Parabolic Subalgebras*

Let I be a subset of S. Let $\boldsymbol{\alpha}_I$ denote the restriction of $\boldsymbol{\alpha}$ to I. Then $\mathcal{H}_A(W_I, I, \boldsymbol{\alpha}_I)$ can be canonically identified with the A-subalgebra of $\mathcal{H} = \mathcal{H}_A(W,S,\boldsymbol{\alpha})$ with A-basis $(T_w^{\boldsymbol{\alpha}})_{w \in W_I}$ through the map $T_w^{\boldsymbol{\alpha}_I} \mapsto T_w^{\boldsymbol{\alpha}}$. Of course, we make this identification and we set $\mathcal{H}_I = \mathcal{H}_A(W_I, I, \boldsymbol{\alpha}_I)$. If $s_1, s_2, \ldots, s_n$ belong to S, we simply write $\mathcal{H}_{s_1,s_2,\ldots,s_n}$ for $\mathcal{H}_{\{s_1,s_2,\ldots,s_n\}}$. The next result is easy (see, for instance, [GePf, §4.4.6]).

Proposition 4.1.11. *$(T_x)_{x \in X_I}$ is an $\mathcal{H}_I$-basis of the **right** $\mathcal{H}_I$-module $\mathcal{H}$.*

Definition 4.1.12. *The subalgebra $\mathcal{H}_I$ is called a **parabolic** subalgebra of $\mathcal{H}$.*

4.1.C. *Anti-involution*

Let

$$\begin{array}{ccc} \mathcal{H}_A(W,S,\boldsymbol{\alpha}) & \longrightarrow & \mathcal{H}_A(W,S,\boldsymbol{\alpha}) \\ h & \longmapsto & h^* \end{array}$$

be the unique A-linear map such that $T_w^* = T_{w^{-1}}$ for all $w \in W$. Then

(4.1.13) *$*$ is an involutive anti-automorphism of the A-algebra $\mathcal{H}$.*

Remark 4.1.14. If W is commutative (i.e. if $m_{st} = 2$ for all $s \neq t$), then $*$ is the identity (since then $w^{-1} = w$ for all $w \in W$). □

4.1.D. Central Form

Let

$$\tau : \mathcal{H} \longrightarrow A$$

be the unique A-linear map such that

$$\tau(T_w) = \begin{cases} 1 & \text{if } w = 1, \\ 0 & \text{if } w \neq 1 \end{cases}$$

for all $w \in W$ (in other words, $\tau(T_w) = \delta_{w=1}$). By Example 3.4.10, the map $\beta : S \to (A, \times)$ can be extended to a unique braided map $\beta : W \to (A, \times)$, $w \mapsto \beta_w$. Recall the following result [GePf, Proposition 8.1.1] (and recall from Appendix A, Definition A.1.1, that a *central form* on $\mathcal{H}$ is a linear form $f : \mathcal{H} \to A$ such that $f(hk) = f(kh)$ for all h, $k \in \mathcal{H}$):

Proposition 4.1.15. *If $x, y \in W$, then*

$$\tau(T_x T_y) = \delta_{xy=1} \beta_x.$$

In particular, τ is a central form and

$$\tau(h^*) = \tau(h)$$

for all $h \in \mathcal{H}$.

4.2. The Case where A is a Group Algebra

From now on, we fix an abelian group $\mathcal{A}$, a commutative ring R, a map $\varphi \in$ Weight$(S, \mathcal{A})$ and we assume that $A = R[\mathcal{A}]$ (we denote it exponentially as in Chapter 2) and $\boldsymbol{\alpha}(s) = (v^{\varphi(s)} - v^{-\varphi(s)}, 1)$ for all $s \in S$. The Hecke algebra $\mathcal{H} = \mathcal{H}_A(W, S, \boldsymbol{\alpha})$ is then denoted by $\mathcal{H}_R(W, S, \varphi)$ and is called the ***Hecke algebra with parameters*** *φ. If $w \in W$, the element T_w of $\mathcal{H}$ can also be denoted by T_w^{φ} if necessary.*

With this assumption, the quadratic relations (Q) become

$$(Q) \qquad \forall\, s \in S,\ (T_s - v^{\varphi(s)})(T_s + v^{-\varphi(s)}) = 0.$$

Also, formulas 4.1.6 and 4.1.7 can be rewritten as

$$(4.2.1) \qquad T_s T_w = \begin{cases} T_{sw} & \text{if } sw > w, \\ T_{sw} + (v^{\varphi(s)} - v^{-\varphi(s)})T_w & \text{if } sw < w \end{cases}$$

or

$$(4.2.2) \qquad T_s T_w = T_{sw} + \delta_{sw<w}(v^{\varphi(s)} - v^{-\varphi(s)})T_w.$$

***Example* 4.2.3.** Assume here that $\mathscr{A} = \mathbb{Z}$ and that $\varphi(s) = 1$ for all $s \in S$. Then $\varphi(w) = \ell(w)$ for all $w \in W$. This case will be referred to as the *split case*. □

4.2.A. *Index, Sign*

As $\mathscr{A}$ is an abelian group, we can (and we will) identify Weight$(S, \mathscr{A})$ with Braid$(W, \mathscr{A})$, according to Example 3.4.10. This means that φ extends uniquely to a braided map still denoted by $\varphi : W \to \mathscr{A}$. Let $\mathrm{ind} : \mathscr{H} \to A$ (resp. $\mathrm{sgn} : \mathscr{H} \to A$) be the unique A-linear map such that $\mathrm{ind}(T_w) = v^{\varphi(w)}$ (resp. $\mathrm{sgn}(T_w) = \varepsilon(w)v^{-\varphi(w)}$) for all $w \in W$. Then

(4.2.4) ind *and* sgn *are morphisms of A-algebras.*

If necessary, ind and sgn will be denoted by ind_φ and sgn_φ, or ind_W and sgn_W, or $\mathrm{ind}_{W,\varphi}$ and $\mathrm{sgn}_{W,\varphi}\cdots$

4.2.B. *Central Form*

The assumption, together with Proposition 4.1.15, implies that

$$(4.2.5) \qquad \tau(T_x T_y) = \delta_{xy=1}$$

for all $x, y \in W$. Consequently,

(4.2.6) *if W is finite, then τ is a symmetrizing form and so $\mathscr{H}$ is symmetric.*

4.2.C. Invertibility

The standard generators T_s are now invertible: in fact,

$$T_s^{-1} = T_s - v^{\varphi(s)} + v^{-\varphi(s)}. \tag{4.2.7}$$

Consequently,

$$T_w \in \mathscr{H}^\times \tag{4.2.8}$$

for all $w \in W$.

4.2.D. Involutions

As the elements of the standard basis are invertible, we can define two new involutions on $\mathscr{H}$ as follows. First, we denote by

$$\begin{aligned} \mathscr{H} &\longrightarrow \mathscr{H} \\ h &\longmapsto h^\dagger \end{aligned}$$

the unique A-linear map such that

$$T_w^\dagger = \varepsilon(w) T_{w^{-1}}^{-1} \tag{4.2.9}$$

for all $w \in W$.

Also, recall that $A = R[\mathscr{A}]$ is endowed with an involutive R-linear automorphism $\bar{\ }$ that sends v^a to v^{-a} (as in Chapter 2). We denote by

$$\begin{aligned} \mathscr{H} &\longrightarrow \mathscr{H} \\ h &\longmapsto \overline{h} \end{aligned}$$

the unique A-*semilinear* map such that

$$\overline{T}_w = T_{w^{-1}}^{-1} \tag{4.2.10}$$

for all $w \in W$.

Proposition 4.2.11. *The map $\dagger : \mathscr{H} \to \mathscr{H}$ (resp. $\bar{\ } : \mathscr{H} \to \mathscr{H}$) is an involutive A-linear (resp. A-semilinear) automorphism of the ring $\mathscr{H}$.*

Moreover, the involutions $*$, $\dagger$ and $\bar{\ }$ commute with each other and all stabilize the parabolic subalgebras $\mathscr{H}_I$, for $I \subset S$.

4.3. Functoriality, Parameters

4.3.A. *Changing the Group*

Let $\mathscr{A}'$ be an abelian group and let $g : \mathscr{A} \to \mathscr{A}'$ be a morphism of groups. Then g induces a morphism of algebras $R[\mathscr{A}] \to R[\mathscr{A}']$ (still denoted by g), $g \circ \varphi : S \to \mathscr{A}'$ belongs to Weight$(S, \mathscr{A}')$ and, by Remark 4.1.5, the map $T_w^{\varphi} \mapsto T_w^{g\circ\varphi}$ induces an isomorphism of algebras

(4.3.1) $$g_* : R[\mathscr{A}'] \otimes_{R[\mathscr{A}]} \mathscr{H}_R(W,S,\varphi) \xrightarrow{\sim} \mathscr{H}_R(W,S,g\circ\varphi).$$

If $\mathscr{A}''$ is another abelian group and $g' : \mathscr{A}' \to \mathscr{A}''$ is a morphism of groups, then

(4.3.2) $$(g' \circ g)_* = g'_* \circ (\mathrm{Id}_{R[\mathscr{A}']} \otimes_{R[\mathscr{A}]} g_*).$$

Also,

(4.3.3) *g_* commutes with the involutions $*$, $^-$ and †*

and

(4.3.4) *g_* respects the canonical symmetrizing form τ.*

Finally, note that the diagrams

(4.3.5) $$\begin{array}{ccc} \mathscr{H}_R(W,S,\varphi) & \xrightarrow{g_*} & \mathscr{H}_R(W,S,g\circ\varphi) \\ \Big\downarrow \mathrm{ind}_{\varphi} & & \Big\downarrow \mathrm{ind}_{g\circ\varphi} \\ R[\mathscr{A}] & \xrightarrow{g_*} & R[\mathscr{A}'] \end{array}$$

and

(4.3.6) $$\begin{array}{ccc} \mathscr{H}_R(W,S,\varphi) & \xrightarrow{g_*} & \mathscr{H}_R(W,S,g\circ\varphi) \\ \Big\downarrow \mathrm{sgn}_{\varphi} & & \Big\downarrow \mathrm{sgn}_{g\circ\varphi} \\ R[\mathscr{A}] & \xrightarrow{g_*} & R[\mathscr{A}'] \end{array}$$

are commutative.

4.3.B. Generic Hecke Algebra

If $s, t \in S$, we write $s \sim t$ if s and t are conjugate in W. We set $\bar{S} = S/\sim$ and we denote by $\bar{s}$ the image of s in $\bar{S}$. Let $\mathbb{Z}[\bar{S}]$ denote the free abelian group with $\mathbb{Z}$-basis $\bar{S}$, i.e. the group of formal sums $\sum_{\bar{s}\in\bar{S}} \lambda_{\bar{s}}\, \bar{s}$ with finite support. Finally, let

$$\begin{aligned} \mathrm{can}: S &\longrightarrow \mathbb{Z}[\bar{S}] \\ s &\longmapsto \bar{s}. \end{aligned}$$

Then $\mathrm{can} \in \mathrm{Weight}(S, \mathbb{Z}[\bar{S}])$.

Definition 4.3.7. *The Hecke algebra $\mathcal{H}_R(W, S, \mathrm{can})$ is called the* **generic Hecke algebra** *(over R) associated with (W,S). It is denoted by $\boldsymbol{\mathcal{H}}_R$ (or $\boldsymbol{\mathcal{H}}_R(W)$, or $\boldsymbol{\mathcal{H}}_R(W,S)$ if necessary). The element T_w^{can} is denoted by $\boldsymbol{T}_w$.*

In more concrete terms, if we denote by $\boldsymbol{v}_{\bar{s}}$ (or $\boldsymbol{v}_s$, for $s \in S$) the element $v^{\bar{s}}$ of the group algebra $R[\mathbb{Z}[\bar{S}]]$, then $(\boldsymbol{v}_{\bar{s}})_{\bar{s}\in\bar{S}}$ is a family of algebraically independent elements over R and

$$R[\mathbb{Z}[\bar{S}]] = R[(\boldsymbol{v}_{\bar{s}}^{\pm 1})_{\bar{s}\in\bar{S}}]. \tag{4.3.8}$$

Moreover, the quadratic relations become

$$\forall\, s \in S,\ (\boldsymbol{T}_s - \boldsymbol{v}_s)(\boldsymbol{T}_s + \boldsymbol{v}_s^{-1}) = 0. \tag{4.3.9}$$

Note that $\boldsymbol{v}_s = \boldsymbol{v}_t$ if $s \sim t$.

The map $\mathrm{can}: S \to \mathbb{Z}[\bar{S}]$ is the *universal weight function* among all weight functions $S \to \mathscr{A}$, where $\mathscr{A}$ is an abelian group: indeed, there exists a *unique* morphism of groups $\varphi^{\mathrm{lin}}: \mathbb{Z}[\bar{S}] \to \mathscr{A}$ such that $\varphi^{\mathrm{lin}}(\bar{s}) = \varphi(s)$ for all $s \in S$. In other words,

$$\varphi = \varphi^{\mathrm{lin}} \circ \mathrm{can}, \tag{4.3.10}$$

and so, by 4.3.1, $\mathcal{H}_R(W,S,\varphi)$ is a specialization of the generic Hecke algebra $\boldsymbol{\mathcal{H}}_R$, i.e. we have an isomorphism of algebras

$$\varphi_*^{\mathrm{lin}}: R[\mathscr{A}] \otimes_{R[\mathbb{Z}[\bar{S}]]} \boldsymbol{\mathcal{H}}_R \xrightarrow{\sim} \mathcal{H}_R(W,S,\varphi). \tag{4.3.11}$$

Finally, as $\mathrm{can} \in \mathrm{Weight}(S, \mathbb{Z}[\bar{S}])$, then, by Matsumoto's Theorem, it can be extended to a weight function $\mathrm{can}: W \to \mathbb{Z}[\bar{S}]$ and it is easily checked that, with the notation $\ell_{\bar{s}}$ introduced at the end of §3.4,

$$\mathrm{can}(w) = \sum_{\bar{s}\in\bar{S}} \ell_{\bar{s}}(w)\, \bar{s} \tag{4.3.12}$$

for all $w \in W$.

4.3.C. Changing Signs

Let us fix here, and only here, a partition $S = S_+ \dot\cup S_-$ of S into W-closed subsets (see Definition 3.4.12). Let

$$\begin{aligned} \varphi_- : S &\longrightarrow \mathscr{A} \\ s &\longmapsto \begin{cases} \varphi(s) & \text{if } s \in S_+, \\ -\varphi(s) & \text{if } s \in S_-. \end{cases} \end{aligned}$$

Then $\varphi_- \in \mathrm{Weight}(S, \mathscr{A})$ and the next lemma is easily checked:

Lemma 4.3.13. *The unique A-linear map $\Theta_{S_-} : \mathscr{H}_R(W,S,\varphi) \longrightarrow \mathscr{H}_R(W,S,\varphi_-)$ which sends T_w^φ to $(-1)^{\ell_{S_-}(w)} T_w^{\varphi_-}$ is an isomorphism of A-algebras.*

Note also that

(4.3.14) $\qquad\Theta_{S_-}$ *commutes with the involutions* $*$, $^-$ *and* †

and

(4.3.15) $\qquad\Theta_{S_-}$ *respects the canonical symmetrizing form* τ.

4.3.D. Vanishing Parameters

We fix here, and only here, a partition $S = I \dot\cup J$ into W-closed subsets such that $I \subset \{s \in S \mid \varphi(s) = 0\}$. Following §B.3 (see Appendix B), we set

$$\widetilde{J} = \{wsw^{-1} \mid w \in W_I \text{ and } s \in J\} \quad \text{and} \quad \widetilde{W} = \langle \widetilde{J} \rangle.$$

Then, by Theorem B.3.1, $(\widetilde{W}, \widetilde{J})$ is a Coxeter system and

$$W = W_I \ltimes \widetilde{W}.$$

As in §B.3, we denote by $\widetilde{\ell} : \widetilde{W} \to \mathbb{N}$ the length function with respect to $\widetilde{J}$ and we denote by $\nu : \widetilde{J} \to J$ the unique map such that $t \in \widetilde{J}$ and $\nu(t) \in J$ are W_I-conjugate. We denote by $\widetilde{\varphi} : \widetilde{W} \to \mathscr{A}$ the restriction of the braided map $\varphi : W \to \mathscr{A}$ extending $\varphi : S \to \mathscr{A}$. Then:

Lemma 4.3.16. *If $a \in W_I$ and $w \in \widetilde{W}$, then $\varphi(aw) = \varphi(wa) = \varphi(w)$ and $\widetilde{\varphi}(awa^{-1}) = \widetilde{\varphi}(w)$.*

Proof. The proof is similar to the proof of Corollary B.2.5. ■

Corollary 4.3.17. *The map $\widetilde{\varphi} : \widetilde{W} \to \mathscr{A}$ is braided and W_I-invariant.*

Now, we set $\widetilde{\mathscr{H}} = \mathscr{H}_R(\widetilde{W}, \widetilde{J}, \widetilde{\varphi})$: as W_I acts on the Coxeter system $(\widetilde{W}, \widetilde{J})$ and stabilizes $\widetilde{\varphi}$, it also acts on $\widetilde{\mathscr{H}}$. Hence we can build the semidirect product $W_I \ltimes \widetilde{\mathscr{H}}$. If $w \in \widetilde{W}$, we denote by $\widetilde{T}_w$ the element $T_w^{\widetilde{\varphi}}$ of $\widetilde{\mathscr{H}}$: then $(x \ltimes \widetilde{T}_w)_{(x,w)\in W_I \times \widetilde{W}}$ is an A-basis of $W_I \ltimes \widetilde{\mathscr{H}}$.

Lemma 4.3.18. *The A-linear map $W_I \ltimes \widetilde{\mathscr{H}} \to \mathscr{H}$ that sends $x \ltimes \widetilde{T}_w$ to T_{xw} is an isomorphism of A-algebras.*

Proof. It is sufficient to check that, if $x \in W_I$ and $w \in W$, then

(4.3.19) $$T_{xw} = T_x T_w \quad \text{and} \quad T_{wx} = T_w T_x,$$

which is obvious. ■

4.4. *r*-Polynomials

For $y \in W$, write
$$\overline{T}_y = \sum_{x \in W} \overline{r}_{x,y}\, T_x,$$
with $r_{x,y} \in A$. If necessary, we write $r_{x,y}^{\varphi}$ for the element $r_{x,y}$. If $\mathscr{A}'$ is another abelian group and $g : \mathscr{A} \to \mathscr{A}'$ is a morphism of groups, then

(4.4.1) $$g_*(r_{x,y}^{\varphi}) = r_{x,y}^{g\circ\varphi}$$

by 4.3.3.

Commentary 4.4.2. Formula 4.4.1 shows that the elements $r_{x,y}$ have a *generic* nature. In other words, if we denote by $\boldsymbol{r}_{x,y}$ the polynomials $r_{x,y}^{\mathrm{can}}$ computed in the generic Hecke algebra, then $r_{x,y}^{\varphi}$ is just the specialization of $\boldsymbol{r}_{x,y}$ through the morphism $\varphi_*^{\mathrm{lin}} : R[\mathbb{Z}[\bar{S}]] \to A, \boldsymbol{v}_s \mapsto v^{\varphi(s)}$. We can even take $R = \mathbb{Z}$.

Note also that, if $x, y \in W_I$ for some $I \subset S$, then the element $r_{x,y}$ computed inside $\mathscr{H}$ equals the one computed inside $\mathscr{H}_I$, because the involution $\bar{\ }$ stabilizes $\mathscr{H}_I$ (see Proposition 4.2.11). □

We now investigate in this subsection some properties of the $r_{x,y}$. The first set of properties is easy:

(4.4.3) $$\begin{cases} \sum_{z\in W} \overline{r}_{x,z}\, r_{z,y} = \delta_{x=y}, \\ r_{y,y} = 1, \\ r_{x,y} \neq 0 \implies x \leqslant y \end{cases}$$

for all $x, y \in W$.

***Remark* 4.4.4.** In other words, the involution $\bar{\ }$ on $\mathscr{H}$ satisfies the same properties (I1) and (I2) as the involution $\bar{\ }$ on the module $\mathscr{M}$ in Chapter 2 (see 2.2.1). □

Proof of 4.4.3. The first property follows immediately from the fact that $\bar{\ }$ is an involution. Let $(s_1, s_2, \ldots, s_r)$ be a reduced expression of y. Then

$$\begin{aligned}\overline{T}_y &= \overline{T}_{s_1}\overline{T}_{s_2}\cdots\overline{T}_{s_r}\\ &= (T_{s_1} - v^{\varphi(s_1)} + v^{-\varphi(s_1)})(T_{s_2} - v^{\varphi(s_2)} + v^{-\varphi(s_2)})\cdots(T_{s_r} - v^{\varphi(s_r)} + v^{-\varphi(s_r)}).\end{aligned}$$

If we expand this product using the multiplication rules in the standard basis, we obtain the two other properties. ■

Let us investigate further properties of the r-polynomials. First of all, the fact that $\bar{\ }$ commutes with $*$ implies that

(4.4.5) $$r_{x^{-1},y^{-1}} = r_{x,y}$$

for all $x, y \in W$. Moreover, as $\bar{\ }$ commutes with †, one gets

(4.4.6) $$\overline{r}_{x,y} = \varepsilon(xy) r_{x,y}.$$

Proof of 4.4.6. Indeed, $T_y^\dagger = \overline{(\overline{T}_y)^\dagger}$. But

$$T_y^\dagger = \varepsilon(y)\overline{T}_y = \sum_{x\in W} \varepsilon(y)\overline{r}_{x,y}\, T_x$$

$$\overline{(\overline{T}_y)^\dagger} = \overline{\sum_{x\in W} \overline{r}_{x,y}\, T_x^\dagger} = \overline{\sum_{x\in W} \overline{r}_{x,y}\varepsilon(x)\overline{T}_x} = \sum_{x\in W} \varepsilon(x) r_{x,y}\, T_x.$$

The result follows. ■

We now provide an inductive algorithm for computing the polynomials $r_{x,y}$:

Lemma 4.4.7. *Let $x, y \in W$ and let $s \in S$ be such that $sy > y$. Then*

$$r_{x,sy} = r_{sx,y} + \delta_{sx>x}(v^{\varphi(s)} - v^{-\varphi(s)}) r_{x,y}.$$

Proof. As $sy > y$, we have $T_{sy} = T_s T_y$. Therefore,

$$\overline{T}_{sy} = \overline{T}_s\overline{T}_y = (T_s - v^{\varphi(s)} + v^{-\varphi(s)}) \sum_{x\in W} \overline{r}_{x,y}\, T_x.$$

But, by 4.2.2,

$$\begin{aligned}(T_s - v^{\varphi(s)} + v^{-\varphi(s)})T_x &= T_{sx} + \delta_{sx<x}(v^{\varphi(s)} - v^{-\varphi(s)})T_x - (v^{\varphi(s)} - v^{-\varphi(s)})T_x \\ &= T_{sx} + (\delta_{sx<x} - 1)(v^{\varphi(s)} - v^{-\varphi(s)})T_x \\ &= T_{sx} - \delta_{sx>x}(v^{\varphi(s)} - v^{-\varphi(s)})T_x.\end{aligned}$$

Therefore,

$$\overline{T}_{sy} = \sum_{x\in W} \big(\overline{r}_{sx,y} - \delta_{sx>x}(v^{\varphi(s)} - v^{-\varphi(s)})\overline{r}_{x,y}\big)\, T_x,$$

and the result follows. ■

Corollary 4.4.8. *If* $x, y \in W$, *then* $\boldsymbol{r}_{x,y} \in R[(\boldsymbol{v}_s - \boldsymbol{v}_s^{-1})_{s\in S}]$.

Proof. This follows immediately from Lemma 4.4.7 by induction on $\ell(y)$. ■

Corollary 4.4.9. $r_{x,y} \in v^{\varphi(y)-\varphi(x)} R[2\mathscr{A}] = \bigoplus_{a\in\mathscr{A}} R\, v^{2a}$.

Proof. We argue by induction on $\ell(y)$, using Lemma 4.4.7. If $\ell(y) = 0$, then $y = 1$ and $\overline{r}_{1,1} = 1$ while $\overline{r}_{x,1} = 0$ if $x \neq 1$. Now assume that $\ell(y) > 0$ and that the result holds for elements of strictly smaller length. Let $s \in S$ be such that $sy < y$. Write $sy = y'$, so that $y' < sy' = y$. So, by Lemma 4.4.7,

$$(v^{\varphi(y)-\varphi(x)})r_{x,y} = v^{\varphi(y)-\varphi(x)} r_{sx,y'} + \delta_{sx>x} v^{\varphi(y)-\varphi(x)}(v^{\varphi(s)} - v^{-\varphi(s)})r_{x,y'}.$$

Now, $\varphi(y) = \varphi(y') + \varphi(s)$ as $\varphi : W \to \mathscr{A}$ is a weight function and, if we set $a = \varphi(sx) - \varphi(x)$, then

$$a \in \{\varphi(s), -\varphi(s)\}.$$

With this notation,

$$(v^{\varphi(y)-\varphi(x)})r_{x,y} = v^{\varphi(s)+a} v^{\varphi(y')-\varphi(sx)} r_{sx,y'} + \delta_{sx>x}(v^{2\varphi(s)} - 1)v^{\varphi(y')-\varphi(x)} r_{x,y'}.$$

As $v^{\varphi(y')-\varphi(sx)} r_{sx,y'} \in R[2\mathscr{A}]$ and $v^{\varphi(y')-\varphi(x)} r_{x,y'} \in R[2\mathscr{A}]$ by the induction hypothesis, and since $\varphi(s) + a \in 2\mathscr{A}$, the result follows. ■

Remark 4.4.10. Recall that $\boldsymbol{r}_{x,y}$ denotes the generic r-polynomial $r_{x,y}^{\mathrm{can}}$. If $\gamma = \sum_{\bar{s}\in\bar{S}} \gamma_{\bar{s}}\, \bar{s}$ and $\gamma' = \sum_{\bar{s}\in\bar{S}} \gamma'_{\bar{s}}\, \bar{s}$ are two elements of $\mathbb{Z}[\bar{S}]$, we write $\gamma \preccurlyeq \gamma'$ if $\gamma_{\bar{s}} \leq \gamma'_{\bar{s}}$ for all $\bar{s} \in \bar{S}$. This defines a partial order on $\mathbb{Z}[\bar{S}]$ which is compatible with the group structure.

Then it follows easily from the same induction argument as in the proof of Corollary 4.4.9 that

$$\boldsymbol{r}_{x,y} \equiv v^{\mathrm{can}(y)-\mathrm{can}(x)} \mod \Big(\bigoplus_{a \prec \mathrm{can}(y)-\mathrm{can}(x)} R\, v^a \Big) \tag{4.4.11}$$

for all $x \leqslant y$ in W. □

Remark 4.4.12 (Verma). If $x < y$, then

$$\sum_{x \leqslant z \leqslant y} \varepsilon(z) = 0.$$

To prove this, assume in this remark, and only in this remark, that $R = \mathbb{Z}$, $\mathscr{A} = \mathbb{Z}$ and $\varphi = \ell$ (so that $\mathbb{Z}[\mathscr{A}] = \mathbb{Z}[v, v^{-1}]$ and $\varphi(s) = 1$ for all $s \in S$). Then 4.4.3 and 4.4.6 imply that

$$(*) \qquad \sum_{x \leqslant z \leqslant y} \varepsilon(xz) r_{x,z} r_{z,y} = 0.$$

But, by 4.4.11 and 4.4.1, $\deg(r_{x,z}) = \ell(z) - \ell(x)$ and the coefficient of $v^{\ell(z)-\ell(x)}$ in $r_{x,z}$ is equal to 1. Therefore, the coefficient of $v^{\ell(y)-\ell(x)}$ in the left-hand side of $(*)$ is equal to

$$\varepsilon(x) \sum_{x \leqslant z \leqslant y} \varepsilon(z).$$

The result follows. □

Example 4.4.13 (Type A_2). As in Example 3.12.7, we assume here that $S = \{s, t\}$ with $m_{st} = 3$, so that (W, S) is of type A_2. We compute in this case the generic r-polynomials. Here is the corresponding Coxeter graph:

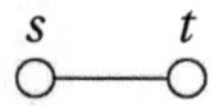

Since s and t are conjugate, we have $\boldsymbol{v}_s = \boldsymbol{v}_t$: we write $\boldsymbol{v} = \boldsymbol{v}_s = \boldsymbol{v}_t$. Let $\breve{\boldsymbol{v}} = \boldsymbol{v}^{-1} - \boldsymbol{v}$: it follows from Corollary 4.4.8 that all the $\boldsymbol{r}_{x,y}$ belong to $\mathbb{Z}[\breve{\boldsymbol{v}}]$. The values of $\boldsymbol{r}_{x,y}$ are given in Table 4.a (where a dot replaces 0).

$x \backslash y$	1	s	t	st	ts	w_0
1	1	$\breve{\boldsymbol{v}}$	$\breve{\boldsymbol{v}}$	$\breve{\boldsymbol{v}}^2$	$\breve{\boldsymbol{v}}^2$	$\breve{\boldsymbol{v}}^3 + \breve{\boldsymbol{v}}$
s	.	1	.	$\breve{\boldsymbol{v}}$	$\breve{\boldsymbol{v}}$	$\breve{\boldsymbol{v}}^2$
t	.	.	1	$\breve{\boldsymbol{v}}$	$\breve{\boldsymbol{v}}$	$\breve{\boldsymbol{v}}^2$
st	.	.	.	1	.	$\breve{\boldsymbol{v}}$
ts	.	.	.	.	1	$\breve{\boldsymbol{v}}$
w_0	.	.	.	.	.	1

Table 4.a. $\boldsymbol{r}$-polynomials in type A_2

Note that, if $\ell(y) - \ell(x) \leq 2$, then $\boldsymbol{r}_{x,y} = \breve{\boldsymbol{v}}^{\ell(y)-\ell(x)}$. □

Example 4.4.14 (Type B_2). As in Example 3.12.8, we assume here that $S = \{s,t\}$ with $m_{st} = 4$, so that (W,S) is of type B_2. Here is the corresponding Coxeter graph:

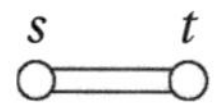

Then $\boldsymbol{v}_s \neq \boldsymbol{v}_t$ and we set $\breve{\boldsymbol{v}}_s = \boldsymbol{v}_s - \boldsymbol{v}_s^{-1}$ and $\breve{\boldsymbol{v}}_t = \boldsymbol{v}_t - \boldsymbol{v}_t^{-1}$. It follows from Corollary 4.4.8 that all the $\boldsymbol{r}_{x,y}$ belong to $\mathbb{Z}[\breve{\boldsymbol{v}}_s, \breve{\boldsymbol{v}}_t]$. The values of $\boldsymbol{r}_{x,y}$ are given in Table 4.b. □

x \ y	1	s	t	st	ts	sts	tst	w_0
1	1	$\breve{\boldsymbol{v}}_s$	$\breve{\boldsymbol{v}}_t$	$\breve{\boldsymbol{v}}_s\breve{\boldsymbol{v}}_t$	$\breve{\boldsymbol{v}}_s\breve{\boldsymbol{v}}_t$	$\breve{\boldsymbol{v}}_t(\breve{\boldsymbol{v}}_s^2+1)$	$\breve{\boldsymbol{v}}_s(\breve{\boldsymbol{v}}_t^2+1)$	$\breve{\boldsymbol{v}}_s^2\breve{\boldsymbol{v}}_t^2+\breve{\boldsymbol{v}}_s^2+\breve{\boldsymbol{v}}_t^2$
s	.	1	.	$\breve{\boldsymbol{v}}_t$	$\breve{\boldsymbol{v}}_t$	$\breve{\boldsymbol{v}}_s\breve{\boldsymbol{v}}_t$	$\breve{\boldsymbol{v}}_s\breve{\boldsymbol{v}}_t$	$\breve{\boldsymbol{v}}_s(\breve{\boldsymbol{v}}_t^2+1)$
t	.	.	1	$\breve{\boldsymbol{v}}_s$	$\breve{\boldsymbol{v}}_s$	$\breve{\boldsymbol{v}}_s\breve{\boldsymbol{v}}_t$	$\breve{\boldsymbol{v}}_s\breve{\boldsymbol{v}}_t$	$\breve{\boldsymbol{v}}_t(\breve{\boldsymbol{v}}_s^2+1)$
st	.	.	.	1	.	$\breve{\boldsymbol{v}}_s$	$\breve{\boldsymbol{v}}_t$	$\breve{\boldsymbol{v}}_s\breve{\boldsymbol{v}}_t$
ts	.	.	.	.	1	$\breve{\boldsymbol{v}}_s$	$\breve{\boldsymbol{v}}_t$	$\breve{\boldsymbol{v}}_s\breve{\boldsymbol{v}}_t$
sts	.	.	.	.	.	1	.	$\breve{\boldsymbol{v}}_t$
tst	.	.	.	.	.	.	1	$\breve{\boldsymbol{v}}_s$
w_0	.	.	.	.	.	.	.	1

Table 4.b. $\boldsymbol{r}$-polynomials in type B_2

4.5. Reflection Representation

> **Hypothesis.** *We assume in this section, and only in this section, that $R = \mathbb{R}$.*

Note that $\mathcal{H} = \mathcal{H}_{\mathbb{R}}(W,S,\varphi)$ and recall that $\mathbb{R}(\mathcal{A})$ denotes the fraction field of $\mathbb{R}[\mathcal{A}]$. It follows from Tits' Deformation Theorem [GePf, Theorem 7.4.6] applied to the special case of Hecke algebras [GePf, Theorem 8.1.7] that, if W is finite, then the $\mathbb{R}(\mathcal{A})$-algebras $\mathbb{R}(\mathcal{A})\mathcal{H} = \mathbb{R}(\mathcal{A}) \otimes_{\mathbb{R}[\mathcal{A}]} \mathcal{H}$ and $\mathbb{R}(\mathcal{A})W$ are isomorphic: this implies that any representation of W can be deformed to a representation of $\mathbb{R}(\mathcal{A})\mathcal{H}$. We recall how the reflection representation E of

W can be deformed to a representation of $\mathscr{H}$, even if W is infinite. We fix a family $(c_{s,t})_{s,t\in S}$ of elements of $\mathbb{R}[\mathscr{A}]$ such that, for all s, $t \in S$, we have:

(R1) $c_{s,s} = v^{\varphi(s)} + v^{-\varphi(s)}$.
(R2) $c_{s,t} = c_{t,s} = 0$ if $m_{st} = 2$.
(R3) $c_{s,t}c_{t,s} = v^{\varphi(s)} + v^{\varphi(t)} + 2\cos(\frac{\pi}{m_{s,t}})$ if $m_{st} \geqslant 3$.

If $s \in S$, we define an endomorphism ρ_s^φ of the free $\mathbb{R}[\mathscr{A}]$-module $\mathbb{R}[\mathscr{A}] \otimes_{\mathbb{R}} E$ by

$$\rho_s^\varphi(e_t^\varphi) = v^{\varphi(s)} e_t^\varphi - c_{s,t} e_s^\varphi$$

for all $t \in S$ (here, e_t^φ denotes the element $1 \otimes_{\mathbb{R}} e_t$ of $\mathbb{R}[\mathscr{A}] \otimes_{\mathbb{R}} E$).

Theorem 4.5.1. *There exists a unique left $\mathscr{H}$-module structure on $\mathbb{R}[\mathscr{A}] \otimes_{\mathbb{R}} E$ such that*

$$T_s \cdot e = \rho_s^\varphi(e)$$

for all $s \in S$ and $e \in E$.

Proof. See, for instance, [GePf, Theorem 8.1.11]. ∎

We denote by E^φ the free $\mathbb{R}[\mathscr{A}]$-module $\mathbb{R}[\mathscr{A}] \otimes_{\mathbb{R}} E$ endowed with the left $\mathscr{H}$-module structure given by Theorem 4.5.1. It is called the *reflection representation* of $\mathscr{H}$. Note that, through the specialization morphism $\mathbb{R}[\mathscr{A}] \to \mathbb{R}$, $v^a \mapsto 1$, we see that $\mathbb{R} \otimes_{\mathbb{R}[\mathscr{A}]} E^\varphi \simeq E$, as an $\mathbb{R}W$-module.

The next elementary result will be used in Chapter 12:

Lemma 4.5.2. *If $s \in S$, then*

$$\{e \in E^\varphi \mid T_s e = -v^{-\varphi(s)} e\} = \mathbb{R}[\mathscr{A}] e_s^\varphi.$$

Notes

All the results gathered in this chapter are classical and may almost all be found, for instance, in [Bou] or [GePf]. Note that Theorem 4.5.1 appeared in Kilmoyer's Thesis [Kil] and was published in [CuIwKij]. More detailed bibliographical comments can be found in [GePf, §8.5].

Exercises

Exercise 4.1 (Type G_2). As in Exercise 3.2, let us assume here that $S = \{s, t\}$ with $m_{st} = 6$, so that (W, S) is of type G_2. Here is the corresponding Coxeter graph:

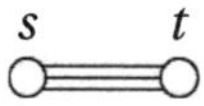

Compute the generic r-polynomials for W.

Exercise 4.2 (Direct products). As in Exercise 3.2, assume that $S = S_1 \dot{\cup} S_2$ is a partition such that, if $s_1 \in S_1$ and $s_2 \in S_2$, then $m_{s_1 s_2} = 2$ (i.e. $s_1 s_2 = s_2 s_1$). Let $W_i = W_{S_i}$ $(i \in \{1,2\})$. Let φ_i denote the restriction of φ to S_i and let $\mathcal{H}_i = \mathcal{H}_A(W_i, S_i, \varphi_i)$. Show that

$$\mathcal{H} = \mathcal{H}_1 \otimes_A \mathcal{H}_2$$

and that, if (x_1, x_2) and (y_1, y_2) are two elements of $W_1 \times W_2$, then

$$\boldsymbol{r}_{x_1 x_2, y_1 y_2} = \boldsymbol{r}_{x_1, y_1} \boldsymbol{r}_{x_2, y_2}.$$

Exercise 4.3. Let $I \subset S$ and let $a \in X_I$, x, $y \in W_I$. Show that $\boldsymbol{r}_{ax,ay} = \boldsymbol{r}_{x,y}$ and $\boldsymbol{r}_{xa^{-1},ya^{-1}} = \boldsymbol{r}_{x,y}$.

Exercise 4.4. The aim of this exercise is to prove the last statement of Example 4.1.8. Let λ and μ be two elements of A and let T be an indeterminate over A. Let $\mathcal{H}_1 = A[T]/\langle (T-\lambda)(T-\mu)\rangle$ and

$$\mathcal{H}_2 = \{(a,b) \in A \times A \mid a \equiv b \mod \lambda - \mu\}.$$

We aim to prove that $\mathcal{H}_1 \simeq \mathcal{H}_2$ whenever $\lambda \neq \mu$ and A is an integral domain.

(a) Show that $\mathcal{H}_2$ is indeed a subring of $A \times A$.
(b) Let $\pi : A[T] \to A \times A$, $P \mapsto (P(\lambda), P(\mu))$. Show that π is a morphism of A-algebras, whose image is $\mathcal{H}_2$.
(c) Assume here that $\lambda \neq \mu$ and that A is an integral domain. Show that $\mathrm{Ker}(\pi)$ is generated by $(T-\lambda)(T-\mu)$. Conclude.

Exercise 4.5. Let I and J be two subsets of S. Show that

$$\mathcal{H} = \bigoplus_{x \in X_{IJ}} \mathcal{H}_I \cdot T_x \cdot \mathcal{H}_J.$$

Exercise 4.6. Let $\mathrm{aug} : \mathcal{A} \to \{1\}$ be the trivial morphism. Then $\mathrm{aug}_* : A \to \mathbb{Z}$ is the augmentation morphism and it induces an isomorphism

$$\mathrm{aug}_* : \mathbb{Z} \otimes_A \mathcal{H} \xrightarrow{\sim} \mathbb{Z}W.$$

Now, let I be a W-closed subset of S.

(a) Show that there is a unique morphism of A-algebras $\mathrm{sgn}_I : \mathcal{H} \to A$ such that

$$\mathrm{sgn}_I(T_s) = \begin{cases} -v^{-\varphi(s)} & \text{if } s \in I, \\ v^{\varphi(s)} & \text{if } s \in S \setminus I. \end{cases}$$

If necessary, aug_* and sgn_I are denoted by aug_*^{φ} and sgn_I^{φ}, respectively.

(b) Show that the diagram

$$\begin{array}{ccc} \mathcal{H} & \xrightarrow{\ \mathrm{sgn}_I\ } & A \\ {\scriptstyle \mathrm{aug}_*}\downarrow & & \downarrow{\scriptstyle \mathrm{aug}_*} \\ \mathbb{Z}W & \xrightarrow{\ \varepsilon_I\ } & \mathbb{Z} \end{array}$$

is commutative (here, $\varepsilon_I : \mathbb{Z}W \to \mathbb{Z}$ is the morphism of algebras induced by the morphism of groups $\varepsilon_I : W \to \{1, -1\}$ defined in Exercise 3.5).

(c) Keep the notation of §4.3.C (S_+, S_-, φ_-,...). Show that the diagram

$$\begin{array}{ccc} \mathcal{H}^{\varphi} & \xrightarrow{\ \Theta_{S_-}\ } & \mathcal{H}^{\varphi^-} \\ {\scriptstyle \mathbb{Z}\otimes_A -}\downarrow & & \downarrow{\scriptstyle \mathbb{Z}\otimes_A -} \\ \mathbb{Z}W & \xrightarrow{\ \varepsilon_{S_-}^{\circ}\ } & \mathbb{Z}W \end{array}$$

is commutative (here, $\varepsilon_I^{\circ} : \mathbb{Z}W \to \mathbb{Z}W$ denotes the automorphism of $\mathbb{Z}W$ such that $\varepsilon_I^{\circ}(w) = \varepsilon_I(w)w$ for all $w \in W$).

Exercise 4.7. Assume that W is finite. If $h \in \mathcal{H}$, let $\det(h, \mathcal{H})$ denote the determinant of the map $\mathcal{H} \to \mathcal{H}$, $h' \mapsto hh'$. Show that

$$\det(T_w, \mathcal{H}) = (-1)^{|W|\ell(w)/2}$$

for all $w \in W$. (***Hint:*** If $w = s \in S$, use the decomposition $\mathcal{H} = \bigoplus_{x \in X_s} (AT_{x^{-1}} \oplus AT_{sx^{-1}})$).

Part III
Kazhdan–Lusztig Cells

In this part, we really start the study of the main objects of this book, namely the Kazhdan–Lusztig basis, Kazhdan–Lusztig cells and their associated modules. Based on the general constructions of Part I (preorders, invariant basis), the particular context of Hecke algebras leads to particularly interesting representations, with connections to many other fields of mathematics (representations of reductive groups over many different types of fields, Lie algebras, geometry of flag varieties, unipotent classes...) that will not be discussed here.

This part is organized as follows. The Kazhdan–Lusztig basis is constructed in Chapter 5 while the Kazhdan–Lusztig cells are defined in Chapter 6. Finally, in Chapter 7, we study how, conjecturally, the Kazhdan–Lusztig cells change whenever the weight function φ changes.

Notation. From now on, and until the end of this book (except in the Appendices), we fix a Coxeter system (W, S), a commutative group $\mathscr{A}$ whose law is denoted additively and a map $\varphi \in \text{Weight}(S, \mathscr{A})$. We denote by $\mathscr{A}$ the group algebra $\mathbb{Z}[\mathscr{A}]$ (still denoted exponentially) and $\mathscr{H}$ the Hecke algebra $\mathscr{H}_{\mathbb{Z}}(W, S, \varphi)$. We denote by K the fraction field of A. If F is a field of characteristic 0, then we denote by $F(\mathscr{A})$ the fraction field of the group algebra $F[\mathscr{A}]$ (for instance, $K = \mathbb{Q}(\mathscr{A})$).

Hypothesis. *We also assume from now on that $\mathscr{A}$ is a totally ordered abelian group and we keep the notation of Chapter 2 ($\mathscr{A}_{\leqslant 0}, \mathscr{A}_{<0}, A_{\leqslant 0}, A_{<0}, \mathscr{H}_{<0}, \mathscr{H}_{anti}$...).*

As we mainly study the Kazhdan–Lusztig theory of cells whenever φ takes only positive values (because we will show that all cases can be reduced to this one), we will use a notation for this: we write $\varphi > 0$ to mean that $\varphi(s) > 0$ for all $s \in S$.

Chapter 5
The Kazhdan–Lusztig Basis

5.1. Degree, Valuation, Properties of A

If $\beta \in A$ and $a \in \mathscr{A}$, we denote by $\mathrm{coef}_a(\beta)$ the coefficient of v^a in β. It belongs to $\mathbb{Z}$. Hence,

$$\beta = \sum_{a \in \mathscr{A}} \mathrm{coef}_a(\beta) v^a.$$

If $\beta \neq 0$, we set

$$\deg(\beta) = \max\{a \in \mathscr{A} \mid \mathrm{coef}_a(\beta) \neq 0\} \quad \text{and} \quad \mathrm{val}(\beta) = \min\{a \in \mathscr{A} \mid \mathrm{coef}_a(\beta) \neq 0\}$$

for the *degree* and the *valuation* of β, respectively. Following the usual (and appropriate) convention, we set $\deg(0) = -\infty$ and $\mathrm{val}(0) = +\infty$.

The first obvious remark is about the group $\mathscr{A}$ and follows immediately from the fact that it is ordered:

(5.1.1) $\mathscr{A}$ *is torsion-free.*

Note also that, if $\alpha, \beta \in A$, then

(5.1.2) $$\deg(\alpha\beta) = \deg(\alpha) + \deg(\beta) \quad \text{and} \quad \mathrm{val}(\alpha\beta) = \mathrm{val}(\alpha) + \mathrm{val}(\beta)$$

while

(5.1.3) $$\begin{cases} \deg(\alpha + \beta) \leqslant \max\{\deg(\alpha), \deg(\beta)\}, \\ \mathrm{val}(\alpha + \beta) \geqslant \min\{\mathrm{val}(\alpha), \mathrm{val}(\beta)\}. \end{cases}$$

Finally, note that

(5.1.4) $$\deg(\overline{\alpha}) = -\mathrm{val}(\alpha) \quad \text{and} \quad \mathrm{val}(\overline{\alpha}) = -\deg(\alpha).$$

Proposition 5.1.5. *The ring A is integral and integrally closed.*

C. Bonnafé, *Kazhdan–Lusztig Cells with Unequal Parameters*,
Algebra and Applications 24, https://doi.org/10.1007/978-3-319-70736-5_5

Proof. The fact that A is integral follows immediately from 5.1.2. Let us now show that A is integrally closed. First, assume that $\mathscr{A}$ is finitely generated. Then it is free over $\mathbb{Z}$ by 5.1.1: let $(a_1,\ldots,a_r)$ be a $\mathbb{Z}$-basis of $\mathscr{A}$ and let $v_i = v^{a_i}$. Then $v_1,\ldots, v_n$ are algebraically independent indeterminates over $\mathbb{Z}$ and $A = \mathbb{Z}[v_1, v_1^{-1}, v_2, v_2^{-1},\ldots, v_r, v_r^{-1}]$, so it is integrally closed. This proves the result whenever $\mathscr{A}$ is finitely generated.

Now, let us come back to the general case. Let α, $\beta \in A$ with $\beta \neq 0$ be such that α/β is integral over A. Let $P = \mathbf{t}^n + \alpha_{n-1}\mathbf{t}^{n-1} + \cdots + \alpha_1\mathbf{t} + \alpha_0 \in A[\mathbf{t}]$ be monic and such that $P(\alpha/\beta) = 0$. Now, there exists a finitely generated subgroup $\mathscr{A}'$ of $\mathscr{A}$ such that $\mathbb{Z}[\mathscr{A}']$ contains the finite family of elements α, $\beta, \alpha_0, \alpha_1,\ldots, \alpha_{n-1}$. Hence α/β is integral over $\mathbb{Z}[\mathscr{A}']$ and belongs to $\mathbb{Q}(\mathscr{A}')$. So $\alpha/\beta \in \mathbb{Z}[\mathscr{A}'] \subset A$ because we have just proved that $\mathbb{Z}[\mathscr{A}']$ is integrally closed. ■

5.2. The Kazhdan–Lusztig Basis

The next theorem is an immediate consequence of Corollary 2.1.6 and property 4.4.3:

Theorem 5.2.1. *Let $y \in W$. Then there exists a unique element $\mathbf{C}_y \in \mathscr{H}$ such that*

$$\begin{cases} \overline{\mathbf{C}}_y = \mathbf{C}_y, \\ \mathbf{C}_y \equiv T_y \mod \mathscr{H}_{<0}. \end{cases}$$

Moreover, $\mathbf{C}_y \equiv T_y \mod \left(\bigoplus_{x<y} A_{<0}\, T_x\right)$.

Corollary 5.2.2. *The family $(\mathbf{C}_y)_{y\in W}$ is an A-basis of $\mathscr{H}$.*

Definition 5.2.3. *The basis $(\mathbf{C}_y)_{y\in W}$ is called the **Kazhdan–Lusztig basis** of $\mathscr{H}$.*

***Example* 5.2.4.** We have $\mathbf{C}_1 = T_1 = 1$ and, if $s \in S$, then

$$\mathbf{C}_s = \begin{cases} T_s + v^{-\varphi(s)} & \text{if } \varphi(s) > 0, \\ T_s & \text{if } \varphi(s) = 0, \\ T_s - v^{\varphi(s)} & \text{if } \varphi(s) < 0. \end{cases}$$

(by 4.2.7). A direct computation shows that

$$\mathbf{C}_s^2 = \begin{cases} (v^{\varphi(s)} + v^{-\varphi(s)})\mathbf{C}_s & \text{if } \varphi(s) > 0, \\ T_1 = \mathbf{C}_1 = 1 & \text{if } \varphi(s) = 0, \\ -(v^{\varphi(s)} + v^{-\varphi(s)})\mathbf{C}_s & \text{if } \varphi(s) < 0. \end{cases}$$

Note that $\mathbf{C}_s$ depends heavily on the choice of φ and on the order on $\mathscr{A}$. Also,

$$\mathbf{C}_s^\dagger = \begin{cases} -T_s + v^{\varphi(s)} & \text{if } \varphi(s) > 0, \\ -T_s & \text{if } \varphi(s) = 0, \\ -T_s - v^{-\varphi(s)} & \text{if } \varphi(s) < 0. \end{cases}$$

Consequently,

$$\mathbf{C}_s\mathbf{C}_s^\dagger = \mathbf{C}_s^\dagger\mathbf{C}_s = \begin{cases} 0 & \text{if } \varphi(s) \neq 0, \\ -1 & \text{if } \varphi(s) = 0. \end{cases}$$

Note that, as expected, $(1, \mathbf{C}_s)$ is an A-basis of $\mathscr{H}_s$, and the structure constants in this basis have been computed above. □

Since the involutions $\overline{}$ and $*$ commute, and since $*$ stabilizes $\mathscr{H}_{<0}$, we get

(5.2.5) $$\mathbf{C}_{y^{-1}} = \mathbf{C}_y^*$$

for all $y \in W$.

5.3. Kazhdan–Lusztig Polynomials

Let us write

$$\mathbf{C}_y = \sum_{x \in W} p_{x,y}\, T_x.$$

Then

(5.3.1) $$\begin{cases} p_{x,y} \in A_{<0} & \text{if } x \neq y, \\ p_{x,y} = \sum_{x \leqslant z \leqslant y} r_{x,z}\overline{p}_{z,y}, \\ p_{x,x} = 1, \\ p_{x,y} \neq 0 \Longrightarrow x \leqslant y \end{cases}$$

for all $x, y \in W$ (see 2.2.2 and 4.4.6). We normalize the elements $p_{x,y}$ in the following way: we set

$$P_{x,y} = v^{\varphi(y)-\varphi(x)} p_{x,y}.$$

By 5.2.5 and 3.4.3, we get that

(5.3.2) $$p_{x^{-1},y^{-1}} = p_{x,y} \quad \text{and} \quad P_{x^{-1},y^{-1}} = P_{x,y}$$

for all $x, y \in W$.

Definition 5.3.3. *If $\varphi > 0$, then the $P_{x,y}$ are called the* ***Kazhdan–Lusztig polynomials*** *of (W, S).*

Remark 5.3.4. Of course, symmetrically, there exists a unique element $\mathbf{C}_y^\# \in \mathscr{H}$ such that $\overline{\mathbf{C}_y^\#} = \mathbf{C}_y^\#$ and $\mathbf{C}_y^\# \equiv T_y \mod \mathscr{H}_{>0}$. In fact,

$$\mathbf{C}_y^\# = \varepsilon(y)\mathbf{C}_y^\dagger.$$

Indeed, as the involutions $^{-}$ and † commute, one gets $\mathbf{C}_y^\dagger = \overline{\mathbf{C}}_y^\dagger = \sum_{x\in W} \overline{p}_{x,y} \overline{T}_x^\dagger$, and so $\mathbf{C}_y^\dagger = \sum_{x\in W} \varepsilon(x)\overline{p}_{x,y} T_x \equiv \varepsilon(y) T_y \mod \mathscr{H}_{>0}$. So the result follows from the uniqueness statement. □

5.4. Functoriality, Parameters

5.4.A. *Changing $\mathscr{A}$*

Example 5.2.4 shows that, in contrast to the standard basis $(T_y)_{y\in W}$ (which is just the specialization of the generic standard basis $(\boldsymbol{T}_y)_{y\in W}$), the Kazhdan–Lusztig basis heavily depends on the order of $\mathscr{A}$ and on the choice of φ: if $\varphi(s)<0$ (resp. $\varphi(s)=0$, resp. $\varphi(s)>0$), then $\mathbf{C}_s$ is the specialization of $\boldsymbol{T}_s - \boldsymbol{v}_s$ (resp. $\boldsymbol{T}_s$, resp. $\boldsymbol{T}_s + \boldsymbol{v}_s^{-1}$). As usual, if we want to emphasize φ, we denote by $\mathbf{C}_y^\varphi$, $p_{x,y}^\varphi$ and $P_{x,y}^\varphi$ the objects $\mathbf{C}_y$, $p_{x,y}$ and $P_{x,y}$.

Therefore, the functoriality properties of the Kazhdan–Lusztig basis must involve the order on $\mathscr{A}$. So, let $\mathscr{A}'$ be another totally ordered abelian group and let $g : \mathscr{A} \to \mathscr{A}'$ be a *strictly increasing* morphism of groups (i.e., if $a_1 < a_2$ in $\mathscr{A}$, then $g(a_1) < g(a_2)$ in $\mathscr{A}'$). In particular, g is injective and

$$g_*(\mathscr{H}_{<0}^\varphi) \subset \mathscr{H}_{<0}^{g\circ\varphi} \tag{5.4.1}$$

so it follows from 4.4.3 that

$$g_*(\mathbf{C}_y^\varphi) = \mathbf{C}_y^{g\circ\varphi} \tag{5.4.2}$$

for all $y \in W$.

5.4.B. *Changing Signs*

As in §4.3.C, let us fix in this subsection a partition $S = S_+ \dot\cup S_-$ of S into W-closed subsets and let

$$\begin{aligned} \varphi_- : S &\longrightarrow \mathscr{A} \\ s &\longmapsto \begin{cases} \varphi(s) & \text{if } s \in S_+, \\ -\varphi(s) & \text{if } s \in S_-. \end{cases} \end{aligned}$$

By Lemma 4.3.13, the A-linear map $\Theta_{S_-} : \mathscr{H}^\varphi \longrightarrow \mathscr{H}^{\varphi_-}$ which sends T_w^φ to $(-1)^{\ell_{S_-}(w)} T_w^{\varphi_-}$ is an isomorphism of A-algebras. Moreover, Θ_{S_-} commutes with the involutions $*$, $^{-}$ and †. Also, by construction,

$$\Theta_{S_-}(\mathscr{H}_{<0}^\varphi) = \mathscr{H}_{<0}^{\varphi_-}. \tag{5.4.3}$$

Therefore,

$$\Theta_{S_-}(\mathbf{C}_w^\varphi) = (-1)^{\ell_{S_-}(w)} \mathbf{C}_w^{\varphi_-}. \tag{5.4.4}$$

5.4.C. Vanishing Parameters

As in §4.3.D, we fix in this subsection a partition $S = I \dot\cup J$ into W-closed subsets such that $\varphi(s) = 0$ for all $s \in I$ and we set

$$\widetilde{J} = \{wsw^{-1} \mid w \in W_I \text{ and } s \in J\} \quad \text{and} \quad \widetilde{W} = \langle \widetilde{J} \rangle.$$

Recall that $(\widetilde{W}, \widetilde{J})$ is a Coxeter system and that $W = W_I \ltimes \widetilde{W}$ (see Theorem B.3.1). By Corollary 4.3.17, the map $\widetilde{\varphi} : \widetilde{W} \to \mathscr{A}$ obtained by restriction from φ is braided and W_I-invariant. Also, as in §4.3.D, we set $\widetilde{\mathscr{H}} = \mathscr{H}_{\mathbb{Z}}(\widetilde{W}, \widetilde{J}, \widetilde{\varphi})$ and we define $\widetilde{T}_w = T_w^{\widetilde{\varphi}}$ and $\widetilde{\mathbf{C}}_w = \mathbf{C}_w^{\widetilde{\varphi}}$ for all $w \in \widetilde{W}$.

By Lemma 4.3.18, the A-linear map

$$\begin{array}{ccc} W_I \ltimes \widetilde{\mathscr{H}} & \longrightarrow & \mathscr{H} \\ x \ltimes \widetilde{T}_w & \longmapsto & T_{xw} \end{array}$$

is an isomorphism of A-algebras. So we view $\widetilde{\mathscr{H}}$ as an A-subalgebra of $\mathscr{H}$.

Proposition 5.4.5. *If $x \in W_I$ and $w \in \widetilde{W}$, then:*

(a) $\mathbf{C}_x = T_x$.
(b) $\mathbf{C}_w = \widetilde{\mathbf{C}}_w$.
(c) $\mathbf{C}_{xw} = \mathbf{C}_x \mathbf{C}_w = T_x \widetilde{\mathbf{C}}_w$.

Proof. (a) If $s \in I$, then $\mathbf{C}_s = T_s$ by Example 5.2.4. So (a) follows easily by induction on the length of x.

(b) follows from the fact that the inclusion $\widetilde{\mathscr{H}} \hookrightarrow \mathscr{H}$ commutes with the involution $\overline{}$ and from the characterization of the Kazhdan–Lusztig basis (see Theorem 5.2.1).

(c) As $T_x T_w = T_{xw}$, we have that

$$\overline{\mathbf{C}_x \mathbf{C}_w} = \mathbf{C}_x \mathbf{C}_w \quad \text{and} \quad \mathbf{C}_x \mathbf{C}_w \equiv T_{xw} \mod \mathscr{H}_{>0}.$$

Again, the statement (c) follows from the characterization of the Kazhdan–Lusztig basis. ■

Corollary 5.4.6. *If $x \in W_I$ and $w \in W$, then $\mathbf{C}_x \mathbf{C}_w = \mathbf{C}_{xw}$.*

Commentary 5.4.7. By 5.4.4, the computation of the Kazhdan–Lusztig basis is reduced to the case where $\varphi(s) \geqslant 0$ for all $s \in S$. By Proposition 5.4.5, the computation of the Kazhdan–Lusztig basis is reduced, possibly after changing the group, to the case where $\varphi(s) \neq 0$ for all $s \in S$. So, theoretically, the computation of the Kazhdan–Lusztig basis is reduced to the case where $\varphi > 0$. However, note that the group $(\widetilde{W}, \widetilde{J})$ might be much more complicated than (W, S): for instance, $\widetilde{J}$ might be infinite even though S is finite (see Example B.3.6). □

Example 5.4.8. If $\varphi(s)=0$ for all $s\in S$, then $\mathscr{H}=AW$ and $\mathbf{C}_w=T_w$ for all $w\in W$. □

5.5. Structure Constants

Let $(h_{x,y,z})_{x,y,z\in W}$ be the family of elements of A such that

$$\mathbf{C}_x\mathbf{C}_y=\sum_{z\in W} h_{x,y,z}\mathbf{C}_z.$$

These are the structure constants of the Hecke algebra $\mathscr{H}$ in the Kazhdan–Lusztig basis. If necessary, $h_{x,y,z}$ is denoted by $h^{\varphi}_{x,y,z}$. Of course, these structure constants obey the functoriality properties of §5.4.A. Since $\mathbf{C}_y=\overline{\mathbf{C}}_y$ and by 5.2.5, we get

$$\overline{h}_{x,y,z}=h_{x,y,z} \quad \text{and} \quad h_{y^{-1},x^{-1},z^{-1}}=h_{x,y,z} \tag{5.5.1}$$

for all $x,y,z\in W$.

In this section, we are particularly interested in the case where x or y belongs to S. We set $\lambda^s_{x,y}=h_{s,y,x}$ and $\rho^s_{x,y}=h_{y,s,x}$ (here, λ^s stands for *left* multiplication by $\mathbf{C}_s$ while ρ^s stands for *right* multiplication by $\mathbf{C}_s$). In [Lus20, §6.2 and Theorem 6.6], $\lambda^s_{x,y}$ is denoted by $\mu^s_{x,y}$ (and defined only if $sx<x<y<sy$). With this notation,

$$\mathbf{C}_s\mathbf{C}_y=\sum_{x\in W}\lambda^s_{x,y}\,\mathbf{C}_x \tag{5.5.2}$$

and

$$\mathbf{C}_y\mathbf{C}_s=\sum_{x\in W}\rho^s_{x,y}\,\mathbf{C}_x. \tag{5.5.3}$$

By 5.5.1, we obtain

$$\lambda^s_{x^{-1},y^{-1}}=\rho^s_{x,y}. \tag{5.5.4}$$

The main result of this section is the following:

Theorem 5.5.5. *Let $s\in S$ be such that $\varphi(s)\neq 0$. Then $(\mathbf{C}_s T_x)_{sx>x}$ and $(\mathbf{C}_y)_{sy<y}$ are two A-bases of the right ideal $\mathbf{C}_s\mathscr{H}$. Moreover,*

$$\begin{aligned}\mathbf{C}_s\mathscr{H}&=\{h\in\mathscr{H}\mid \mathbf{C}_s^{\dagger}h=0\}\\ &=\begin{cases}\{h\in\mathscr{H}\mid \mathbf{C}_s h=(v^{\varphi(s)}+v^{-\varphi(s)})h\} & \textit{if } \varphi(s)>0,\\[2ex] \{h\in\mathscr{H}\mid \mathbf{C}_s h=-(v^{\varphi(s)}+v^{-\varphi(s)})h\} & \textit{if } \varphi(s)<0.\end{cases}\end{aligned}$$

Proof. By applying the involution $*$ to Proposition 4.1.11, we get that $\mathcal{H}$ is a free left $\mathcal{H}_s$-module with basis $(T_x)_{sx>x}$. But, by Example 5.2.4, $\mathbf{C}_s\mathcal{H}_s = A\mathbf{C}_s$. This shows that $(\mathbf{C}_s T_x)_{sx>x}$ is an A-basis of $\mathbf{C}_s\mathcal{H}$.

Let $(\mathbf{C}_s\mathcal{H})_{<0} = \bigoplus_{sx>x} A_{<0}\mathbf{C}_s T_x$. As $\mathbf{C}_s T_x = T_{sx} + p_{1,s} T_x$ whenever $sx>x$, we get that

$$(\flat) \qquad \mathbf{C}_s T_x \equiv T_{sx} \mod \mathcal{H}_{<0}$$

and

$$(\sharp) \qquad (\mathbf{C}_s\mathcal{H})_{<0} \subset \mathcal{H}_{<0}.$$

On the other hand, the involution $^-$ stabilizes $\mathbf{C}_s\mathcal{H}$ so it is clear that, if $sy>y$, then

$$\overline{\mathbf{C}_s T_y} \equiv \mathbf{C}_s T_y \mod \Big(\bigoplus_{\substack{sx>x\\ x<y}} A\mathbf{C}_s T_x\Big).$$

Hence, we can apply Corollary 2.1.6 to the A-module $\mathbf{C}_s\mathcal{H}$: it implies that there exists a unique element $C \in (\mathbf{C}_s\mathcal{H})_{<0}$ such that

$$\begin{cases} \overline{C} = C, \\ C \equiv \mathbf{C}_s T_y \mod (\mathbf{C}_s\mathcal{H})_{<0}. \end{cases}$$

Therefore, by $(\flat)$ and $(\sharp)$, we get

$$\begin{cases} \overline{C} = C, \\ C \equiv T_{sy} \mod \mathcal{H}_{<0}. \end{cases}$$

In other words, by Theorem 5.2.1, $C = \mathbf{C}_{sy}$. As in Corollary 2.1.6, $(\mathbf{C}_{sy})_{sy>y}$ is an A-basis of $\mathbf{C}_s\mathcal{H}$. In other words, as $W = X_s^{-1} \dot\cup sX_s^{-1}$, $(\mathbf{C}_x)_{sx<x}$ is an A-basis of $\mathbf{C}_s\mathcal{H}$, as desired.

Let us now prove the last statements. Since $\mathcal{H}$ is a free left $\mathcal{H}_s$-module with basis $(T_x)_{sx>x}$, it is sufficient to show the result for $\mathcal{H} = \mathcal{H}_s$. In this case, this follows from the elementary computations in Example 5.2.4. ■

Corollary 5.5.6. *Let $s \in S$ be such that $\varphi(s) \neq 0$ and let $x, y \in W$. Then:*

(a) *If $sy<y$, then*

$$\mathbf{C}_s\mathbf{C}_y = \begin{cases} (v^{\varphi(s)} + v^{-\varphi(s)})\mathbf{C}_y & \text{if } \varphi(s)>0, \\ -(v^{\varphi(s)} + v^{-\varphi(s)})\mathbf{C}_y & \text{if } \varphi(s)<0. \end{cases}$$

(b) *If $sy>y$, then*

$$\lambda_{x,y}^s = \begin{cases} 1 & \text{if } x=sy, \\ 0 & \text{if } sx>x, \\ 0 & \text{if } x \neq sy \text{ and } x \not< y. \end{cases}$$

Proof. (a) If $sy<y$, then $\mathbf{C}_y \in \mathbf{C}_s\mathcal{H}$ by Theorem 5.5.5. So the result follows from the last statement of Theorem 5.5.5.

(b) Since $\mathbf{C}_s\mathbf{C}_y \in \mathbf{C}_s\mathcal{H}$, it follows from 5.5.2 and Theorem 5.5.5 that $\lambda_{x,y}^s=0$ whenever $sx>x$. Also,

$$\mathbf{C}_s\mathbf{C}_y \equiv T_{sy} \mod \Big(\bigoplus_{z<sy} A\, T_z\Big).$$

This shows that

$$\mathbf{C}_s\mathbf{C}_y - \mathbf{C}_{sy} \in \Big(\bigoplus_{z<sy} A\, T_z\Big) = \Big(\bigoplus_{z<sy} A\mathbf{C}_z\Big),$$

and so $\lambda_{sy,y}^s=1$.

Now, write $\mathbf{C}_y = \sum_{x\leqslant y} p_{x,y} T_x$. Then $\mathbf{C}_s\mathbf{C}_y \equiv T_s\mathbf{C}_y \mod \mathcal{H}_{<0}$ and

$$T_s\mathbf{C}_y = T_{sy} + \sum_{x<y} p_{x,y} T_{sx} + \sum_{sx<x<y} p_{x,y}(v^{\varphi(s)} - v^{-\varphi(s)})T_x.$$

Therefore,

$$\mathbf{C}_s\mathbf{C}_y - \mathbf{C}_{sy} \in \Big(\bigoplus_{x<y} A\, T_x\Big) + \mathcal{H}_{<0}.$$

So it follows from 5.3.1 and Corollary 2.1.5 that

$$\mathbf{C}_s\mathbf{C}_y - \mathbf{C}_{sy} \in \bigoplus_{x<y} A\, T_x = \bigoplus_{x<y} A\mathbf{C}_x.$$

This shows that $\lambda_{x,y}^s=0$ whenever $x\neq sy$ and $x\not< y$. ∎

The involutive anti-automorphism $*: T_w \mapsto T_{w^{-1}}$ allows one to switch from left to right multiplication, and so to deduce from Theorem 5.5.5 and Corollary 5.5.6 the following results:

Theorem 5.5.7. *Let $s\in S$ be such that $\varphi(s)\neq 0$. Then $(T_x\mathbf{C}_s)_{xs>x}$ and $(\mathbf{C}_y)_{ys<y}$ are two A-bases of the left ideal $\mathcal{H}\mathbf{C}_s$. Moreover,*

$$\begin{aligned}\mathcal{H}\mathbf{C}_s &= \{h\in\mathcal{H} \mid h\mathbf{C}_s^\dagger=0\}\\ &= \begin{cases}\{h\in\mathcal{H} \mid h\mathbf{C}_s=(v^{\varphi(s)}+v^{-\varphi(s)})h\} & \textit{if } \varphi(s)>0,\\[2ex] \{h\in\mathcal{H} \mid h\mathbf{C}_s=-(v^{\varphi(s)}+v^{-\varphi(s)})h\} & \textit{if } \varphi(s)<0.\end{cases}\end{aligned}$$

Corollary 5.5.8. *Let $s\in S$ be such that $\varphi(s)\neq 0$ and let $x, y\in W$. Then:*

(a) *If $ys<y$, then*

$$\mathbf{C}_y\mathbf{C}_s = \begin{cases}(v^{\varphi(s)}+v^{-\varphi(s)})\mathbf{C}_y & \textit{if } \varphi(s)>0,\\ -(v^{\varphi(s)}+v^{-\varphi(s)})\mathbf{C}_y & \textit{if } \varphi(s)<0.\end{cases}$$

(b) *If* $ys > y$, *then*

$$\rho^s_{x,y} = \begin{cases} 1 & \text{if } x = ys, \\ 0 & \text{if } xs > x, \\ 0 & \text{if } x \neq ys \text{ and } x \not< y. \end{cases}$$

***Commentary* 5.5.9.** In Chapter 8, we investigate the relations between the cells of W and those of its parabolic subgroups. The results of this section are just special cases of the Geck induction theorem (see Theorem 8.2.1), applied to the case where the parabolic subgroup is generated by a single simple reflection s. □

5.6. An Algorithm

The aim of this section is to describe a basic algorithm, based only on the elementary properties proved in the previous section 5.5, for computing *simultaneously* the Kazhdan–Lusztig polynomials $p_{x,y}$ and the structure constants $\lambda^s_{x,y}$. This algorithm was sufficient for computing the full Kazhdan–Lusztig basis of a finite Coxeter group of rank $\leqslant 5$ (except for type H_4) on a standard computer in 2010. More sophisticated algorithms are available, relying on finer properties of the Kazhdan–Lusztig basis.

In all the results of this section, we assume that φ takes positive values on some (or all) elements of S: it has been explained in Commentary 5.4.7 how to handle the general case from this one.

The algorithm computes the polynomials $p_{x,y}$ and the structure constants $\lambda^s_{x,y}$ by induction on the pair $(\ell(y), \ell(y) - \ell(x))$, ordered lexicographically. It is based on the following observations:

Proposition 5.6.1. *Let* x, $y \in W$ *and let* $s \in S$ *be such that* $sy > y$ *and* $\varphi(s) > 0$. *Then:*

(a) $$p_{x,sy} = \begin{cases} p_{sx,y} + v^{\varphi(s)} p_{x,y} - \displaystyle\sum_{\substack{x \leqslant z < y \\ (sz < z)}} p_{x,z} \lambda^s_{z,y} & \text{if } sx < x, \\ v^{-\varphi(s)} p_{sx,sy} & \text{if } sx > x. \end{cases}$$

(b) *If* $sx < x < y < sy$, *then* $\lambda^s_{x,y}$ *satisfies the following two properties:*

$$\begin{cases} \lambda^s_{x,y} \equiv v^{\varphi(s)} p_{x,y} - \displaystyle\sum_{\substack{x < z < y \\ (sz < z)}} p_{x,z} \lambda^s_{z,y} \mod A_{<0}, \\ \overline{\lambda}^s_{x,y} = \lambda^s_{x,y}. \end{cases}$$

***Remark* 5.6.2.** In the above formulas, the condition $sz < z$ has been written in parentheses, meaning that it may be added or removed according to our needs: indeed, if $sz \not< z$, then $\lambda^s_{z,y} = 0$ and so the contribution of the corresponding term in the sum is zero. □

Proof. (a) Since $sy > y = ssy$ and $\varphi(s) > 0$, we have

$$T_s\mathbf{C}_{sy} = v^{\varphi(s)}\mathbf{C}_{sy}$$

(see Example 5.2.4 and Corollary 5.5.6(a)). Therefore,

$$\sum_{x\in W} p_{x,sy}\left(T_{sx} + \delta_{sx<x}(v^{\varphi(s)} - v^{-\varphi(s)})T_x\right) = \sum_{x\in W} p_{x,sy}\, v^{\varphi(s)}\, T_x.$$

In other words,

$$\sum_{x\in W}\left(p_{sx,sy} + (\delta_{sx<x}(v^{\varphi(s)} - v^{-\varphi(s)}) - v^{\varphi(s)})p_{x,sy}\right)T_x = 0.$$

Thus, if $sx > x$, then $p_{sx,sy} = v^{\varphi(s)}p_{x,sy}$: this shows the second equality in (a).

Let us now prove the first one. By 5.5.2, we have $\mathbf{C}_s\mathbf{C}_y = \sum_{z\in W}\lambda^s_{z,y}\mathbf{C}_z$. Therefore,

$$\sum_{x\in W} p_{x,y}\left(T_{sx} + \delta_{sx<x}(v^{\varphi(s)} - v^{-\varphi(s)})T_x + v^{-\varphi(s)}T_x\right) = \sum_{x,z\in W} p_{x,z}\lambda^s_{z,y}\, T_x.$$

By comparing the coefficient of T_x on both sides, one gets, if $sx < x$,

$$p_{sx,y} + v^{\varphi(s)}p_{x,y} = \sum_{z\in W} p_{x,z}\lambda^s_{z,y}.$$

As $\lambda^s_{sy,y} = 1$ and $\lambda^s_{z,y} = 0$ if $z \not< y$ and $z \neq sy$ (see Corollary 5.5.6(c)), we get

$$p_{x,sy} = p_{sx,y} + v^{\varphi(s)}p_{x,y} - \sum_{x\leqslant z<y\in W} p_{x,z}\lambda^s_{z,y},$$

as desired.

(b) Assume now that $sx < x < y < sy$. By (a), and since $p_{x,x} = 1$, one gets

$$\lambda^s_{x,y} = -p_{x,sy} + p_{sx,y} + v^{\varphi(s)}p_{x,y} - \sum_{x<z<y\in W} p_{x,z}\lambda^s_{z,y}.$$

But, since $x < sy$, and $sx < y$, we have that $p_{x,sy} \in A_{<0}$ and $p_{sx,y} \in A_{<0}$ by 5.3.1. Consequently,

$$\lambda^s_{x,y} \equiv v^{\varphi(s)}p_{x,y} - \sum_{x<z<y\in W} p_{x,z}\lambda^s_{z,y} \mod A_{<0},$$

as desired. Finally, the fact that $\lambda^s_{x,y}$ is invariant under the involution $\bar{}$ follows from 5.5.1. ■

Let us now explain how Proposition 5.6.1 provides an algorithm to compute the $p_{x,y}$ and the $\lambda^s_{x,y}$. In the easy example where $\ell(y) - \ell(x) = 1$, the algorithm is developed in Example 5.7.3. So let $w \in W$: we want to compute the $p_{x,y}$ and the $\lambda^s_{x,y}$ for all $x \leqslant y \leqslant w$ and all $s \in S$.

- First, by Corollary 5.5.6(b), if $x \not\leqslant y$, then $p_{x,y} = 0$ and

$$\lambda_{x,y}^{s} = \begin{cases} 1 & \text{if } x = sy \geqslant y, \\ 0 & \text{otherwise.} \end{cases}$$

- The first step shows that we may, and we will, now assume that $x \leqslant y$. The algorithm works by induction on the pair $(\ell(y), \ell(y) - \ell(x)) \in \mathbb{N} \times \mathbb{N}$, ordered lexicographically. Note that, if $\ell(y) = \ell(x)$, that is, if $x = y$, then $p_{y,y} = 1$ and, by Corollary 5.5.6(a),

$$\lambda_{y,y}^{s} = \begin{cases} 0 & \text{if } sy \geqslant y, \\ v^{\varphi(s)} + v^{-\varphi(s)} & \text{if } sy \leqslant y. \end{cases}$$

- So assume now that $\ell(y) - \ell(x) \geqslant 1$ and that the $p_{x',y'}$ and the $\lambda_{x',y'}^{s'}$ have been computed if $(\ell(y'), \ell(y') - \ell(x'))$ is strictly smaller than $(\ell(y), \ell(y) - \ell(x))$ for the lexicographic order. Then:
  - As $\ell(y) \geqslant 1$, there exists an $s \in S$ such that $sy < y$. Then:

$$p_{x,y} = \begin{cases} p_{sx,sy} + v^{\varphi(s)} p_{x,sy} - \displaystyle\sum_{\substack{x \leqslant z < sy \\ (sz < z)}} p_{x,z} \lambda_{z,sy}^{s} & \text{if } sx < x, \\ v^{-\varphi(s)} p_{sx,y} & \text{if } sx > x \end{cases}$$

    (see Proposition 5.6.1(a)). Hence $p_{x,y}$ can be computed using this formula, since all the terms that are involved in the right-hand side are known by the induction hypothesis.
  - Now let $s' \in S$. If $s'y < y$, then $\lambda_{x,y}^{s'} = 0$ by Corollary 5.5.6(b). If $s'y > y$, then $v^{\varphi(s')} p_{x,y} - \sum_{\substack{x < z < y \\ (sz < z)}} p_{x,z} \lambda_{z,y}^{s'}$ is known thanks to the previous step and the induction hypothesis. By Proposition 5.6.1(b), $\lambda_{x,y}^{s'}$ is now easily determined by the two conditions (see Remark 2.1.4):

$$\begin{cases} \lambda_{x,y}^{s'} \equiv v^{\varphi(s')} p_{x,y} - \displaystyle\sum_{\substack{x < z < y \\ (sz < z)}} p_{x,z} \lambda_{z,y}^{s'} \mod A_{<0}, \\ \overline{\lambda}_{x,y}^{s'} = \lambda_{x,y}^{s'}. \end{cases}$$

Commentary 5.6.3. The previous algorithm is more efficient than the naive algorithm based on the proof of Corollary 2.1.6 which consists in computing first the polynomials $r_{x,y}$ and then deducing from the proof how to compute the $p_{x,y}$. Moreover, it also gives the structure constants $\lambda_{x,y}^s$ which are necessary if one wants to compute the Kazhdan–Lusztig cells, as is explained in the next Chapter 6.

However, if one is only interested in computing some Kazhdan–Lusztig polynomials $p_{x,y}^{\varphi}$ (with x and y fixed) for different values of φ, then the algorithm based on the relations 5.3.1 between Kazhdan–Lusztig polynomials and r-polynomials becomes more efficient, as one can use the fact that the r-polynomials do not depend on φ.

Much more efficient algorithms are due to F. du Cloux [duCl2] or M. Geck [Gec9]. □

We now gather properties that can be proved by an induction argument (induction on the pair $(\ell(y), \ell(y)-\ell(x))$) following the principles described in the previous algorithm (the easy proof is omitted):

Proposition 5.6.4. *Let x and y be two elements of W such that $x \leqslant y$. We also assume that $\varphi > 0$. Then:*

(a) $P_{x,y} \equiv 1 \mod \mathbb{Z}[2\mathscr{A}_{>0}]$ *(recall that $P_{x,y} = v^{\varphi(y)-\varphi(x)} p_{x,y}$).*
(b) $\mathrm{val}(p_{x,y}) = \varphi(x) - \varphi(y)$.
(c) *If $s \in S$ is such that $sx < x < y < sy$, then $\deg(\lambda_{x,y}^s) < \varphi(s)$ and $v^{\varphi(y)-\varphi(x)} \lambda_{x,y}^s \in \mathbb{Z}[2\mathscr{A}]$.*

Corollary 5.6.5. *Let x and y be two elements of W such that $x \leqslant y$. Then the following properties hold, without assumption on φ:*

(a) $P_{x,y} \in \mathbb{Z}[2\mathscr{A}]$.
(b) *If $s \in S$ is such that $sx < x < y < sy$, then $v^{\varphi(y)-\varphi(x)} \lambda_{x,y}^s \in \mathbb{Z}[2\mathscr{A}]$.*

Proof. Using Proposition 5.4.5, we may assume that $\varphi(s) \neq 0$ for all $s \in S$. Now, let $S_- = \{s \in S \mid \varphi(s) < 0\}$ and $S_+ = \{s \in S \mid \varphi(s) > 0\}$ and let us apply the results of §5.4.B in this context. Then $p_{x,y}^{\varphi} = (-1)^{\ell_{S_-}(y)-\ell_{S_-}(x)} p_{x,y}^{\varphi_-}$. But $\varphi_-(s) > 0$ for all $s \in S$, so $P_{x,y}^{\varphi_-} = v^{\varphi_-(y)-\varphi_-(x)} p_{x,y}^{\varphi_-} \in \mathbb{Z}[2\mathscr{A}]$. On the other hand, $\varphi_-(w) - \varphi(w) \in 2\mathscr{A}$ for all $w \in W$: this shows (a).

(b) follows from the same argument. ■

Again, by using the involutive anti-automorphism $* : T_w \longmapsto T_{w^{-1}}$, one deduces immediately from the previous results the following:

Proposition 5.6.6. *Let x, $y \in W$ and let $s \in S$ be such that $ys > y$ and $\varphi(s) > 0$. Then:*

$$\text{(a) } p_{x,ys} = \begin{cases} p_{xs,y} + v^{\varphi(s)} p_{x,y} - \displaystyle\sum_{\substack{x \leqslant z < y \\ (zs < z)}} p_{x,z} \rho_{z,y}^s & \text{if } xs < x, \\ v^{-\varphi(s)} p_{xs,ys} & \text{if } xs > x. \end{cases}$$

(b) *If* $xs < x < y < ys$, *then* $\rho^s_{x,y}$ *satisfies the following two properties:*

$$\begin{cases} \rho^s_{x,y} \equiv v^{\varphi(s)} p_{x,y} - \displaystyle\sum_{\substack{x<z<y \\ (zs<z)}} p_{x,z} \rho^s_{z,y} \mod A_{<0}, \\ \overline{\rho}^s_{x,y} = \rho^s_{x,y}. \end{cases}$$

Corollary 5.6.7. *Let* x *and* y *be two elements of* W *such that* $x \leqslant y$. *Then:*

(a) *If* $s \in S$ *is such that* $xs < x < y < ys$, *then* $v^{\varphi(y)-\varphi(x)} \rho^s_{x,y} \in \mathbb{Z}[2\mathscr{A}]$.

(b) *If, moreover,* $\varphi > 0$ *and if* $s \in S$ *is such that* $xs < x < y < ys$, *then* $\deg(\rho^s_{x,y}) < \varphi(s)$ *and* $v^{\varphi(y)-\varphi(x)} \rho^s_{x,y} \in \mathbb{Z}[2\mathscr{A}]$.

5.7. Examples

Example 5.7.1 (Parabolic subgroups). Let I be a subset of S and let $\varphi_I : I \to \mathscr{A}$ denote the restriction of φ. Let $w \in W_I$: we denote by $\mathbf{C}^{\varphi_I}_w$ the element of the Kazhdan–Lusztig basis computed in the parabolic subalgebra $\mathscr{H}_I$ and by $\mathbf{C}^{\varphi}_w$ the element of the Kazhdan–Lusztig basis computed in $\mathscr{H}$. Then

$$\mathbf{C}^{\varphi_I}_w = \mathbf{C}^{\varphi}_w.$$

Indeed, the involution $\overline{}$ stabilizes $\mathscr{H}_I$ and the result follows from the characterization of the Kazhdan–Lusztig basis.

If x, y, $z \in W_I$, this shows that the elements $p_{x,y}$, $P_{x,y}$ and the structure constants $h_{x,y,z}$ can be computed in $\mathscr{H}$ as well as in $\mathscr{H}_I$.

Note that all the facts established here could have also been proved using an induction argument based on the algorithm described in the previous section 5.6. □

Example 5.7.2 (Disjoint support). Let I and J be two subsets of S such that $I \cap J = \varnothing$ and let $x \in W_I$ and $y \in W_J$. Then

$$\mathbf{C}_x \mathbf{C}_y = \mathbf{C}_{xy}.$$

Indeed, if $x' \leqslant x$ and $y' \leqslant y$, then $x' \in W_I$ and $y' \in W_J$ (see 3.5.3) and so $T_{x'} T_{y'} = T_{x'y'}$ (see the equality (♡) in Example 3.6.10). Therefore,

$$\mathbf{C}_x \mathbf{C}_y = \sum_{\substack{x' \leqslant x \\ y' \leqslant y}} p_{x',x} p_{y',y} T_{x'y'} \equiv T_{xy} \mod \mathscr{H}_{<0}.$$

Since $\overline{\mathbf{C}_x \mathbf{C}_y} = \mathbf{C}_x \mathbf{C}_y$, the result follows from the characterization of the Kazhdan–Lusztig basis. Note also that the above argument shows that

$$p_{x'y',xy} = p_{x',x} p_{y',y} \qquad \text{and} \qquad P_{x'y',xy} = P_{x',x} P_{y',y}$$

for all $x' \leqslant x \in W_I$ and $y' \leqslant y \in W_J$. □

Example 5.7.3 (Length difference 1). Here we apply the algorithm to deal with the easy case when the length difference is equal to 1. So, let x and y in W be such that $\ell(y) = \ell(x)+1$ and $x<y$. Let $(s_1, s_2, \ldots, s_r)$ be a reduced expression of y and let i be such that $x = s_1 s_2 \cdots s_{i-1} s_{i+1} \cdots s_r$. *We assume in this example that* $\varphi(s_i) > 0$. Then

$$(\clubsuit) \qquad \begin{cases} \boldsymbol{r}_{x,y} = \boldsymbol{v}_{s_i} - \boldsymbol{v}_{s_i}^{-1}, \\ r_{x,y} = v^{\varphi(s_i)} - v^{-\varphi(s_i)}, \\ p_{x,y} = v^{-\varphi(s_i)}. \end{cases}$$

Let us prove the first and third equalities (note that the second one follows immediately from the first one) by induction on r: note that the case $r=1$ is obvious. Two cases may occur.

• If $i=1$, then $s_1 x = y > x$ and it follows from Lemma 4.4.7 that

$$\boldsymbol{r}_{x,y} = \boldsymbol{r}_{x,s_1 x} = \boldsymbol{r}_{s_1 x,x} + (\boldsymbol{v}_{s_1} - \boldsymbol{v}_{s_1}^{-1})\boldsymbol{r}_{x,x}.$$

But $\boldsymbol{r}_{x,x} = 1$ and $\boldsymbol{r}_{s_1 x,x} = 0$ since $s_1 x > x$. On the other hand, $p_{x,y} = p_{x,s_1 x} = v^{-\varphi(s_1)} p_{s_1 x, s_1 x} = v^{-\varphi(s_1)}$ (see Proposition 5.6.1(a)).

• If $i \geqslant 2$, then $s_1 x < x$ and we set $y' = s_1 y < y$. It now follows from Lemma 4.4.7 that

$$\boldsymbol{r}_{x,y} = \boldsymbol{r}_{x,s_1 y'} = \boldsymbol{r}_{s_1 x,y'} = \boldsymbol{v}_{s_i} - \boldsymbol{v}_{s_i}^{-1}$$

by the induction hypothesis. On the other hand, $x \not\leqslant y'$ (indeed, as $\ell(x) = \ell(y')$, $x \leqslant y'$ would imply that $x = y'$, but this is impossible as $s_1 x < x$ and $s_1 y' > y'$). So $p_{x,y} = p_{x,s_1 y'} = p_{s_1 x, y'}$ by Proposition 5.6.1(a), and $p_{s_1 x,y'} = v^{-\varphi(s_1)}$ by the induction hypothesis.

Now, let $s \in S$ be such that

$$sx < x < y < sy.$$

Then

$$(\diamondsuit) \qquad \lambda_{x,y}^s = \begin{cases} v^{\varphi(s)-\varphi(s_i)} + v^{\varphi(s_i)-\varphi(s)} & \text{if } \varphi(s) > \varphi(s_i) > 0, \\ 1 & \text{if } \varphi(s) = \varphi(s_i) > 0, \\ 0 & \text{if } \varphi(s_i) > \varphi(s) > 0. \end{cases}$$

Indeed, by Proposition 5.6.1(b), we have $\lambda_{x,y}^s \equiv v^{\varphi(s)-\varphi(s_i)} \bmod A_{<0}$. Since $\overline{\lambda}_{x,y}^s = \lambda_{x,y}^s$, the result follows. □

Example 5.7.4 ($\varphi = \ell$, the split case). Assume in this example, and only in this example, that φ is constant and let $a \in \mathscr{A}$ denote this constant value. The case where $a = 0$ has been treated in §5.4.C, so we assume moreover in this example that $a \neq 0$. By 5.4.4, we may assume that $a > 0$. Then the map $g : \mathbb{Z} \to \mathscr{A}$, $n \mapsto na$ is strictly increasing and $\varphi = g \circ \ell$. So, by 5.4.2, we may, and we will, assume in this example, and only in this example, that $\mathscr{A} = \mathbb{Z}$ and $\varphi = \ell$: this is called the *split case* throughout this book (see Example 4.2.3).

If $x, y \in W$, we follow tradition and denote by $\mu(x,y)$ the coefficient of v^{-1} in $p_{x,y}$ (recall that $\deg(p_{x,y}) \leqslant -1$ if $x \neq y$). In other words, $\mu(x,y) = \mathrm{coef}_{-1}(p_{x,y})$. Note that $\mu(x,y) = 0$ if $\ell(y) - \ell(x)$ is even (see Proposition 5.6.4). Now, let $s \in S$. Let us prove the following statements:

(5.7.5) *If $x \neq y$, then $\lambda^s_{x,y}, \rho^s_{x,y} \in \mathbb{Z}$.*

(5.7.6) *If $x \neq y$ and $\lambda^s_{x,y} \neq 0$ or $\rho^s_{x,y} \neq 0$, then $\ell(y) - \ell(x)$ is odd.*

(5.7.7) *If $sx < x < y < sy$, then $\lambda^s_{x,y} = \mu(x,y)$.*

(5.7.8) *If $xs < x < y < ys$, then $\rho^s_{x,y} = \mu(x,y)$.*

Note first that, to prove these four statements, it is sufficient to prove only those about the $\lambda^s_{x,y}$. Also, by Corollary 5.5.6(b), we may assume that $sx < x < y < sy$. Therefore, by Proposition 5.6.4(c), we have $\lambda^s_{x,y} \in \mathbb{Z}$. In this case, Proposition 5.6.1(b) implies that $\lambda^s_{x,y} \equiv v p_{x,y} \mod A_{<0}$. This proves 5.7.7, while 5.7.5 and 5.7.6 follow from 5.7.7. □

Example 5.7.9 (Type A_2). As in Example 3.12.7, we assume here that $S = \{s, t\}$ with $m_{st} = 3$ so that (W, S) is of type A_2. Let $a = \varphi(s) = \varphi(t)$ (since s and t are conjugate in W). Here is the corresponding Coxeter graph, together with the values of φ:

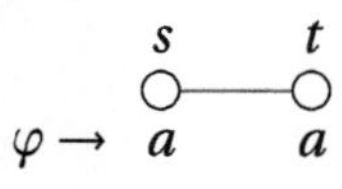

Recall from Example 3.12.7 that $W = \{1, s, t, st, ts, tst = sts = w_0\}$. By Example 5.4.8, we may assume that $a \neq 0$ and, by Commentary 5.4.7, we may assume that $a > 0$.

An easy application of the algorithm described in §5.6 shows that

$$p_{x,y} = v^{-a(\ell(y)-\ell(x))} \text{ for all } x \leqslant y.$$

In other words,

$$P_{x,y} = 1 \textit{ for all } x \leqslant y.$$

Also, the structure constants $\lambda^{?}_{x,y}$ are determined by the general facts (see Corollary 5.5.6) and the following special cases (which follow from Example 5.7.3). We only write the result of the multiplication $\mathbf{C}_s\mathbf{C}_x$ (or $\mathbf{C}_t\mathbf{C}_x$) whenever $sx > x$ (or $tx > x$).

$$\begin{cases} \mathbf{C}_s\mathbf{C}_1 = \mathbf{C}_s, \\ \mathbf{C}_s\mathbf{C}_t = \mathbf{C}_{st}, \\ \mathbf{C}_s\mathbf{C}_{ts} = \mathbf{C}_{w_0} + \mathbf{C}_s, \end{cases} \quad \text{and} \quad \begin{cases} \mathbf{C}_t\mathbf{C}_1 = \mathbf{C}_t, \\ \mathbf{C}_t\mathbf{C}_s = \mathbf{C}_{ts}, \\ \mathbf{C}_t\mathbf{C}_{st} = \mathbf{C}_{w_0} + \mathbf{C}_t. \end{cases}$$

Among other things, note that $\mathbf{C}_{w_0} = \sum_{w \in W} v^{\varphi(w)-\varphi(w_0)} T_w$ and

$$\mathbf{C}_{w_0}^2 = v^{-a\ell(w_0)}\Big(\sum_{w \in W} v^{2a\ell(w)}\Big)\mathbf{C}_{w_0}.$$

Note that $\sum_{w \in W} v^{2a\ell(w)} = (1+v^{2a})(1+v^{2a}+v^{4a})$ and $\ell(w_0) = 3$. □

Example 5.7.10 (Type B_2). As in Example 3.12.8, we assume that $S = \{s,t\}$ with $m_{st} = 4$, so that (W,S) is of type B_2. We write $a = \varphi(s)$ and $b = \varphi(t)$. Here is the corresponding Coxeter graph, together with the values of φ:

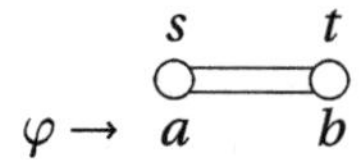

As in the previous example 5.7.9, using Commentary 5.4.7, we may, and we will, assume that $a, b > 0$. Again, the structure constants $\lambda^{?}_{x,y}$ are determined by the general facts (see Corollary 5.5.6) and the following special cases (which follow from Example 5.7.3).

$$\begin{cases} \mathbf{C}_s\mathbf{C}_1 = \mathbf{C}_s, \\ \mathbf{C}_s\mathbf{C}_t = \mathbf{C}_{st}, \\ \mathbf{C}_s\mathbf{C}_{ts} = \begin{cases} \mathbf{C}_{sts} & \text{if } a < b, \\ \mathbf{C}_{sts} + \mathbf{C}_s & \text{if } a = b, \\ \mathbf{C}_{sts} + (v^{b-a} + v^{a-b})\mathbf{C}_s & \text{if } a > b, \end{cases} \\ \mathbf{C}_s\mathbf{C}_{tst} = \begin{cases} \mathbf{C}_{w_0} & \text{if } a < b, \\ \mathbf{C}_{w_0} + \mathbf{C}_{st} & \text{if } a = b, \\ \mathbf{C}_{w_0} + (v^{b-a} + v^{a-b})\mathbf{C}_{st} & \text{if } a > b, \end{cases} \end{cases}$$

$$\text{and}\quad \begin{cases} \mathbf{C}_t\mathbf{C}_1 = \mathbf{C}_t, \\ \mathbf{C}_t\mathbf{C}_s = \mathbf{C}_{ts}, \\ \mathbf{C}_t\mathbf{C}_{st} = \begin{cases} \mathbf{C}_{tst} + (v^{b-a}+v^{a-b})\mathbf{C}_t & \text{if } a<b, \\ \mathbf{C}_{tst} + \mathbf{C}_t & \text{if } a=b, \\ \mathbf{C}_{tst} & \text{if } a>b, \end{cases} \\ \mathbf{C}_t\mathbf{C}_{sts} = \begin{cases} \mathbf{C}_{w_0} + (v^{b-a}+v^{a-b})\mathbf{C}_{ts} & \text{if } a<b, \\ \mathbf{C}_{w_0} + \mathbf{C}_{ts} & \text{if } a=b, \\ \mathbf{C}_{w_0} & \text{if } a>b. \end{cases} \end{cases}$$

We also obtain the values of the Kazhdan–Lusztig polynomials $P_{x,y}$: they are given in Table 5.a.

x \ y	1	s	t	st	ts	sts	tst	w_0
1	1	1	1	1	1	$\begin{cases} 1+v^{2a} & \text{if } a<b \\ 1 & \text{if } a=b, \\ 1-v^{2b} & \text{if } a>b, \end{cases}$	$\begin{cases} 1-v^{2a} & \text{if } a<b \\ 1 & \text{if } a=b, \\ 1+v^{2b} & \text{if } a>b, \end{cases}$	1
s	.	1	.	1	1	$\begin{cases} 1+v^{2a} & \text{if } a<b \\ 1 & \text{if } a=b, \\ 1-v^{2b} & \text{if } a>b, \end{cases}$	1	1
t	.	.	1	1	1	1	$\begin{cases} 1-v^{2a} & \text{if } a<b \\ 1 & \text{if } a=b, \\ 1+v^{2b} & \text{if } a>b, \end{cases}$	1
st	.	.	.	1	.	1	1	1
ts	.	.	.	.	1	1	1	1
sts	.	.	.	.	.	1	.	1
tst	.	.	.	.	.	.	1	1
w_0	.	.	.	.	.	.	.	1

Table 5.a. Kazhdan–Lusztig polynomials in type B_2

It follows from this table that, if φ is constant (i.e. if $a=b$), then $P_{x,y}=1$ for all $x \leqslant y$. Finally, note that $\mathbf{C}_{w_0} = \sum_{w\in W} v^{\varphi(w)-\varphi(w_0)} T_w$ and

$$\mathbf{C}_{w_0}^2 = v^{-\varphi(w_0)}\Big(\sum_{w\in W} v^{\varphi(w)}\Big)\mathbf{C}_{w_0}$$

for all $a, b \in \mathscr{A}_{>0}$. □

Notes

Almost all the results of this chapter have been proved by Kazhdan and Lusztig [KaLu1] in the *equal parameter case* (i.e. $\mathcal{A} = \mathbb{Z}$ and $\varphi(s) = 1$ for all $s \in S$) and have been extended by Lusztig to the general case [Lus9], [Lus20]. However, we have sometimes given slightly different proofs, based on a systematic use of Corollaries 2.1.3 and 2.1.6 (especially for Theorem 5.5.5 and its consequences).

Note also that the basis $(\mathbf{C}_w)_{w\in W}$ was denoted by $(C'_w)_{w\in W}$ in [KaLu1] and $(c_w)_{w\in W}$ in [Lus20]. The structure constants $\lambda^s_{x,y}$ were generally denoted by $\mu^s_{x,y}$ or $M^s_{x,y}$ in the literature (see, for instance, [Lus9], [Lus20], [Shi5]).

Exercises

Exercise 5.1 (Type G_2). As in Exercise 3.1, assume here that $S = \{s,t\}$ with $m_{st} = 6$, so that (W,S) is of type G_2. Let $a = \varphi(s)$ and $b = \varphi(t)$ (recall that s and t are not conjugate in W) and assume that $a, b > 0$. Here is the corresponding Coxeter graph, together with the values of φ:

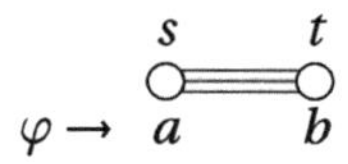

Compute the Kazhdan–Lusztig polynomials according to the values of a and b. Check that, if $a = b$, then all the Kazhdan–Lusztig polynomials $P_{x,y}$ and all the structure constants $h_{x,y,z}$ have non-negative coefficients.

Show that, in all cases, $\mathbf{C}_{w_0} = \sum_{w\in W} v^{\varphi(w)-\varphi(w_0)} T_w$ and

$$\mathbf{C}_{w_0}^2 = v^{-\varphi(w_0)}\Big(\sum_{w\in W} v^{2\varphi(w)}\Big)\mathbf{C}_{w_0}.$$

Extend, if possible, all of the above results to the case where $a, b \geqslant 0$.

Exercise 5.2 (Direct products). As in Exercises 3.2 and 4.2, assume that $S = S_1 \dot\cup S_2$ is a partition such that, if $s_1 \in S_1$ and $s_2 \in S_2$, then $m_{s_1 s_2} = 2$ (i.e. $s_1 s_2 = s_2 s_1$). Let $W_i = W_{S_i}$ $(i \in \{1,2\})$. Let φ_i denote the restriction of φ to S_i and let $\mathcal{H}_i = \mathcal{H}_A(W_i, S_i, \varphi_i)$. If $(w_1, w_2) \in W_1 \times W_2$, then show that

$$\mathbf{C}^{\varphi}_{w_1 w_2} = \mathbf{C}^{\varphi_1}_{w_1}\mathbf{C}^{\varphi_2}_{w_2}$$

through the isomorphism $\mathcal{H} = \mathcal{H}_1 \otimes \mathcal{H}_2$ of Exercise 4.2.

Deduce the corresponding properties of Kazhdan–Lusztig polynomials and structure constants.

Exercise 5.3. Write $T_y = \sum_{x \in W} q_{x,y} \mathbf{C}_x$, with $q_{x,y} \in A$. Show that:

(a) $q_{x,x} = 1$.
(b) If $q_{x,y} \neq 0$, then $x \leqslant y$.
(c) If $x \neq y$, then $q_{x,y} \in A_{<0}$.

Exercise 5.4. Let $x, y \in W$ and $s \in S$ be such that $s \in \mathscr{R}(x) \cap \mathscr{L}(y)$. Show that $\mathbf{C}_x \mathbf{C}_y \in (v^{\varphi(s)} + v^{-\varphi(s)})\mathscr{H}$.

Exercise 5.5. Let I be a spherical subset of S of cardinality $\geqslant 2$. Recall that E^{φ} denotes the vector space $\mathbb{R}(\mathscr{A})E$ endowed with the left $\mathbb{R}(\mathscr{A})\mathscr{H}$-module structure given by Theorem 4.5.1. We wish to prove that $\mathbf{C}^{\dagger}_{w_I}$ acts by 0 on E^{φ}. So let $e \in E^{\varphi}$ and let $f = \mathbf{C}^{\dagger}_{w_I} e$.

(a) Show that $T_s f = -v^{-\varphi(s)} f$ for all $s \in I$.
(b) Using Lemma 4.5.2 and the fact that $|I| \geqslant 2$, show that $f = 0$.

Chapter 6
Kazhdan–Lusztig Cells

This chapter is devoted to the definition and the very first elementary properties of *Kazhdan–Lusztig cells,* defined using the Kazhdan–Lusztig basis $(\mathbf{C}_w)_{w\in W}$ and the constructions developed in Chapter 1.

6.1. Preorders on W

We keep the notation and notions developed in Chapter 1, but here we assume that $X = W$ and $\mathscr{C} = (\mathbf{C}_w)_{w\in W}$. In this case, the elements $a_{x,y,z}$ of Chapter 1 are denoted $h_{x,y,z}$ in the context of Hecke algebras endowed with the Kazhdan–Lusztig basis. For instance, $\overset{L}{\twoheadleftarrow}$ denotes the relation defined by

$$x \overset{L}{\twoheadleftarrow} y \Longleftrightarrow \exists w \in W,\ h_{w,y,x} \neq 0. \tag{6.1.1}$$

Then $\leqslant_L$ denotes the transitive closure of $\overset{L}{\twoheadleftarrow}$. Similarly, we define the relations $\sim_L$, $\overset{R}{\twoheadleftarrow}$, $\leqslant_R$, $\sim_R$, $\overset{LR}{\twoheadleftarrow}$, $\leqslant_{LR}$ and $\sim_{LR}$ as in §1.1.

It must be noticed that, as well as the Kazhdan–Lusztig basis depending on φ, the above relations also heavily depend on φ: we will see this in Example 6.1.11 (see also Chapter 21). Therefore, if we need to emphasize the weight function φ, we use, for instance, the notation $\overset{L,\varphi}{\twoheadleftarrow}$, $\leqslant_R^\varphi$ and $\sim_{LR}^\varphi$ instead of $\overset{L}{\twoheadleftarrow}$, $\leqslant_R$ and $\sim_{LR}$, respectively.

Example 6.1.2. Let x and $y \in W$ be such that $\ell(xy) = \ell(x) + \ell(y)$. Then

$$h_{x,y,xy} = 1.$$

C. Bonnafé, *Kazhdan–Lusztig Cells with Unequal Parameters*,
Algebra and Applications 24, https://doi.org/10.1007/978-3-319-70736-5_6

Indeed, let $C = \mathbf{C}_x\mathbf{C}_y - \mathbf{C}_{xy}$. An easy computation shows that $C \in (\bigoplus_{z<xy} AT_z)$. So it follows from Corollary 2.1.5 that $C \in \bigoplus_{z<xy} A\mathbf{C}_z$. This proves that $h_{x,y,xy} = 1$.

So, by definition, $xy \leqslant_L y$ and $xy \leqslant_R x$. As an example, take $y = 1$, so that $x \leqslant_L 1$ (and so $x \leqslant_R 1$) for all $x \in W$. □

Remark 6.1.3. Let $\mathbf{h} = (\mathbf{C}_s)_{s\in S}$: as $(T_s)_{s\in S}$ generates $\mathcal{H}$, $\mathbf{h}$ also generates $\mathcal{H}$. We can then apply Remark 1.1.5 to compute the relations $\leqslant_?$ and $\sim_?$: we denote by $\overset{L}{\longleftarrow}$ and $\overset{R}{\longleftarrow}$ the relations $\overset{\mathbf{h},L}{\twoheadleftarrow}$ and $\overset{\mathbf{h},R}{\twoheadleftarrow}$, respectively. Computing the structure constants $\lambda_{x,y}^s$ and $\rho_{x,y}^s$ (or at least determining when they vanish or not) allows us to determine the relations $\overset{L}{\longleftarrow}$ and $\overset{R}{\longleftarrow}$, respectively, and incidently allows us to compute $\leqslant_L$ and $\leqslant_R$, respectively. □

6.1.A. Cells

As in §1.1.2, we define cells in W as follow:

Definition 6.1.4 (Cells). *A **(Kazhdan–Lusztig) left cell** (resp. **right cell**, resp. **two-sided cell**) is an equivalence class in W for the relation $\sim_L$ (resp. $\sim_R$, resp. $\sim_{LR}$). If one wants to emphasize the weight function φ, they are called left, right or two-sided **φ-cells**. The set of these cells is denoted by* $\mathrm{Cell}_L(W)$ *(resp.* $\mathrm{Cell}_R(W)$, *resp.* $\mathrm{Cell}_{LR}(W)$*) or* $\mathrm{Cell}_L^\varphi(W)$ *(resp.* $\mathrm{Cell}_R^\varphi(W)$, *resp.* $\mathrm{Cell}_{LR}^\varphi(W)$*).*

Using the anti-involution $*$ and the fact that $\mathbf{C}_w^* = \mathbf{C}_{w^{-1}}$ (see 5.2.5), we immediately obtain the following links between the "left" and the "right" relations:

Lemma 6.1.5. *Let x and y be two elements of W. Then:*

(a) *$x \overset{L}{\twoheadleftarrow} y$ if and only if $x^{-1} \overset{R}{\twoheadleftarrow} y^{-1}$.*
(b) *$x \leqslant_L y$ if and only if $x^{-1} \leqslant_R y^{-1}$.*
(c) *$x \sim_L y$ if and only if $x^{-1} \sim_R y^{-1}$.*

Corollary 6.1.6. *If C is a left cell, then C^{-1} is a right cell.*

6.1.B. Cell Modules

As in §1.1.C, we define *left, right or two-sided ideals* of W as well as *left, right or two-sided convex* subsets of W (see Definition 1.1.13) and, if C is such a subset, we define the subsets $W^{\leqslant_? C}$ and $W^{<_? C}$ of W, as well as the modules

$\mathcal{H}^{\leqslant_? C}$, $\mathcal{H}^{<_? C}$ and $\mathcal{H}^?[C]$ (these are left modules if $? = L$, right modules if $? = R$ and $(\mathcal{H},\mathcal{H})$-bimodules if $? = LR$). Note that $W^{\leqslant_? C}$ and $W^{<_? C}$ are left, right or two-sided ideals (according to the situation) of W.

Remark 6.1.7. Note that, of course, the notions of left, right or two-sided ideals and the notions of left, right or two-sided convex subset of W depend on φ. If we need to emphasize the weight function φ, we will talk about ideals *with respect to* φ or convex subsets *with respect to* φ. In this case, we add an exponent φ in the notation (for instance, $W^{\leqslant_?^\varphi C}$, $\mathcal{H}^{\varphi,?}[C]$,...). □

Question 6.1.8. *Does any left (resp. right, resp. two-sided) ideal of W contain a maximal left (resp. right, resp. two-sided ideal) cell?*

To our knowledge, the answer to this question is unknown in general. We will see in Chapter 16 that some conjectures by Lusztig imply that the answer to this question is positive (whenever S is finite); see Proposition 16.1.3. Of course, it holds if W is finite, or at least if the number of left cells is finite (which happens to be the case in affine Weyl groups [Lus20, Theorem 18.2] but which is not the case in general, as we will see in Chapter 24).

Definition 6.1.9 (Cell modules). *If $? \in \{L, R, LR\}$ and if $C \in \mathrm{Cell}_?(W)$, the module $\mathcal{H}^?[C]$ is called the* **(Kazhdan–Lusztig) cell module** *associated with C.*

> **Notation.** *If C is a left, right or two-sided convex subset of W and if $w \in C$, we denote by $\mathbf{c}_w$ (or $\mathbf{c}_w^?$ if necessary) its image in $\mathcal{H}^?[C]$. Note that*
> $$\mathcal{H}^?[C] = \bigoplus_{w \in C} A\mathbf{c}_w$$
> *and that $(\mathbf{c}_w)_{w\in C}$ is an A-basis of $\mathcal{H}^?[C]$.*

If M is a left $\mathcal{H}$-module, we denote by M_* the right $\mathcal{H}$-module whose underlying A-module is M and such that the right action of $h \in \mathcal{H}$ on M_* is given by the left action of h^* on M. Let C be a left convex subset of W. Then C^{-1} is right convex by Lemma 6.1.5(b) and it is easily checked from 5.5.4 that we have an isomorphism of right $\mathcal{H}$-modules

$$\mathcal{H}^L[C]_* \simeq \mathcal{H}^R[C^{-1}]. \tag{6.1.10}$$

Example 6.1.11. Assume here that $S = \{s\}$ and let $? \in \{L, R, LR\}$. Then, it follows from Example 5.2.4 that two cases may occur:

- If $\varphi(s) = 0$, then $s \leqslant_? 1$ and $1 \leqslant_? s$, so that $1 \sim_? s$.

- If $\varphi(s) \neq 0$, then $s \leqslant_? 1$ but $1 \not\leqslant_? s$; hence, $1 \not\sim_? s$. Therefore,

$$\mathrm{Cell}_?(W) = \begin{cases} \{W\} & \text{if } \varphi(s) = 0, \\ \{\{1\},\{s\}\} & \text{if } \varphi(s) \neq 0. \end{cases}$$

This shows that the partition into cells depends on φ.

Finally, let us describe the left cell modules. If $\varphi(s) = 0$, then W is the unique cell and the associated left cell module is $\mathcal{H} \simeq AW$ itself. So assume that $\varphi(s) \neq 0$. The structure of the left cell modules $\mathcal{H}^L[w]$ is determined by the action of T_s on $\mathbf{c}_w$. By Example 5.2.4, it is given by:

$$T_s \mathbf{c}_1 = \begin{cases} -v^{-\varphi(s)}\mathbf{c}_1 & \text{if } \varphi(s) > 0, \\ v^{\varphi(s)}\mathbf{c}_1 & \text{if } \varphi(s) < 0, \end{cases} \qquad \text{and} \qquad T_s \mathbf{c}_s = \begin{cases} v^{\varphi(s)}\mathbf{c}_s & \text{if } \varphi(s) > 0, \\ -v^{-\varphi(s)}\mathbf{c}_s & \text{if } \varphi(s) < 0. \end{cases}$$

Recall that $\mathrm{ind}(T_s) = v^{\varphi(s)}$ and $\mathrm{sgn}(T_s) = -v^{-\varphi(s)}$ (see §4.2.A). □

6.2. Functoriality, Parameters

6.2.A. *Changing the Group*

As in §5.4.A, let $\mathcal{A}'$ be another totally ordered abelian group and let $g : \mathcal{A} \to \mathcal{A}'$ be a *strictly increasing* morphism of groups. As the map $g_* : \mathcal{H}^\varphi \longrightarrow \mathcal{H}^{g\circ\varphi}$ is injective and since $g_*(\mathbf{C}_w^\varphi) = \mathbf{C}_w^{g\circ\varphi}$ by 5.4.2, we immediately obtain the following results:

Proposition 6.2.1. *Let x and y be two elements of W and let $? \in \{L, R, LR\}$. Then:*

(a) $x \overset{?,\varphi}{\twoheadleftarrow} y$ *if and only if* $x \overset{?,g\circ\varphi}{\twoheadleftarrow} y$.
(b) $x \leqslant_?^\varphi y$ *if and only if* $x \leqslant_?^{g\circ\varphi} y$.
(c) $x \sim_?^\varphi y$ *if and only if* $x \sim_?^{g\circ\varphi} y$.

Corollary 6.2.2. $\mathrm{Cell}_?^\varphi(W) = \mathrm{Cell}_?^{g\circ\varphi}(W)$.

At the level of modules, it is easily checked that a subset C of W is left, right or two-sided convex with respect to φ if and only if it is left, right or two-sided convex with respect to $g \circ \varphi$, and that in this case, we have

(6.2.3) $$\mathcal{H}^{g\circ\varphi,?}[C] \simeq \mathbb{Z}[\mathcal{A}'] \otimes_{\mathbb{Z}[\mathcal{A}]} \mathcal{H}^{\varphi,?}[C].$$

6.2.B. Changing Signs

As in §4.3.C and 5.4.B, we fix here a partition $S = S_+ \dot\cup S_-$ of S into W-closed subsets and we set

$$\begin{aligned} \varphi_- : S &\longrightarrow \mathscr{A} \\ s &\longmapsto \begin{cases} \varphi(s) & \text{if } s \in S_+, \\ -\varphi(s) & \text{if } s \in S_-. \end{cases} \end{aligned}$$

As $\Theta_{S_-}(\mathbf{C}_w^\varphi) = (-1)^{\ell_{S_-}(w)} \mathbf{C}_w^{\varphi_-}$ (see 5.4.4), we get immediately that:

Proposition 6.2.4. *Let x and y be two elements of W and let $? \in \{L, R, LR\}$. Then:*

(a) $x \overset{?,\varphi}{\twoheadleftarrow} y$ *if and only if* $x \overset{?,\varphi_-}{\twoheadleftarrow} y$.
(b) $x \leqslant_?^\varphi y$ *if and only if* $x \leqslant_?^{\varphi_-} y$.
(c) $x \sim_?^\varphi y$ *if and only if* $x \sim_?^{\varphi_-} y$.

Corollary 6.2.5. $\mathrm{Cell}_?^\varphi(W) = \mathrm{Cell}_?^{\varphi_-}(W)$.

At the level of modules, it is easily checked that a subset C of W is left, right or two-sided convex with respect to φ if and only if it is left, right or two-sided convex with respect to φ_-, and that in this case, we have

$$\mathscr{H}^{\varphi,?}[C] \simeq \Theta_{S_-}^* \mathscr{H}^{\varphi_-,?}[C], \tag{6.2.6}$$

where $\Theta_{S_-}^* M$ is the $\mathscr{H}^\varphi$-module (or $(\mathscr{H}^\varphi, \mathscr{H}^\varphi)$-bimodule) obtained from the $\mathscr{H}^\varphi$-module (or $(\mathscr{H}^\varphi, \mathscr{H}^\varphi)$-bimodule) M by precomposing the action with the isomorphism Θ_{S_-}.

If $\gamma : W \to \{1, -1\}$ is a linear character, we denote by $\mathbb{Z}_\gamma^L$ (resp. $\mathbb{Z}_\gamma^R$) the left (resp. right) $\mathbb{Z}W$-module whose underlying $\mathbb{Z}$-module is $\mathbb{Z}$ but on which an element $w \in W$ acts by multiplication by $\gamma(w)$. If $\gamma' : W \to \{1, -1\}$ is another linear character, we denote by $\mathbb{Z}_{\gamma \boxtimes \gamma'}^{LR}$ the $(\mathbb{Z}W, \mathbb{Z}W)$-bimodule whose underlying $\mathbb{Z}$-module is $\mathbb{Z}$ but on which an element $(w, w') \in W \times W$ acts by multiplication by $\gamma(w)\gamma'(w')$. Now, it follows from Exercise 4.6(c) that the isomorphism 6.2.6 induces an isomorphism of $\mathbb{Z}W$-modules

$$\mathbb{Z} \otimes_A \mathscr{H}^{\varphi,?}[C] \simeq \begin{cases} \mathbb{Z}_{\varepsilon_{S_-}}^? \otimes_{\mathbb{Z}} (\mathbb{Z} \otimes_A \mathscr{H}^{\varphi_-,?}[C]) & \text{if } ? \in \{L, R\}, \\ \mathbb{Z}_{\varepsilon_{S_-} \boxtimes \varepsilon_{S_-}}^{LR} \otimes_{\mathbb{Z}} (\mathbb{Z} \otimes_A \mathscr{H}^{\varphi_-,?}[C]) & \text{if } ? = LR. \end{cases} \tag{6.2.7}$$

6.2.C. Vanishing Parameters

As in §4.3.D and 5.4.C, we fix here a partition $S = I \dot\cup J$ of S into W-closed subsets such that $\varphi(s) = 0$ for all $s \in I$, we set

$$\widetilde{J} = \{wsw^{-1} \mid w \in W_I \text{ and } s \in J\} \quad \text{and} \quad \widetilde{W} = \langle \widetilde{J} \rangle,$$

we denote by $\tilde{\varphi}$ the restriction to $\widetilde{J}$ of the braided map $\varphi : W \to \mathscr{A}$ extending φ and we set $\tilde{\mathscr{H}} = \mathscr{H}_{\mathbb{Z}}(\widetilde{W}, \widetilde{J}, \tilde{\varphi}) \subset \mathscr{H}$.

Proposition 6.2.8. *Let x, $x' \in W_I$ and w, $w' \in \widetilde{W}$. Then:*

(a) $xw \leqslant_L^{\varphi} x'w'$ *if and only if* $w \leqslant_L^{\tilde{\varphi}} w'$.
(b) $wx \leqslant_R^{\varphi} w'x'$ *if and only if* $w \leqslant_R^{\tilde{\varphi}} w'$.
(c) $xw \leqslant_{LR}^{\varphi} x'w'$ *if and only if there exists a* $\xi \in W_I$ *such that* $w \leqslant_{LR}^{\tilde{\varphi}} \xi w' \xi^{-1}$.

Proof. By Lemma 6.1.5, the statement (b) follows from (a). Moreover, (c) follows from (a) and (b). So let us prove (a). By an easy induction argument, it is sufficient to show that $xw \overset{L,\varphi}{\twoheadleftarrow} x'w'$ if and only if $w \overset{L,\tilde{\varphi}}{\twoheadleftarrow} w'$.

So assume that $xw \overset{L,\varphi}{\twoheadleftarrow} x'w'$. Then there exist an $a \in W_I$ and a $z \in \widetilde{W}$ such that $h_{az,x'w',xw}^{\varphi} \neq 0$. But, by Corollary 5.4.6, $h_{az,x'w',xw}^{\varphi} = \delta_{ax'=x} h_{x'^{-1}zx',w',w}^{\tilde{\varphi}}$. So $h_{x'^{-1}zx',w',w}^{\tilde{\varphi}} \neq 0$: this shows that $w \overset{L,\tilde{\varphi}}{\twoheadleftarrow} w'$. The converse follows by a similar argument. ■

Corollary 6.2.9. *Let x, $x' \in W_I$ and w, $w' \in \widetilde{W}$. Then:*

(a) $xw \sim_L^{\varphi} x'w'$ *if and only if* $w \sim_L^{\tilde{\varphi}} w'$.
(b) $wx \sim_R^{\varphi} w'x'$ *if and only if* $w \sim_R^{\tilde{\varphi}} w'$.
(c) $xw \sim_{LR}^{\varphi} x'w'$ *if and only if there exists a* $\xi \in W_I$ *such that* $w \sim_{LR}^{\tilde{\varphi}} \xi w' \xi^{-1}$.

In other words, if C is a left (resp. right, resp. two-sided) $\tilde{\varphi}$-cell of $\widetilde{W}$, then $W_I \cdot C$ (resp. $C \cdot W_I$, resp. $W_I \cdot C \cdot W_I$) is a left (resp. right, resp. two-sided) φ-cell of W.

At the level of modules, the consequence of the previous results can be expressed as follows. If C is a left convex subset of $\widetilde{W}$, then $W_I \cdot C$ is a left convex subset of W and

$$\mathscr{H}^L[W_I \cdot C] \simeq \mathscr{H} \otimes_{\tilde{\mathscr{H}}} \tilde{\mathscr{H}}^L[C]. \tag{6.2.10}$$

An analogous result holds for right convex subsets, but the obvious generalisation to two-sided convex subsets does not hold in general.

Example 6.2.11. As a particular case, we assume here that $\varphi(s) = 0$ for all $s \in S$, so that we can take $I = S$ (and $J = \widetilde{J} = \varnothing$). Then $\mathscr{H} = AW$ and Proposition 6.2.8 implies that

W is the only left, right or two-sided cell,

and 6.2.10 can be rewritten as

$$\mathscr{H}^L[W] = \mathscr{H} = AW.$$

In fact, we even have

$$\mathscr{H}^{LR}[W] = AW,$$

as an $(\mathscr{H}, \mathscr{H})$-bimodule. This case was already treated in Example 1.1.19. □

Commentary 6.2.12. As in Commentary 5.4.7, if we want to compute cells or to describe cell modules then, at least theoretically, we may assume that $\varphi > 0$. Going back to the general case is possible by using Propositions 6.2.4 and 6.2.8, Corollaries 6.2.5 and 6.2.9 and Equations 6.2.7 and 6.2.10.

Therefore, in most of the examples in this book, we assume that $\varphi > 0$. Note, however, that considering vanishing parameters can be interesting, as will be explained in Chapter 7. □

6.3. Descent Sets

The first invariant of cells is given by descent sets (see the next proposition): it shows that the partition into left or right cells is not completely trivial (at least in the interesting case where $\varphi(s) \neq 0$ for all $s \in S$).

Proposition 6.3.1. *Let $S_{\neq} = \{s \in S \mid \varphi(s) \neq 0\}$. Let x and y be two elements of W. Then:*

(a) *If $x \leqslant_L y$, then $\mathscr{R}(y) \cap S_{\neq} \subseteq \mathscr{R}(x) \cap S_{\neq}$.*
(b) *If $x \sim_L y$, then $\mathscr{R}(x) \cap S_{\neq} = \mathscr{R}(y) \cap S_{\neq}$.*
(a′) *If $x \leqslant_R y$, then $\mathscr{L}(y) \cap S_{\neq} \subseteq \mathscr{L}(x) \cap S_{\neq}$.*
(b′) *If $x \sim_R y$, then $\mathscr{L}(x) \cap S_{\neq} = \mathscr{L}(y) \cap S_{\neq}$.*

Proof. It is sufficient to prove (a), since (a′) follows from (a) by applying Lemma 6.1.5 and 3.7.4, and (b) and (b′) then follow easily. By an easy induction argument, we may assume that $x \overset{L}{\twoheadleftarrow} y$ (in other words, there exists an $h \in \mathscr{H}$ such that $\mathbf{C}_x$ appears with a non-zero coefficient in the decomposition of $h\mathbf{C}_y$ in the Kazhdan–Lusztig basis). So, let $s \in \mathscr{R}(y) \cap S_{\neq}$. Since $\varphi(s) \neq 0$, it follows from Theorem 5.5.7 that

$$\mathscr{H}\mathbf{C}_s = \bigoplus_{s \in \mathscr{R}(w)} A\mathbf{C}_w.$$

Hence $\mathbf{C}_y \in \mathscr{H}\mathbf{C}_s$, and so $h\mathbf{C}_y \in \mathscr{H}\mathbf{C}_s$. Therefore $s \in \mathscr{R}(x)$, as desired. ■

Corollary 6.3.2. *Let x and y be two elements of W and assume that $\varphi(s) \neq 0$ for all $s \in S$. Then:*

(a) *If $x \leqslant_L y$, then $\mathscr{R}(y) \subseteq \mathscr{R}(x)$.*
(b) *If $x \sim_L y$, then $\mathscr{R}(x) = \mathscr{R}(y)$.*
(a′) *If $x \leqslant_R y$, then $\mathscr{L}(y) \subseteq \mathscr{L}(x)$.*
(b′) *If $x \sim_R y$, then $\mathscr{L}(x) = \mathscr{L}(y)$.*

6.4. About the Structure Constants $\lambda^s_{x,y}$ and $\rho^s_{x,y}$

As explained in Remark 6.1.3, the knowledge of the set of triples $(s,x,y) \in S \times W \times W$ such that $\lambda^s_{x,y} \neq 0$ determines the preorder $\leqslant_L$ and so the partition into left cells. The knowledge of the precise values of the $\lambda^s_{x,y}$ determines moreover the structure of the left cell modules. Using 5.5.4, this information also determines the preorders $\leqslant_R$ and $\leqslant_{LR}$, the partition into right and two-sided cells, as well as the structures of the right and two-sided cell modules. In this section, we give some first elementary results in this direction, focusing on the vanishing of the $\lambda^s_{x,y}$: according to Corollary 5.5.6, we may restrict to the case where $sx < x < y < sy$.

Proposition 6.4.1. *Assume that $\varphi > 0$. Let $s, t \in S$ and $x, y \in W$ be such that $sx < x < y < sy$, $\lambda^s_{x,y} \neq 0$ and $t \in \mathscr{L}(y) \setminus \mathscr{L}(x)$. Then:*

(a) $\mathscr{R}(y) \subseteq \mathscr{R}(x)$.
(b) $st \neq ts$.
(c) $\varphi(s) \geqslant \varphi(t)$.
(d) *If, moreover, $\varphi(s) = \varphi(t)$, then $y = tx$ and $\lambda^s_{x,y} = 1$.*

Proof. Note that (a) is just a consequence of Corollary 6.3.2(a). We now prove (b), (c) and (d) by contradiction. Let $s, t \in S$ and $x, y \in W$ be such that $sx < x < y < sy$, $t \in \mathscr{L}(y) \setminus \mathscr{L}(x)$ and $\varphi(s), \varphi(t) > 0$.

(b) Assume that $st = ts$. Then $\mathbf{C}_t \mathbf{C}_s \mathbf{C}_y = \mathbf{C}_s \mathbf{C}_t \mathbf{C}_y = (v^{\varphi(t)} + v^{-\varphi(t)}) \mathbf{C}_s \mathbf{C}_y$. Hence, $\mathbf{C}_s \mathbf{C}_y \in \mathbf{C}_t \mathscr{H} = \bigoplus_{t \in \mathscr{L}(w)} A\,\mathbf{C}_w$ by Theorem 5.5.5. So $\lambda^s_{x,y} = 0$.

(c) and (d) Assume that $\varphi(s) \leqslant \varphi(t)$. We now show by induction on $\ell(y) - \ell(x)$ that $\lambda^s_{x,y} = 0$ or that $y = tx$, $\varphi(s) = \varphi(t)$ and $\lambda^s_{x,y} = 1$. If $\ell(y) - \ell(x) = 1$, then this follows from Example 5.7.3. So assume that $\ell(y) - \ell(x) > 1$. Recall from Proposition 5.6.1(b) that

$$\begin{cases} \lambda^s_{x,y} \equiv v^{\varphi(s)} p_{x,y} - \displaystyle\sum_{\substack{x<z<y \\ (sz<z)}} p_{x,z} \lambda^s_{z,y} \mod A_{<0}, \\ \overline{\lambda}^s_{x,y} = \lambda^s_{x,y}. \end{cases}$$

But it follows from Proposition 5.6.1(a) that $p_{x,y} = v^{-\varphi(t)} p_{tx,y}$. Since $y \neq tx$, this implies that $v^{\varphi(s)} p_{x,y} = v^{\varphi(s) - \varphi(t)} p_{tx,y} \in A_{<0}$. On the other hand, if $x < z < y$ and $sz < z$, then two cases may occur:

• If $tz < z$, then $p_{x,z} = v^{-\varphi(t)} p_{tx,z}$ by Proposition 5.6.1(a) and so $p_{x,z} \lambda^s_{z,y} = v^{-\varphi(t)} p_{tx,z} \lambda^s_{z,y} \in A_{<0}$ by Proposition 5.6.4(c).

• If $tz > z$, then it follows from the induction hypothesis that $\lambda^s_{z,y} = 0$ or that $y = tz$ and $\lambda^s_{z,y} = 1$. In this last case, we have again $p_{x,z} \lambda^s_{z,y} = p_{x,z} \in A_{<0}$.

This shows that $\lambda^s_{x,y} \equiv 0 \mod A_{<0}$, and so $\lambda^s_{x,y} = 0$, contrary to our assumption. ■

Corollary 6.4.2. *Assume that* $\varphi > 0$. *Let* $s, t \in S$ *and* $x, y \in W$ *be such that* $xs<x<y<ys$, $\rho^s_{x,y} \neq 0$ *and* $t \in \mathscr{R}(y) \setminus \mathscr{R}(x)$. *Then:*

(a) $\mathscr{L}(y) \subseteq \mathscr{L}(x)$.
(b) $st \neq ts$.
(c) $\varphi(s) \geqslant \varphi(t)$.
(d) *If, moreover,* $\varphi(s)=\varphi(t)$, *then* $y=xt$ *and* $\rho^s_{x,y}=1$.

6.5. Some Conjectures

During its development, the Kazhdan–Lusztig theory of cells has led to several questions and conjectures about the partition of W into cells. Most of these conjectures are due to Lusztig: they concern the properties of the partition once a weight function φ is fixed. We present them in this section. Some others, due to the author, are concerned with how this partition changes whenever φ varies: they will be explained in Chapter 7.

We first need some definitions.

Definition 6.5.1. *A subset* $\mathscr{X}$ *of* W *is said to be* **right-connected** *if, for all* $x, y \in \mathscr{X}$, *there exists a sequence* $s_1, s_2, \dots, s_r$ *of elements of* S *such that* $y=xs_1s_2\cdots s_r$ *and* $xs_1s_2\cdots s_i \in \mathscr{X}$ *for all* i. *It is said to be* **left-connected** *if* $\mathscr{X}^{-1}$ *is right-connected.*

The maximal left-connected subsets of $\mathscr{X}$ *are called the* **left-connected components** *of* $\mathscr{X}$.

Of course, every subset of W is the disjoint union of its left-connected components.

Conjectures 6.5.2 (Lusztig).

(L1) *Every left cell contains an involution.*
(L2) *If* $w \in W$, *then* $w \sim_{LR} w^{-1}$.
(L3) *The equivalence relation* $\sim_{LR}$ *is generated by* $\sim_L$ *and* $\sim_R$.
(L4) *If* $x, y \in W$ *are such that* $x \leqslant_L y$ *and* $x \sim_{LR} y$, *then* $x \sim_L y$.
(L5) *Every left cell is left-connected.*
(L6) *If* Γ *is a two-sided cell, then the left-connected components of* Γ *are the left cells of* W *contained in* Γ.
(L7) *Every two-sided cell meets a finite (i.e. spherical) standard parabolic subgroup.*
(L8) *If* S *is finite, then the number of two-sided cells is finite.*

Note that (L1) implies (L2), as if $d \in W$ is an involution lying in the left cell of w, then $w \sim_L d$ and $w^{-1} \sim_R d^{-1}=d$, so $w \sim_{LR} w^{-1}$. Also, (L7) implies (L8): indeed, if S is finite, then the number of elements of W contained in a spherical standard parabolic subgroup is finite and, according to (L7), this gives an upper bound for the number of two-sided cells.

As can already be seen from this set of conjectures, and as will be shown in the rest of this book, involutions play an important role in the theory of cells. Also, as Kazhdan–Lusztig cells depend heavily on the choice of the set S of simple reflections, it should be difficult to relate them to conjugacy classes in W. However, with M. Geck, we have proposed a new conjecture, checked for several examples, relating cells and conjugacy classes *of involutions*:

Conjectures 6.5.3 (Bonnafé–Geck). *Let Γ be a two-sided cell, let C be a left cell contained in Γ and let $\mathscr{C}$ be a conjugacy class* ***of involutions****. Then $\mathscr{C} \cap \Gamma \neq \varnothing$ if and only if $\mathscr{C} \cap C \neq \varnothing$.*

6.6. Drawing Kazhdan–Lusztig Cells

In this book, we will sometimes draw pictures describing the left and two-sided cells of some particular Coxeter groups of rank 2 or 3. In rank 2, we will use the Tits cone introduced in Section 3.13, and represent the elements of the group as in §3.14.A. In rank 3, we will use the Coxeter complex introduced in Section 3.13, and represent the elements of the group as in §3.14.B. Therefore, to each element w will be assigned a triangle $\mathfrak{T}_w$ (which is equal to $\overline{\mathfrak{C}}_w$ in rank 2 or $\overline{\mathfrak{C}}_w^{\Sigma}$ in rank 3, and might be planar, spherical or hyperbolic).

Fact. *A subset $\mathscr{X}$ of W is left-connected if and only if $\bigcup_{w \in \mathscr{X}} \mathfrak{T}_w$ is connected.*

The Kazhdan–Lusztig cells will be drawn as follows (this is a standard convention, see, for instance, [Lus12], [Gui4], [Gun2],...):

Convention for pictures.

- *In all the cases for which we will draw a picture, S is finite and Conjecture (L8) holds, so that there are only finitely many two-sided cells: each triangle associated with an element $w \in W$ will be assigned a true colour (red, blue, green,...), and two elements w and w' are in the same two-sided cell if and only if their associated triangles have the same colour.*
- *In all the cases for which we will draw a picture, Conjecture (L6) holds: in other words, the left cells are the connected components of the two-sided cells, and so can be obtained as the connected components of the regions of the same colour.*

6.7. Examples

***Example 6.7.1** ($\varphi = 0$).* We assume in this example, and only in this example, that $\varphi(s) = 0$ for all $s \in S$. Then we can apply the results of §6.2.C with $I = S$ and we obtain easily that $W = W_I$ is the unique φ-cell (left, right or two-sided) and that the associated $\mathscr{H}$-module (or $(\mathscr{H}, \mathscr{H})$-bimodule) is $\mathscr{H}$ itself (recall that $\mathscr{H} \simeq AW$).

One can easily check that Lusztig's Conjectures 6.5.2 and Conjecture 6.5.3 hold in this case. □

***Example 6.7.2** (**Cells containing 1**).* We denote by C_1 (resp. Γ_1) the left (resp. two-sided) cell of W containing 1. Note that C_1^{-1} is the right cell containing 1. It follows from Example 6.1.2 that $1 \leqslant_L w$ and $1 \leqslant_R w$ for all $w \in W$. Therefore, C_1 (resp. C_1^{-1}, resp. Γ_1) is the unique maximal left (resp. right, resp. two-sided) cell of W. We aim to describe C_1 and Γ_1.

• First, let us assume that $\varphi(s) \neq 0$ for all $s \in S$. It then follows from Corollary 5.5.6(b) that $\lambda_{1,w}^s = 0$ for all $w \in W$. So, by Remark 6.1.3, $w <_L 1$ for all $w \in W \setminus \{1\}$. This shows that

(6.7.3) *If $\varphi(s) \neq 0$ for all $s \in S$, then $C_1 = C_1^{-1} = \Gamma_1 = \{1\}$.*

Let us now describe the structure of the $(\mathscr{H}, \mathscr{H})$-bimodule $\mathscr{H}^{LR}[1]$. First, $\mathscr{H}^{LR}[1] = A\mathbf{c}_1$ and $\mathbf{C}_s \mathbf{c}_1 = \mathbf{c}_1 \mathbf{C}_s = 0$ for all $s \in S$ (as $\mathbf{C}_s \mathbf{C}_1 = \mathbf{C}_s$). Now, recall that $\mathbf{C}_s$ has been determined in Example 5.2.4, so

$$T_s \mathbf{c}_1 = \mathbf{c}_1 T_s = \begin{cases} -v^{-\varphi(s)} \mathbf{c}_1 & \text{if } \varphi(s) > 0, \\ v^{\varphi(s)} \mathbf{c}_1 & \text{if } \varphi(s) < 0. \end{cases}$$

Now, let $S_+ = \{s \in S \mid \varphi(s) > 0\}$. Then

(6.7.4) *If $\varphi(s) \neq 0$ for all $s \in S$, then $h \mathbf{c}_1 h' = \mathrm{sgn}_{S_+}(hh') \mathbf{c}_1$*

for all $h, h' \in \mathscr{H}$.

• Let us come back to the general case. Let $I = \{s \in S \mid \varphi(s) = 0\}$, let $J = S \setminus I$ and, as in §6.2.C, let

$$\widetilde{J} = \{wsw^{-1} \mid w \in W_I \text{ and } s \in J\} \qquad \text{and} \qquad \widetilde{W} = \langle \widetilde{J} \rangle.$$

We denote by $\tilde{\varphi}$ the restriction to $\widetilde{J}$ of the unique braided map $\varphi : W \to \mathscr{A}$ extending φ and we set $\tilde{\mathscr{H}} = \mathscr{H}(\widetilde{W}, \widetilde{J}, \tilde{\varphi})$. Then, by Corollary 6.2.9 and 6.7.3 applied to $\widetilde{W}$, we get

(6.7.5) $$C_1 = C_1^{-1} = \Gamma_1 = W_I.$$

Then the structure of the $(\mathscr{H}, \mathscr{H})$-bimodule $\mathscr{H}^{LR}[W_I]$ can be described easily from 6.7.4 and 6.2.10. □

Example 6.7.6 (Type A_2). As in Example 3.12.7 (see also Examples 4.4.13 and 5.7.9), we assume here that $S=\{s,t\}$ and $m_{st}=3$, so that (W,S) is of type A_2. We set $\varphi(s)=\varphi(t)=a$ and we assume, following Commentary 6.2.12 and as in Example 5.7.9, that $a>0$. Here is the corresponding Coxeter graph, together with the values of φ:

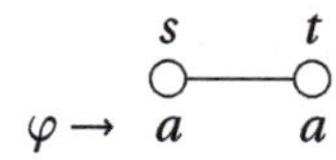

Then it follows from the computation of the left multiplication by $\mathbf{C}_s$ and $\mathbf{C}_t$ given in Example 5.7.9 that the left and two-sided cells are given by Figure 6.a, according to the convention described in Section 6.6.

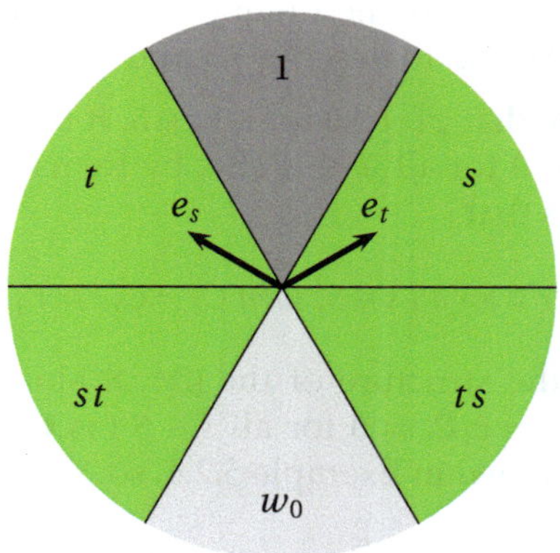

Fig. 6.a. Cells in type A_2

In other words, the two-sided cells are

$$\{1\},\{s,t,st,ts\} \text{ and } \{w_0\},$$

while the left cells are

$$\{1\},\{s,ts\},\{t,st\} \text{ and } \{w_0\}.$$

It is straightforward that

$$\{w_0\} \leqslant_L \{s,ts\} \leqslant_L \{1\}, \quad \text{and} \quad \{w_0\} \leqslant_L \{t,st\} \leqslant_L \{1\},$$

but the left cells $\{s,ts\}$ and $\{t,st\}$ are not comparable. Also,

$$\{w_0\} \leqslant_{LR} \{s,t,st,ts\} \leqslant_{LR} \{1\}.$$

One can easily check that Lusztig's Conjectures 6.5.2 and Conjecture 6.5.3 hold in this case.

Let us now describe the left cell modules. A direct computations shows that, if $h \in \mathcal{H}$, then

$$h\mathbf{c}_1 = \mathrm{ind}(h)\mathbf{c}_1 \qquad \text{and} \qquad h\mathbf{c}_{w_0} = \mathrm{sgn}(h)\mathbf{c}_{w_0}.$$

This describes the structure of $\mathcal{H}^L[1]$ and $\mathcal{H}^L[w_0]$. Now, setting $\boldsymbol{v} = v^a$, the actions of T_s and T_t are given by the matrices

$$T_s \longmapsto \begin{pmatrix} \boldsymbol{v} & 1 \\ 0 & -\boldsymbol{v}^{-1} \end{pmatrix}, \quad T_t \longmapsto \begin{pmatrix} -\boldsymbol{v}^{-1} & 0 \\ 1 & \boldsymbol{v} \end{pmatrix} \quad \text{on } \mathcal{H}^L[s,ts] = A\mathbf{c}_s \oplus A\mathbf{c}_{ts},$$

$$\text{and} \quad T_s \longmapsto \begin{pmatrix} -\boldsymbol{v}^{-1} & 0 \\ 1 & \boldsymbol{v} \end{pmatrix}, \quad T_t \longmapsto \begin{pmatrix} \boldsymbol{v} & 1 \\ 0 & -\boldsymbol{v}^{-1} \end{pmatrix} \quad \text{on } \mathcal{H}^L[t,st] = A\mathbf{c}_t \oplus A\mathbf{c}_{st},$$

where $\boldsymbol{v} = v^a$. It must be noticed that the A-linear map $\mathcal{H}^L[s,ts] \to \mathcal{H}^L[t,st]$, $\mathbf{c}_s \mapsto \mathbf{c}_{st}$, $\mathbf{c}_{ts} \mapsto \mathbf{c}_t$ is an isomorphism of left $\mathcal{H}$-modules. Also, the $K\mathcal{H}$-module $K\mathcal{H}^L[s,ts]$ is irreducible. Note that $\mathbb{R}(\mathcal{A})\mathcal{H}^L[s,ts]$ is isomorphic to the reflection representation of $\mathbb{R}(\mathcal{A})\mathcal{H}$ defined in Section 4.4. □

***Example 6.7.7* (*Type B_2*).** As in Example 3.12.8 (see also Examples 4.4.14 and 5.7.10), we assume here that $S = \{s,t\}$ with $m_{st} = 4$, so that (W,S) is of type B_2. We set $\varphi(s) = a$ and $\varphi(t) = b$, and assume, following Commentary 6.2.12 and as in Example 5.7.10, that $a,\ b > 0$. Here is the corresponding Coxeter graph, together with the values of φ:

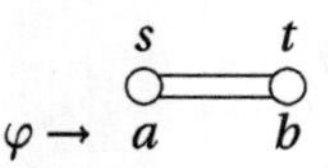

The computations made in Example 5.7.10 show that the partitions into cells are given by the Figure 6.b, according to the convention of Section 6.6.

These figures show that the partitions into cells depend only on whether $a < b$, $a = b$ or $a > b$. If $? \in \{L, R, LR\}$, let $\sim_?^<$ (resp. $\sim_?^=$, resp. $\sim_?^>$) denote the equivalence relations $\sim_?$ whenever $a < b$ (resp. $a = b$, resp $a > b$). Then it is easily checked that

(6.7.8) *the relation* $\sim_?^=$ *is generated by* $\sim_?^<$ *and* $\sim_?^>$.

Note that the Figure 6.b do not give the preorders $\leqslant_L$ and $\leqslant_{LR}$: in fact, in all cases, $\leqslant_L$ is just generated by the elementary relations $uw \leqslant_L w$ whenever $w \in W$ and $u \in S$ are such that $uw > w$, and $\leqslant_{LR}$ can be easily deduced from this. Note that $\leqslant_L$ is not a total order on left cells, while $\leqslant_{LR}$ is a total order on two-sided cells.

One can also easily check that Lusztig's Conjectures 6.5.2 and Conjecture 6.5.3 hold in this case.

Let us now describe the left cell modules. We set $\boldsymbol{v} = v^a$ and $\boldsymbol{V} = v^b$. Since there is an automorphism ω of the Coxeter system (W,S) which exchanges s and t, it is readily seen that the case $a > b$ can be deduced from the case $a < b$. This means that we only consider the following two cases: $a = b$ or $a < b$. First, note that $\{1\}$ and $\{w_0\}$ are alone in their cell, and that, as for type A_2 (see Example 6.7.6),

$$h\mathbf{c}_1 = \mathrm{ind}(h)\mathbf{c}_1 \qquad \text{and} \qquad h\mathbf{c}_{w_0} = \mathrm{sgn}(h)\mathbf{c}_{w_0}$$

for all $h \in \mathscr{H}$.

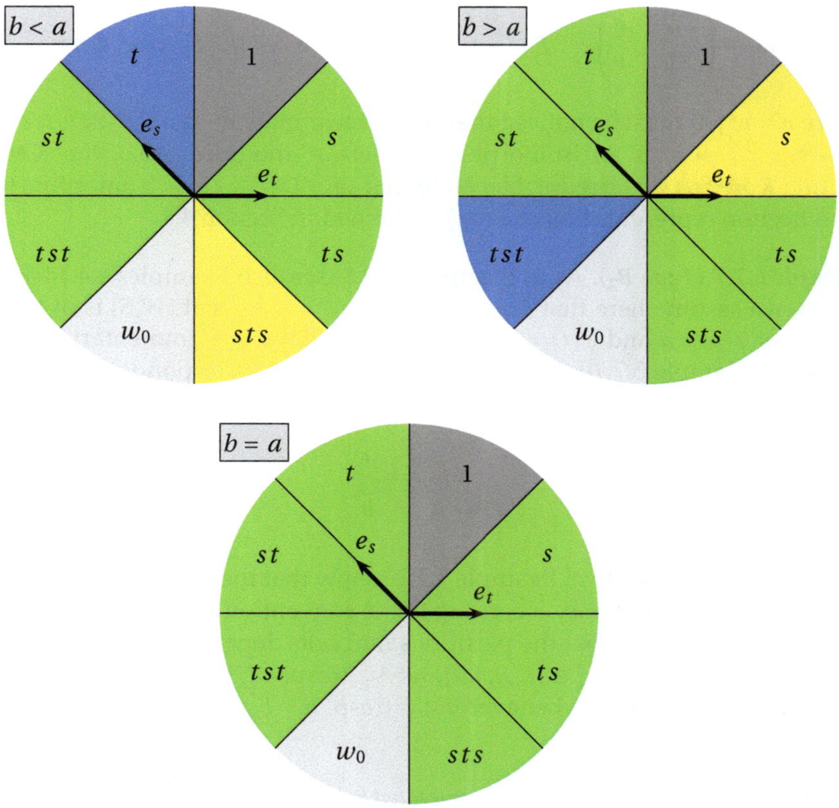

Fig. 6.b. Cells in type B_2

• Assume first that $a = b$ (so that $\boldsymbol{v} = \boldsymbol{V}$): then there are two further left cells $C(s) = \{s, ts, sts\}$ and $C(t) = \{t, st, tst\}$. The action of T_s and T_t on $\mathscr{H}^L[C(s)]$ are given by the matrices

$$T_s \mapsto \begin{pmatrix} v & 1 & 0 \\ 0 & -v^{-1} & 0 \\ 0 & 1 & v \end{pmatrix}, T_t \mapsto \begin{pmatrix} -v^{-1} & 0 & 0 \\ 1 & v & 1 \\ 0 & 0 & -v^{-1} \end{pmatrix} \text{ on } \mathcal{H}^L[C(s)] = A\mathbf{c}_s \oplus A\mathbf{c}_{ts} \oplus A\mathbf{c}_{sts}.$$

As $C(s)$ and $C(t)$ are interchanged by the automorphism ω of (W,S), we deduce that the actions of T_s and T_t on $\mathcal{H}^L[C(t)]$ are given by the matrices

$$T_s \mapsto \begin{pmatrix} -v^{-1} & 0 & 0 \\ 1 & v & 1 \\ 0 & 0 & -v^{-1} \end{pmatrix}, T_t \mapsto \begin{pmatrix} v & 1 & 0 \\ 0 & -v^{-1} & 0 \\ 0 & 1 & v \end{pmatrix} \text{ on } \mathcal{H}^L[C(t)] = A\mathbf{c}_t \oplus A\mathbf{c}_{st} \oplus A\mathbf{c}_{tst}.$$

We deduce that

$$\mathrm{Tr}(T_s, \mathcal{H}^L[C(s)]) = 2v - v^{-1} \neq v - 2v^{-1} = \mathrm{Tr}(T_s, \mathcal{H}^L[C(t)]),$$

so $\mathcal{H}^L[C(s)]$ and $\mathcal{H}^L[C(t)]$ are not isomorphic, even after extension to the fraction field of A.

Note also that $\mathbb{Q}(\mathscr{A})\otimes_A \mathcal{H}^L[C(s)]$ and $\mathbb{Q}(\mathscr{A})\otimes_A \mathcal{H}^L[C(t)]$ are not irreducible $\mathbb{Q}(\mathscr{A})$-modules. Indeed, the subspace of $\mathbb{Q}(\mathscr{A})\otimes_A \mathcal{H}^L[C(s)]$ generated by $\mathbf{c}_s - \mathbf{c}_{sts}$ is a one-dimensional submodule (the other case follows by using the automorphism ω).

• Assume now that $a < b$: then there are four further left cells, two of cardinality 1 and two of cardinality 2. Let us first describe the action of $\mathcal{H}$ on the left cell modules associated with cells of cardinality 1:

$$h\mathbf{c}_s = \mathrm{sgn}_t(h)\mathbf{c}_s \qquad \text{and} \qquad h\mathbf{c}_{tst} = \mathrm{sgn}_s(h)\mathbf{c}_{tst},$$

where $\mathrm{sgn}_I : \mathcal{H} \to A$ has been defined in Exercise 4.6 (for I a W-closed subset of S).

Now, let $C = \{t, st\}$ and $C' = \{ts, sts\}$ be the two remaining left cells. From the computations in Example 5.7.10, it is immediate that the A-linear map $\mathcal{H}^L[C] \to \mathcal{H}^L[C']$, $\mathbf{c}_x \mapsto \mathbf{c}_{xs}$, is an isomorphism of $\mathcal{H}$-modules. Finally, the actions of T_s and T_t on $\mathcal{H}^L[C]$ are given by the matrices

$$T_s \mapsto \begin{pmatrix} v & 0 \\ 1 & -v^{-1} \end{pmatrix}, \quad T_t \mapsto \begin{pmatrix} -V^{-1} & Vv^{-1} + V^{-1}v \\ 0 & V \end{pmatrix} \quad \text{on } \mathcal{H}^L[C] = A\mathbf{c}_t \oplus A\mathbf{c}_{st}.$$

One can check that $K\mathcal{H}^L[C]$ is irreducible and that $\mathbb{R}(\mathscr{A})\mathcal{H}^L[C]$ is isomorphic to the reflection representation of $\mathbb{R}(\mathscr{A})\mathcal{H}$ defined in Section 4.5. □

Remark 6.7.9. Also, in Chapter 21, we will study groups of type $I_2(m)$, with $m \geqslant 3$. If m is odd, then the two simple reflections are conjugate, and the results we will obtain in this case generalize Example 6.7.6. On the other hand, if m is even, then the two simple reflections are not conjugate, and the results we will obtain generalize Example 6.7.7. □

Notes

All the notions defined in this chapter were introduced by Kazhdan and Lusztig [KaLu1] in the equal parameter case and were extended to the general case by Lusztig [Lus9], [Lus20]. Note, however, that Proposition 6.4.1 is due to Shi [Shi5, Proposition 2.3] (see also [Gui4, Lemma 5.1] for a partial result).

Almost all of Lusztig's Conjectures 6.5.2 have been stated in [Lus20]: (L1) is implied by [Lus20, Conjectures 14.2, P6 and P13], (L2) is exactly [Lus20, Conjecture 14.2, P14], (L3) is implied by [Lus20, Conjectures 14.2, P9, P10 and P13], (L7) and (L8) are stated in [Lus20, §13.12]. However, (L5) can be found in [Lus10], while (L6) is implied by (L5) and [Lus20, Conjecture 14.2, P9]. Note that Lusztig's Conjecture (L1) has been proved by M. Geck whenever W is finite [Gec10, Theorem 1.1].

Finally, Conjecture 6.5.3 can be found in [BoGe, Introduction].

Exercises

Exercise 6.1 (Type G_2). As in Exercise 3.1 (see also Exercises 4.1 and 5.1), we assume here that $S = \{s,t\}$ with $m_{st} = 6$, so that (W,S) is of type G_2. We set $\varphi(s) = a$ and $\varphi(t) = b$, and assume first that $a, b > 0$. Here is the corresponding Coxeter graph, together with the values of φ:

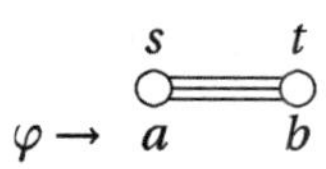

Compute the partition of W into left, right and two-sided cells (according to the values of a and b) and compute the structure of the left cell modules. Also:

(a) Check that if $a \neq b$, then there exist two different left cells C and C' of cardinality 4, and that the left cell modules associated with C and C' are isomorphic. Check also that their extensions to the fraction field $\mathbb{Q}(\mathscr{A})$ of A are not irreducible.

(b) Show that the equivalence relations $\sim_?^{\varphi}$ depend only on whether $a < b$, $a = b$ or $a > b$, and not on the exact values of a and b (we denote by $\sim_?^{<}$, $\sim_?^{=}$ or $\sim_?^{>}$ the relation $\sim_?^{\varphi}$ according to the situation).

(c) Check that $\sim_?^{=}$ is generated by $\sim_?^{<}$ and $\sim_?^{>}$.

Extend, when possible, all of the above results to the case where $a, b \geq 0$.

Finally, check that Lusztig's Conjectures 6.5.2 and Conjecture 6.5.3 hold in type G_2.

Exercise 6.2. Let $\leqslant'_L$ denote the preorder on W defined as in Chapter 1 by using the standard basis $(T_w)_{w\in W}$ instead of the Kazhdan–Lusztig basis $(\mathbf{C}_w)_{w\in W}$. Let $\sim'_L$ be the associated equivalence relation. Show that there is only one equivalence class for $\sim'_L$.

Exercise 6.3. Let I be a subset of S and let $\leqslant_?^I$ denote the preorder $\leqslant_?$ defined inside W_I (using the weight function φ_I obtained by restriction from φ). We define $\sim_?^I$ similarly. Let x and y be two elements of W_I. Show that if $x \leqslant_?^I y$ (resp. $x \sim_?^I y$), then $x \leqslant_? y$ (resp. $x \sim_? y$).

Remark. We will show in Chapter 8 that the converse also holds if $? \in \{L,R\}$ (see Corollary 8.2.3). However, the converse does not hold in general if $? = LR$ (see Example 8.2.5).

Exercise 6.4. Using Examples 6.7.6 and 6.7.7 as well as Exercise 6.1, check that the following two assertions hold whenever (W,S) is of type A_2, B_2 or G_2:

(a) If C is a left cell, then the length of the $K\mathscr{H}$-module $K\mathscr{H}^L[C]$ is equal to the number of involutions in C.

(b) If C and C' are two left cells, then $\dim(\mathrm{Hom}_{K\mathscr{H}}(K\mathscr{H}^L[C], K\mathscr{H}^L[C'])) = |C^{-1} \cap C'|$.

Exercise 6.5 (***W*-graphs**)**.** The notion of a W-graph was introduced by Kazhdan and Lusztig in their original paper [KaLu1], in the equal parameter case. This exercise deals with the general case. Let K' be a field extension of $K = \mathbb{Q}(\mathscr{A})$. A *$W$-graph* (with values in K') is a triple (C, I, μ) where:

- C is a set;
- $I : C \to \mathscr{P}(S)$ is a map;
- μ is a map $\{(s,x,y) \in S \times C \times C \mid s \in I(x) \setminus I(y)\} \to K'$, $(s,x,y) \mapsto \mu^s_{x,y}$,

subject to the following condition: if V_C denotes a K'-vector space with basis $(e_x)_{x\in C}$ and if, for any $s \in S$, $\mathfrak{T}_s : V_C \to V_C$ denotes the unique linear map defined by

$$\mathfrak{T}_s(e_y) = \begin{cases} -v^{-\varphi(s)} e_y + \displaystyle\sum_{\substack{x\in C \\ s\in I(x)}} \mu^s_{x,y}\, e_x & \text{if } s \notin I(y), \\ v^{\varphi(s)} e_y & \text{if } s \in I(y) \end{cases}$$

(for $y \in C$), then the map $T_s \mapsto \mathfrak{T}_s$ extends to a representation of $K'\mathscr{H}$ (on V_C).

A W-graph (C, I, μ) is represented as follows. The elements of the set C are vertices and represented by a circle; inside the circle associated with x, the elements of $I(x)$ are written; for each $s \in S$, we draw an edge between x and y if $s \in I(x) \setminus I(y)$ and $\mu^s_{x,y} \neq 0$; if $\mu^s_{x,y} \neq 1$, then the edge is coloured by $\mu^s_{x,y}$ (otherwise, it is not coloured); if there are several edges between x

and y, another colour is added to each of these edges recalling from which element $s \in S$ it has been constructed.

Now, let C be a left convex subset of W, recall that $\mathscr{L} : C \longrightarrow \mathscr{P}(S)$ is the left descent set map, and let us still write λ for the map $\{(s,x,y) \in S \times C \times C \mid s \in \mathscr{L}(x) \text{ and } s \notin \mathscr{L}(y)\} \longrightarrow K$, $(s,x,y) \mapsto \lambda_{x,y}^s$ (where $\lambda_{x,y}^s = h_{s,y,x}$).

(a) Show that $(C, \mathscr{L}, \lambda)$ is a W-graph (with $K' = K$) and that the left $K\mathscr{H}$-module V_C is isomorphic to $K\mathscr{H}^L[C]$.

(b) If (W,S) is of type A_2 and $\varphi \neq 0$, show that the two W-graphs associated with the left cells of cardinality 2 are isomorphic to

(c) Compute the W-graphs of the cells of cardinality > 1 in types B_2 and G_2.

Exercise 6.6. Assume here that $\varphi(s) \neq 0$ for all $s \in S$ (so that 1 is alone in its left or two-sided cell by Example 6.7.2). Let C and Γ be respectively a left and a two-sided cell of W not containing 1. Show that

$$\mathrm{Hom}_{\mathscr{H}}(\mathscr{H}^L[1], \mathscr{H}^L[C]) = \mathrm{Hom}_{\mathscr{H}}(\mathscr{H}^L[C], \mathscr{H}^L[1]) = 0$$

and $\quad \mathrm{Hom}_{(\mathscr{H},\mathscr{H})}(\mathscr{H}^{LR}[1], \mathscr{H}^{LR}[\Gamma]) = \mathrm{Hom}_{(\mathscr{H},\mathscr{H})}(\mathscr{H}^{LR}[\Gamma], \mathscr{H}^{LR}[1]) = 0.$

Chapter 7
Semicontinuity

The aim of this chapter is to state a conjecture about the behaviour of the sets of cells $\mathrm{Cell}_?^{\varphi}(W)$ whenever φ varies. For simplicity, we assume that φ takes *real values* and that S is finite, so that φ lies in the vector space V of weight functions $S \to \mathbb{R}$. Roughly speaking, our conjectures say that there exists a finite set of rational hyperplanes (which determines a fan) in V such that:

- If φ and φ' lie in the same face of the fan, then $\mathrm{Cell}_?^{\varphi}(W) = \mathrm{Cell}_?^{\varphi'}(W)$.
- If $\varphi(s) \neq 0$ for all $s \in S$, φ lies on some face $\mathscr{F}$ of the fan and if $\mathscr{C}_1, \ldots, \mathscr{C}_r$ denote the open faces whose closure contains φ (and if $\varphi_i \in \mathscr{C}_i$), then the relation $\sim_?^{\varphi}$ is generated by $\sim_?^{\varphi_1}, \ldots, \sim_?^{\varphi_r}$.

If $\varphi(s) = 0$ for some elements $s \in S$, then there is also a description analogous to this, but this requires some more notation and is described rigorously in this chapter.

Hypothesis and notation. *In this chapter, and only in this chapter, we assume that S is* ***finite****. We denote by $V = \mathrm{Weight}(S, \mathbb{R})$ the* ***real*** *vector space of weight functions $S \to \mathbb{R}$. Of course, $\mathbb{R}$ is viewed as an ordered abelian group, with respect to its usual order.*

An $\mathbb{R}$-subspace of V is called ***rational*** *if it is the scalar extension of a $\mathbb{Q}$-subspace in $\mathrm{Weight}(S, \mathbb{Q})$. If $s \in S$, we set*

$$H_s = \{\varphi \in V \mid \varphi(s) = 0\}.$$

It is a rational hyperplane in V. We denote by $\tau_s : V \to V$ the linear map such that, for all $t \in S$,

$$\tau_s(\varphi)(t) = \begin{cases} \varphi(t), & \textit{if } s \textit{ is not conjugate to } t, \\ -\varphi(t) = -\varphi(s), & \textit{if } s \textit{ is conjugate to } t. \end{cases}$$

Then τ_s is a reflection in V, with reflecting hyperplane H_s.

C. Bonnafé, *Kazhdan–Lusztig Cells with Unequal Parameters*,
Algebra and Applications 24, https://doi.org/10.1007/978-3-319-70736-5_7

The dimension of the vector space V is equal to the number of conjugacy classes of reflections in W.

7.1. Faces, Chambers

Each hyperplane H in V divides the open set $V \setminus H$ into two connected components U and U'. We denote by $\smile_H$ the equivalence relation on V whose equivalence classes are U, U' and H.

We fix in this section a *finite* set $\mathfrak{A}$ of hyperplanes in V. We then define the equivalence relation $\smile_{\mathfrak{A}}$ on V by:

$$\varphi \smile_{\mathfrak{A}} \varphi' \text{ if and only if } \varphi \smile_{\mathrm{H}} \varphi' \text{ for all } \mathrm{H} \in \mathfrak{A}.$$

Definition 7.1.1. *The equivalence classes of* $\smile_{\mathfrak{A}}$ *are called the* ***$\mathfrak{A}$-faces****. The open $\mathfrak{A}$-faces are called the* ***$\mathfrak{A}$-chambers****.*

Definition 7.1.2. *The set $\mathfrak{A}$ is called* **complete** *if it contains all the hyperplanes H_s, $s \in S$.*

If $\mathfrak{A}$ is complete and if $\mathscr{F}$ is an $\mathfrak{A}$-face, we set

$$I(\mathscr{F}) = \{s \in S \mid \mathscr{F} \subset H_s\}$$

$$W_{\mathscr{F}} = W_{I(\mathscr{F})}.$$

Corollary 6.2.9 shows that if $\varphi \in \mathscr{F}$, then left (resp. right, resp. two-sided) cells in W are stable under left (resp. right, resp. two-sided) translations by elements of $W_{\mathscr{F}}$.

7.2. Cells

If G is a subgroup of W, we define three equivalence relations $trans_L^G$, $trans_R^G$, $trans_{LR}^G$ by:

- $x \; trans_L^G \; y$ if $Gx = Gy$.
- $x \; trans_R^G \; y$ if $xG = yG$.
- $x \; trans_{LR}^G \; y$ if $GxG = GyG$.

In other words, the set of equivalence classes for $trans_L^G$ (resp. $trans_R^G$, resp. $trans_{LR}^G$) is $G \backslash W$ (resp. W/G, resp. $G \backslash W/G$). We can now state here the Semicontinuity Conjecture.

Semicontinuity Conjecture 7.2.1. *Assume that S is finite. Then there exists a **complete** finite set $\mathfrak{A}$ of **rational** hyperplanes in V such that, for $? \in \{L, R, LR\}$, the following holds:*

(a) *If φ and φ' belong to the same face of $\mathfrak{A}$, then $\sim_?^{\varphi}$ and $\sim_?^{\varphi'}$ coincide.*
(b) *If $\mathscr{F}$ is an $\mathfrak{A}$-face, if $\varphi \in \mathscr{F}$, if $\mathscr{C}_1, \mathscr{C}_2, \ldots, \mathscr{C}_r$ are the $\mathfrak{A}$-chambers such that $\mathscr{F} \subset \overline{\mathscr{C}}_i$ and if $\varphi_i \in \mathscr{C}_i$, then $\sim_?^{\varphi}$ is the equivalence relation generated by the equivalence relations $\sim_?^{\varphi_1}, \sim_?^{\varphi_2}, \ldots, \sim_?^{\varphi_r}$ and $\mathrm{trans}_?^{W_{\mathscr{F}}}$.*

It is easily checked that if $\mathfrak{A}$ and $\mathfrak{A}'$ are two complete finite sets of rational hyperplanes satisfying the conditions of Conjecture 7.2.1, then $\mathfrak{A} \cap \mathfrak{A}'$ also does. Therefore, if Conjecture 7.2.1 holds, then there exists a *minimal* set of hyperplanes $\mathfrak{A}_W$ which satisfies the conditions of Conjecture 7.2.1.

Definition 7.2.2. *If Conjecture 7.2.1 holds, the complete finite set $\mathfrak{A}_W$ of rational hyperplanes is called the **essential hyperplane arrangement** of W. Its elements are called the **essential hyperplanes** of W.*

If $\varphi \in V$, $\lambda \in \mathbb{R}^*$ and $s \in S$, then

(7.2.3) $$\mathrm{Cell}_?^{\lambda\varphi}(W) = \mathrm{Cell}_?^{\varphi}(W) \quad \text{and} \quad \mathrm{Cell}_?^{\tau_s(\varphi)}(W) = \mathrm{Cell}_?^{\varphi}(W)$$

(see Corollaries 6.2.2 and 6.2.5).

Corollary 7.2.4. *If Conjecture 7.2.1 holds, then $\tau_s(\mathfrak{A}_W) = \mathfrak{A}_W$ for all $s \in S$.*

7.3. Examples

Recall that $\bar{S}$ denotes the set of equivalence classes of elements of S under the relation of being conjugate in W (see §4.3.B).

Example 7.3.1. Assume here that $|\bar{S}| = 1$. Let $s \in S$ and let $a = \varphi(s)$. Then V can be identified with $\mathbb{R}$ (through the parameter a) and it follows from Proposition 6.2.1 that the partition of W into cells depends only on whether a is equal to 0 or not. So $\{0\}$ is the only essential hyperplane of W and Conjecture 7.2.1 is trivially verified. □

Example 7.3.2. Assume in this example, and only in this example, that $|\bar{S}| = 2$. Let s, $t \in S$ be such that s is not conjugate to t in W and let $a = \varphi(s)$ and $b = \varphi(t)$. By Propositions 6.2.1 and 6.2.4 and Corollary 7.2.4, the essential hyperplane arrangement (if it exists) is completely determined if we are able to determine the partition into cells whenever $a, b > 0$. Since this partition does not change if we replace (a,b) by $(\lambda a, \lambda b)$ with $\lambda > 0$ (see Proposition 6.2.1 and Corollary 6.2.2), this means that it is enough to determine the values of b/a for which the partition into cells changes (these values are called *essential values*). The next Table 7.a provides examples for which Conjecture 7.2.1 has been checked and gives the list of essential values. In this Table, we give the Coxeter graph of (W,S) (together with its classical name) and the elements of S which are conjugate to s (resp. t) are denoted by s, s_i (resp. t, t_i).

Type	Coxeter graph	Essential values
$I_2(m)$, m even	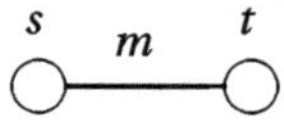	1
F_4	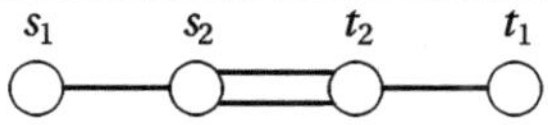	$1/2, 1, 2$
$\tilde{G}_2$	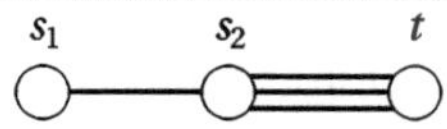	$1, 3/2, 2$

Table 7.a. Essential values for some "small" rank Coxeter groups

The proof for the dihedral case will be given in Chapter 21. The type F_4 case was proved by Geck [Gec4], partly based on computer computations. The proof of Conjecture 7.2.1 in the affine type $\tilde{G}_2$ case is due to Guilhot [Gui4]: the partition into left, right and two-sided cells will be given in Chapter 25. □

Example 7.3.3 (Type B). In this example, and only in this example, we fix $n \geqslant 2$ and we assume that (W,S) is of type B_n, and that its Dynkin diagram is given by

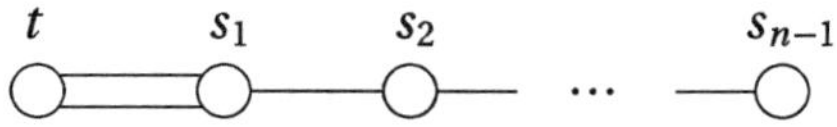

As in the previous Example 7.3.2, we set $a = \varphi(s_1) = \cdots = \varphi(s_{n-1})$ and $b = \varphi(t)$.

Conjecture 7.3.4. *The essential values for* b/a *are* $1, 2, \ldots, n-1$.

Conjecture 7.2.1 for type B_n and Conjecture 7.3.4 have been checked for $n \in \{2,3,4,5,6\}$ thanks to computer calculations. Conjecture 7.3.4 about type B_n is due to Geck, Iancu, Lam and the author [BGIL]. □

Notes

The semicontinuity conjecture was stated by the author in [Bon2].

Exercises

Exercise 7.1. Check Conjecture 7.2.1 whenever W is of type B_2 or G_2.

Exercise 7.2. We aim to justify the terminology "Semicontinuity conjectures". If $\mathscr{X}$ is a topological space and if $(\mathscr{E},\leqslant)$ is a poset, then a map $f:\mathscr{X}\to\mathscr{E}$ is called *lower semi-continuous* if, for all $e\in\mathscr{E}$, the set

$$\{x\in\mathscr{X}\mid f(x)\leqslant e\}$$

is closed in $\mathscr{X}$. We define analogously the notion of an *upper semi-continuous* map.

Let $? \in \{L,R,LR\}$ and assume that Conjecture 7.2.1 holds. We fix $w\in W$ and we denote by $\gamma_w^? : V\to\mathscr{P}(W)$ the map that sends $\varphi\in V$ to the Kazhdan–Lusztig left, right or two-sided (according to the value of ?) φ-cell of w. Show that $\gamma_w^?$ is lower semi-continuous (for the order $\subset$ on $\mathscr{P}(W)$).

Exercise 7.3. Let $s\in S$ and x, $y\in W$ be such that $sx<x<y<sy$ and $\ell(y)-\ell(x)=1$. Show that $\{\varphi\in V\mid h_{s,y,x}^{\varphi}\neq 0\}$ is closed in V.

Part IV
General Properties of Cells

Hypothesis. *From now on, and until the end of this book, we assume that* $\varphi > 0$.

Even though many of the results that are stated below do not require this hypothesis or could be easily adapted to hold in full generality, this will considerably smooth the exposition. Going back to the general case can be done using the results of Sections 4.3, 5.4 or 6.2, as is explained in Commentaries 5.4.7 and 6.2.12.

In Chapter 8, we study the link between the partition into cells of W and of its standard parabolic subgroups (induction, restriction of cells, cellular maps). We then apply these ideas in Chapter 9 to construct lower and upper bounds for the partition into cells. In Chapter 10, we assume that W is finite and explain why translation by w_0 sends a cell to a cell. Finally, Chapter 11 explains the Guilhot induction process (which generalizes the parabolic induction of cells constructed in Chapter 8) and gives an application to the *asymptotic case*.

Chapter 8
Cells and Parabolic Subgroups

Kazhdan–Lusztig cells have been defined in Chapter 6: it is a natural question to see how one can compute these cells, as well as the associated cell modules. A first idea would be to proceed by induction on $|S|$, starting from the knowledge of cells for proper parabolic subgroups. It is the purpose of this chapter to develop tools for relating cells of W and cells of its parabolic subgroups.

Hypothesis and notation. *We fix in this chapter a subset I of S and we denote by $\varphi_I : I \to \mathscr{A}$ the restriction of φ. To avoid confusion, the preorders $\leqslant_L$, $\leqslant_R$, $\leqslant_{LR}$ on W_I associated with the weight function φ_I are denoted by $\leqslant_L^I$, $\leqslant_R^I$ and $\leqslant_{LR}^I$, respectively. Similarly, we denote by $\sim_L^I$, $\sim_R^I$ and $\sim_{LR}^I$ the equivalence relations associated with the preorders $\leqslant_L^I$, $\leqslant_R^I$ and $\leqslant_{LR}^I$, respectively.*

We denote by $\mathrm{pr}_L^I : W \to W_I$, $\mathrm{pr}_R^I : W \to W_I$, $\mathrm{rep}_L^I : W \to X_I$ and $\mathrm{rep}_R^I : W \to X_I$ the unique maps such that $w = \mathrm{rep}_L^I(w)\mathrm{pr}_L^I(w) = \mathrm{pr}_R^I(w)\mathrm{rep}_R^I(w)^{-1}$ for all $w \in W$. If $\delta : W_I \to W_I$ is any map, we denote by $\delta^L : W \to W$ (resp. $\delta^R : W \to W$) the map defined by $\delta^L(xw) = x\delta(w)$ (resp. $\delta^R(wx^{-1}) = \delta(w)x^{-1}$) for all $x \in X_I$ and $w \in W_I$. Then δ^L (resp. δ^R) is called the ***left extension*** *(resp. the* ***right extension****) of δ to W.*

We set $\boldsymbol{\mu}_2 = \{1, -1\}$ and, if $\mu : W_I \to \boldsymbol{\mu}_2$ is any map, we denote by $\mu_L : W \to \boldsymbol{\mu}_2$ (resp. $\mu_R : W \to \boldsymbol{\mu}_2$) the map $\mu \circ \mathrm{pr}_L^I$ (resp. $\mu \circ \mathrm{pr}_R^I$).

All the results of this chapter are concerned with left or right cells. It turns out that, apart from the obvious property that $\leqslant_{LR}^I$ refines the restriction of $\leqslant_{LR}$ to W_I, there seems to be no nice result relating two-sided cells of W and of its parabolic subgroups. Throughout this chapter, the detailed exposition and the proofs are given in the context of left cells: using the anti-involution $*$, everything is then easily translated to the context of right cells.

C. Bonnafé, *Kazhdan–Lusztig Cells with Unequal Parameters*,
Algebra and Applications 24, https://doi.org/10.1007/978-3-319-70736-5_8

8.1. The Kazhdan–Lusztig–Geck Basis

Recall from Proposition 4.1.11 that $(T_x)_{x\in X_I}$ is an $\mathcal{H}_I$-basis of the *right* $\mathcal{H}_I$-module $\mathcal{H}$ (in particular, $\mathcal{H}$ is a flat right $\mathcal{H}_I$-module). It follows that, if $\mathfrak{a}$ is a left ideal of $\mathcal{H}_I$, then

(8.1.1) $$\mathcal{H}\otimes_{\mathcal{H}_I}\mathfrak{a}=\bigoplus_{x\in X_I} T_x\cdot\mathfrak{a}\ \textit{is a left ideal of}\ \mathcal{H}.$$

It is the left ideal of $\mathcal{H}$ generated by $\mathfrak{a}$.

Let us now introduce a new basis of $\mathcal{H}$, which mixes the standard basis and the Kazhdan–Lusztig basis. If $w\in W$, then we set

$$\mathbf{L}_w^I=T_{\mathrm{rep}_L^I(w)}\mathbf{C}_{\mathrm{pr}_L^I(w)}.$$

We now introduce a new order relation on W, which mixes the Bruhat order on W and the preorder $\leqslant_L^I$ on W_I.

Definition 8.1.2. *Let x and $y\in W$. We write $x\sqsubset_L^I y$ if the following three conditions are satisfied:*

(1) $\mathrm{rep}_L^I(x)<\mathrm{rep}_L^I(y)$,
(2) $x<y$,
(3) $\mathrm{pr}_L^I(x)\leqslant_L^I\mathrm{pr}_L^I(y)$.

We write $x\sqsubseteq_L^I y$ if $x\sqsubset_L^I y$ or $x=y$.

Note that

(8.1.3) $$\textit{if}\ x\sqsubseteq_L^I y,\ \textit{then}\ x\leqslant y.$$

Note also that

(8.1.4) $$\mathbf{L}_y^I\equiv T_y^I\equiv\mathbf{C}_y \mod \bigoplus_{x<y} A_{<0}\,T_x.$$

Therefore,

(8.1.5) $$(\mathbf{L}_w^I)_{w\in W}\ \textit{is an A-basis of}\ \mathcal{H}.$$

We call it the *Kazhdan–Lusztig–Geck basis*. The next result, despite its simplicity, is the key to the results of this chapter:

Lemma 8.1.6. *If $y\in W$, then*

$$\overline{\mathbf{L}}_y^I\equiv\mathbf{L}_y^I \mod \bigoplus_{x\sqsubset_L^I y} A\,\mathbf{L}_x^I.$$

Proof. Let $b \in X_I$ and $y_1 \in W_I$ be such that $y = by_1$. Then

$$\overline{\mathbf{L}}_y^I = T_{b^{-1}}^{-1}\mathbf{C}_{y_1} = \mathbf{L}_y^I + \sum_{a<b} \overline{r}_{a,b} T_a \mathbf{C}_{y_1}.$$

Now, let $a < b$. Write $a = a'u$ with $a' \in X_I$ and $u \in W_I$. Then $a' \leqslant a < b$ and $T_a \mathbf{C}_{y_1} = T_{a'} T_u \mathbf{C}_{y_1}$. This shows that

$$T_a \mathbf{C}_{y_1} \in \bigoplus_{x_1 \leqslant_L^I y_1} A\, T_{a'} \mathbf{C}_{x_1}.$$

Therefore,

$$\overline{\mathbf{L}}_y^I - \mathbf{L}_y^I \in \bigoplus_{\substack{a' \in X_I, a' < b \\ x_1 \leqslant_L^I y_1}} A\mathbf{L}_{a'x_1}^I.$$

Moreover, by 8.1.4,

$$\overline{\mathbf{L}}_y^I - \mathbf{L}_y^I \in \bigoplus_{x<y} A\, T_x = \bigoplus_{x<y} A\mathbf{L}_x^I.$$

The result follows from these last two facts. ■

It then follows from Corollary 2.1.6 that there exists a unique element $\mathbf{C}_y^I \in \mathscr{H}$ such that

$$\begin{cases} \overline{\mathbf{C}}_y^I = \mathbf{C}_y^I, \\ \mathbf{C}_y^I \equiv \mathbf{L}_y^I \mod \bigoplus_{x \in W} A_{<0}\, \mathbf{L}_x^I. \end{cases}$$

But, by 8.1.4, this implies that

$$\mathbf{C}_y^I = \mathbf{C}_y.$$

So, if we write

$$\mathbf{C}_y = \sum_{x \in W} p_{x,y}^I \mathbf{L}_x^I,$$

then, by 2.2.2,

(8.1.7)
$$\begin{cases} p_{x,y}^I \in A_{<0}, & \text{if } x \neq y, \\ p_{x,x}^I = 1, & \\ p_{x,y}^I \neq 0 \Longrightarrow x \sqsubseteq_L^I y. & \end{cases}$$

An easy consequence of 8.1.7 is that

(8.1.8)
$$p_{ax,ay}^I = p_{x,y}^I$$

for all $a \in X_I$ and $x, y \in W_I$.

8.2. Induction Theorems

The aim of this section is to show that the induction functor $\mathscr{H} \otimes_{\mathscr{H}_I} -$ on representations of Hecke algebras has a version at the level of Kazhdan–Lusztig cells.

Theorem 8.2.1. *Let $\mathfrak{I}$ be a subset of W_I. If $\mathfrak{I}$ is a left ideal of W_I, then $X_I \cdot \mathfrak{I}$ is a left ideal of W and*

$$\mathscr{H}(X_I \cdot \mathfrak{I}) = \mathscr{H} \otimes_{\mathscr{H}_I} \mathscr{H}_I(\mathfrak{I}).$$

Proof. Let $\mathfrak{I}$ be a left ideal of W_I. By definition,

$$\mathscr{H}(X_I \cdot \mathfrak{I}) = \bigoplus_{\substack{x \in X_I \\ w \in \mathfrak{I}}} A\, \mathbf{C}_{xw}.$$

Since $\mathfrak{I}$ is a left ideal of W_I, it follows from 8.1.1 and 8.1.7 that

$$\mathscr{H}(X_I \cdot \mathfrak{I}) = \bigoplus_{\substack{x \in X_I \\ w \in \mathfrak{I}}} A\, T_x \mathbf{C}_w.$$

So $\mathscr{H}(X_I \cdot \mathfrak{I}) = \mathscr{H} \otimes_{\mathscr{H}_I} \mathscr{H}_I(\mathfrak{I})$. This shows the second statement, and shows also that $\mathscr{H}(X_I \cdot \mathfrak{I})$ is a left ideal of $\mathscr{H}$. So $X_I \cdot \mathfrak{I}$ is a left ideal of W by Proposition 1.1.12. ■

Corollary 8.2.2. *If C is a left convex subset of W_I, then $X_I \cdot C$ is a left convex subset of W and we have an isomorphism of left $\mathscr{H}$-modules*

$$\mathscr{H}^L[X_I \cdot C] \simeq \mathscr{H} \otimes_{\mathscr{H}_I} \mathscr{H}_I^L[C].$$

In particular, $X_I \cdot C$ is a union of left cells of W.

Proof. This follows from Theorem 8.2.1 and Remark 1.1.17, taking into account that the functor $\mathscr{H} \otimes_{\mathscr{H}_I} -$ is exact. ■

Corollary 8.2.3. *Let $x, y \in W$. Then:*

(a) *If $x \leqslant_L y$, then $\mathrm{pr}_L^I(x) \leqslant_L^I \mathrm{pr}_L^I(y)$.*
(b) *If $x \sim_L y$, then $\mathrm{pr}_L^I(x) \sim_L^I \mathrm{pr}_L^I(y)$.*

In particular, the relations $\leqslant_L^I$ and $\sim_L^I$ are the restriction to W_I of $\leqslant_L$ and $\sim_L$ respectively:

Corollary 8.2.4. *Let $x, y \in W_I$. Then:*

(a) *$x \leqslant_L y$ if and only if $x \leqslant_L^I y$.*
(b) *$x \sim_L y$ if and only if $x \sim_L^I y$.*

The following example shows that the analogue of Corollary 8.2.4 for the relations $\leqslant_{LR}$ and $\sim_{LR}$ does not hold.

Example 8.2.5. Assume in this example, and only in this example, that (W,S) is of type A_3, that $S=\{s,t,u\}$, that the Coxeter graph is given by

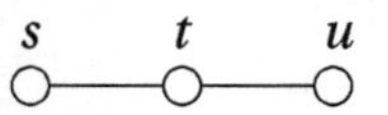

and that $\varphi(s) \neq 0$. Let $I = \{s,u\}$. Then, by Example 6.7.6 and Exercise 6.3, $s \sim_L ts \sim_R t \sim_L ut \sim_R u$. So $s \sim_{LR} u$, but $s \not\sim_{LR}^I u$, as $W_I \simeq W_s \times W_u$. □

We now investigate the properties of the involution $\bar{\ }$ in this context. First, note that if C is a left convex subset of W, then $\mathscr{H}^{\leqslant_L C}$ and $\mathscr{H}^{<_L C}$ are stable under the involution $\bar{\ }$, so this implies that $\mathscr{H}^L[C]$ inherits an A-semilinear involution still denoted by $\bar{\ }$. We have

$$\overline{\mathbf{c}}_w = \mathbf{c}_w.$$

We set

$$\mathscr{H}^L[C]_{<0} = \bigoplus_{w \in C} A_{<0}\,\mathbf{c}_w.$$

Corollary 8.2.6. *Let C be a left convex subset of W_I and, if $w \in X_I \cdot C$, let $\mathbf{l}_w^I$ denote the image of $\mathbf{L}_w^I$ in $\mathscr{H}^L[X_I \cdot C]$. Then:*

(a) $\mathscr{H}^L[X_I \cdot C]_{<0} = \bigoplus_{w \in X_I \cdot C} A_{<0}\,\mathbf{l}_w^I.$

(b) $\overline{\mathbf{l}}_w^I \equiv \mathbf{l}_w^I \mod \mathscr{H}^L[X_I \cdot C]_{<0}.$

(c) *Let $w \in X_I \cdot C$ and let $c \in \mathscr{H}^L[X_I \cdot C]$ be such that*

$$\begin{cases} \overline{c} = c, \\ c \equiv \mathbf{l}_w^I \mod \mathscr{H}^L[X_I \cdot C]_{<0}. \end{cases}$$

Then $c = \mathbf{c}_w$.

Proof. This follows directly from 8.1.7, Lemma 8.1.6 and Corollary 2.1.3. ■

Using the anti-involution $*$, Theorem 8.2.1 and Corollaries 8.2.2, 8.2.3, and 8.2.4 become:

Theorem 8.2.7. *Let $\mathfrak{I}$ be a subset of W_I. If $\mathfrak{I}$ is a right ideal of W_I, then $\mathfrak{I} \cdot X_I^{-1}$ is a right ideal of W and*

$$\mathscr{H}(\mathfrak{I} \cdot X_I^{-1}) = \mathscr{H}_I(\mathfrak{I}) \otimes_{\mathscr{H}_I} \mathscr{H}.$$

Corollary 8.2.8. *If C is a right convex subset of W_I, then $\mathfrak{I} \cdot X_I^{-1}$ is a right convex subset of W and we have an isomorphism of right $\mathscr{H}$-modules*

$$\mathscr{H}^R[C \cdot X_I^{-1}] \simeq \mathscr{H}_I^R[C] \otimes_{\mathscr{H}_I} \mathscr{H}.$$

In particular, $\mathfrak{I} \cdot X_I^{-1}$ *is a union of right cells of* W.

Corollary 8.2.9. *Let* $x, y \in W$. *Then:*

(a) *If* $x \leqslant_R y$, *then* $\mathrm{pr}_R^I(x) \leqslant_R^I \mathrm{pr}_R^I(y)$.
(b) *If* $x \sim_R y$, *then* $\mathrm{pr}_R^I(x) \sim_R^I \mathrm{pr}_R^I(y)$.

Corollary 8.2.10. *Let* $x, y \in W_I$. *Then:*

(a) $x \leqslant_R y$ *if and only if* $x \leqslant_R^I y$.
(b) $x \sim_R y$ *if and only if* $x \sim_R^I y$.

8.3. The Restriction Theorem

As well as the induction functor $\mathscr{H} \otimes_{\mathscr{H}_I} -$ on modules having a cell version, the restriction functor $\mathrm{Res}_{\mathscr{H}_I}^{\mathscr{H}}$ has an analogue at the level of Kazhdan–Lusztig cells. It is the purpose of this section to explain this analogy.

Theorem 8.3.1. *Let* $a \in X_I$ *and* $x, y, z \in W_I$. *Then*

$$h_{ax,y,az} = h_{x,y,z} \qquad \textit{and} \qquad h_{x,ya^{-1},za^{-1}} = h_{x,y,z}.$$

Proof. Let $(q_{u,v}^I)_{u,v \in W}$ denote the inverse matrix of $(p_{u,v}^I)_{u,v \in W}$. Then it follows from 8.1.7 that

$$\begin{cases} q_{u,u}^I = 1, \\ q_{u,v}^I \neq 0 \Longrightarrow u \sqsubseteq_L^I v. \end{cases}$$

Now,

$$\begin{aligned}
\mathbf{C}_{ax}\mathbf{C}_y &= T_a \mathbf{C}_x \mathbf{C}_y + \sum_{\substack{b<a \\ u \in W_I}} p_{bu,ax}^I T_b \mathbf{C}_u \mathbf{C}_y \\
&= \sum_{z \in W_I} h_{x,y,z} T_a \mathbf{C}_z + \sum_{\substack{b<a \\ u \in W_I \\ v \in W_I}} p_{bu,ax}^I h_{u,y,v} T_b \mathbf{C}_v \\
&= \sum_{z \in W_I} h_{x,y,z} \mathbf{C}_{az} + \sum_{\substack{z \in W_I \\ b<a \\ u \in W_I}} h_{x,y,z} q_{bu,az}^I \mathbf{C}_{bu} \\
&\qquad + \sum_{\substack{b<a \\ u \in W_I}} p_{bu,ax}^I h_{u,y,v} \Big(\sum_{\substack{c<b \\ w \in W_I}} q_{cw,bv}^I \mathbf{C}_{cw} \Big).
\end{aligned}$$

On the other hand,

$$\mathbf{C}_{ax}\mathbf{C}_y = \sum_{w \in W} h_{ax,y,w} \mathbf{C}_w.$$

So the first equality follows from the comparison of the coefficient of $\mathbf{C}_{az}$ in both expressions.

The second equality follows from the first one (where x, y and z have been replaced by x^{-1}, y^{-1} and z^{-1}, respectively) by applying 5.5.1. ■

Corollary 8.3.2. *Let $a \in X_I$ and $x, y \in W_I$. Then:*

(a) *If $x \leqslant_L^I y$, then $xa^{-1} \leqslant_L ya^{-1}$.*
(b) *If $x \sim_L^I y$, then $xa^{-1} \sim_L ya^{-1}$.*
(a') *If $x \leqslant_R^I y$, then $ax \leqslant_R ay$.*
(b') *If $x \sim_R^I y$, then $ax \sim_R ay$.*

Proof. The statements (b), (a') and (b') follow easily from (a). So let us prove (a). For this, we may assume that there exists a $u \in W_I$ such that $h_{u,y,x} \neq 0$. Then, by Theorem 8.3.1, we have $h_{u,ya^{-1},xa^{-1}} = h_{u,y,x} \neq 0$. So $xa^{-1} \overset{L}{\twoheadleftarrow} ya^{-1}$, as desired. ■

Corollary 8.3.3. *Let $a \in X_I$. Then:*

(a) *If C is a left convex subset of W, then $Ca \cap W_I$ is a left convex subset of W_I.*
(a') *If C is a right convex subset of W, then $a^{-1}C \cap W_I$ is a right convex subset of W_I.*

Proof. Again, it is sufficient to prove (a). So let C be a left convex subset of W. Let $x, z \in Ca \cap W_I$ and let $y \in W_I$ be such that $x \leqslant_L^I y \leqslant_L^I z$. Then $xa^{-1} \leqslant_L ya^{-1} \leqslant_L za^{-1}$ by Corollary 8.3.2(a), so $ya^{-1} \in C$ because C is left convex. So $y \in Ca \cap W_I$, as desired. ■

At the level of modules, we obtain a rough description of the restriction to $\mathcal{H}_I$ of the left module $\mathcal{H}^L[C]$. For this, we need to define, for $l \in \mathbb{N}$,

$$X_I^{\leqslant l} = \{a \in X_I \mid \ell(a) \leq l\} \qquad \text{and} \qquad X_I^l = \{a \in X_I \mid \ell(a) = l\}.$$

Corollary 8.3.4. *Let C be a left convex subset of W. For $l \in \mathbb{N}$, we define*

$$\mathcal{H}^L[C](I,l) = \bigoplus_{a \in X_I^{\leqslant l}} \Big(\bigoplus_{w \in C \cap W_I a^{-1}} A\mathbf{c}_w \Big).$$

Then $0 \subset \mathcal{H}^L[C](I,0) \subset \mathcal{H}^L[C](I,1) \subset \mathcal{H}^L[C](I,2) \subset \cdots$ is a filtration of the left $\mathcal{H}_I$-module $\operatorname{Res}_{\mathcal{H}_I}^{\mathcal{H}} \mathcal{H}^L[C]$ and

$$\mathcal{H}^L[C](I,l)/\mathcal{H}^L[C](I,l-1) \simeq \bigoplus_{a \in X_I^l} \mathcal{H}_I^L[Ca \cap W_I]$$

as $\mathcal{H}_I$-modules. Moreover,

$$\bigcup_{l \geqslant 0} \mathcal{H}^L[C](I,l) = \mathcal{H}^L[C].$$

Proof. This follows immediately from Theorem 8.3.1 and its proof (especially the first sequence of equalities). ■

Remark 8.3.5. Again, there is no analogue of such results for the relations $\leqslant_{LR}$ and $\sim_{LR}$. □

8.4. Induction of Isomorphic Cells

It is a fact that can already be checked in dihedral groups (see Chapter 21) that there are different cells which give rise to isomorphic cell modules, the isomorphism respecting the Kazhdan–Lusztig basis (a precise definition is given below): we explain in this section how this property is preserved by the induction process.

8.4.A. Definition, Examples

Definition 8.4.1. *Let $\sigma : W \to W$ and $\mu : W \to \boldsymbol{\mu}_2$ be two maps. Then the pair (σ,μ) is called* **left cellular** *if the following two conditions are satisfied for every left cell C of W:*

(LC1) *$\sigma(C)$ is a left cell.*
(LC2) *The A-linear map $(\sigma,\mu)_C : \mathscr{H}^L[C] \to \mathscr{H}^L[\sigma(C)]$, $\mathbf{c}_w \mapsto \mu_w \mathbf{c}_{\sigma(w)}$ is an isomorphism of left $\mathscr{H}$-modules.*

It is called **strongly left cellular** *if it is left cellular and, moreover, satisfies the following condition:*

(LC3) *If $w \in W$, then $\sigma(w) \sim_R w$.*

If μ is constant and (σ,μ) satisfies (LC1) *and* (LC2) *(resp.* (LC1), (LC2) *and* (LC3)*), then we say that σ is a* **left cellular** *map (resp. a* **strongly left cellular** *map).*

We similarly define the notions of **right cellular** *and* **strongly right cellular** *pair or map, as well as the notion of* **two-sided cellular** *pair or map.*

Note that, if (σ,μ) is a left (resp. right, resp. two-sided) cellular pair, then σ is not necessarily bijective: however, its restriction to a left (resp. right, resp. two-sided) cell is injective by (LC2). The next examples show that there are non-trivial cellular maps already in type A_2 and B_2. This will be extended to general dihedral groups in Chapter 21. However, in all these examples, the sign function μ is constant: we will see in Chapter 17 that there exist left cellular pairs (σ,μ) where μ is not constant on cells (see Example 17.3.4).

Example 8.4.2 **(*Type* A_2).** As in Examples 3.12.7, 4.4.13, 5.7.9 and 6.7.6, we assume here that $S = \{s,t\}$ and $m_{st} = 3$, so that (W,S) is of type A_2. We set $\varphi(s) = \varphi(t) = a$ and we assume, following Commentary 6.2.12 and as in Examples 5.7.9 and 6.7.6, that $a > 0$. Here is the corresponding Coxeter graph, together with the values of φ:

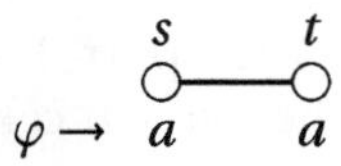

Then it follows from the computation of left cells and left cell modules in Example 6.7.6 that the unique involution $\sigma : W \to W$ such that

$$\begin{cases} \sigma(1) = 1, \\ \sigma(sts) = sts, \\ \sigma(s) = st, \\ \sigma(t) = ts, \end{cases}$$

is a left cellular map. □

Example 8.4.3 (***Type*** B_2). As in Examples 3.12.8, 4.4.14, 5.7.10 and 6.7.7, we assume here that $S = \{s,t\}$ with $m_{st} = 4$, so that (W,S) is of type B_2. We set $\varphi(s) = a$ and $\varphi(t) = b$, and assume, following Commentary 6.2.12 and as in Examples 5.7.10 and 6.7.7, that $a, b > 0$. Here is the corresponding Coxeter graph, together with the values of φ:

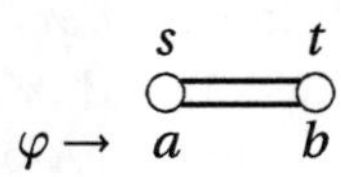

Let us define two involutions $\sigma_=$ and $\sigma_<$ of W by

$$\begin{cases} \sigma_=(1) = 1, \\ \sigma_=(stst) = stst, \\ \sigma_=(s) = sts, \\ \sigma_=(ts) = ts, \\ \sigma_=(t) = tst, \\ \sigma_=(st) = st, \end{cases} \qquad \begin{cases} \sigma_<(1) = 1, \\ \sigma_<(s) = s, \\ \sigma_<(tst) = tst, \\ \sigma_<(stst) = stst, \\ \sigma_<(t) = ts, \\ \sigma_<(st) = sts. \end{cases}$$

We also define an involution $\sigma_>$ by $\sigma_> = \omega \circ \sigma_< \circ \omega^{-1}$, where $\omega : W \to W$ is the unique automorphism such that $\omega(s) = t$ and $\omega(t) = s$. Then it follows from the computation of left cells and left cell modules in Example 6.7.7 that

$$\begin{cases} \textit{If } a < b, \textit{ then } \sigma_< \textit{ is left cellular,} \\ \textit{If } a = b, \textit{ then } \sigma_= \textit{ is left cellular,} \\ \textit{If } a > b, \textit{ then } \sigma_> \textit{ is left cellular.} \end{cases}$$

Note the following facts:

- If $b = a$, then $\sigma_=$ stabilizes all the left cells but induces a non-trivial automorphism of the left cell modules associated with left cells of cardinality 3.
- If $a < b$ (resp. $a > b$), then $\sigma_<$ (resp. $\sigma_>$) stabilizes the four left cells of cardinality 1, but exchanges the two left cells of cardinality 2.

Finally, note that $\sigma_=$ commutes with the automorphism ω. □

8.4.B. Induction

The main result of this section is that the left extension of a left cellular pair is still left cellular.

Theorem 8.4.4. *Recall that I is a subset of S and let (σ,μ) be a left cellular pair for (W_I, I, φ_I). Then (σ^L, μ_L) is a left cellular pair for (W,S,φ). If, moreover, (σ,μ) is strongly left cellular, then (σ^L, μ_L) is strongly left cellular.*

Proof. The proof is divided into four steps:

• *First step: construction and properties of an isomorphism of left $\mathcal{H}$-modules.* Let C be a left cell of W_I. We denote by $\mathfrak{I}$ (resp. $\mathfrak{I}^\#$) the set of elements w in W_I such that $w \leqslant_L C$ (resp. $w <_L C$). By 8.1.1, the families $(\mathbf{L}_w^I)_{w \in X_I \cdot \mathfrak{I}}$ and $(\mathbf{L}_w^I)_{w \in X_I \cdot \mathfrak{I}^\#}$ are A-bases of the left ideals $\mathcal{H}\mathcal{H}_I^{\leqslant_L C}$ and $\mathcal{H}\mathcal{H}_I^{<_L C}$, respectively. By Theorem 8.2.1, $\mathcal{H}\mathcal{H}_I^{\leqslant_L C} = \mathcal{H}(X_I \cdot \mathfrak{I})$ and $\mathcal{H}\mathcal{H}_I^{<_L C} = \mathcal{H}(X_I \cdot \mathfrak{I}^\#)$ and so the families $(\mathbf{C}_w)_{w \in X_I \cdot \mathfrak{I}}$ and $(\mathbf{C}_w)_{w \in X_I \cdot \mathfrak{I}^\#}$ are also A-bases of $\mathcal{H}\mathcal{H}_I^{\leqslant_L C}$ and $\mathcal{H}\mathcal{H}_I^{<_L C}$, respectively.

Recall from Corollary 8.2.2 that $X_I \cdot C$ is left convex and that $\mathcal{H}^L[X_I \cdot C] \simeq \mathcal{H} \otimes_{\mathcal{H}_I} \mathcal{H}^L[C] \simeq \mathcal{H}(X_I \cdot \mathfrak{I})/\mathcal{H}(X_I \cdot \mathfrak{I}^\#)$, so that the images $\mathbf{l}_w^I$ and $\mathbf{c}_w^L$ of $\mathbf{L}_w^I$ and $\mathbf{C}_w$, respectively, in the left $\mathcal{H}$-module $\mathcal{H}^L[X_I \cdot C]$ are well-defined. Again by 8.1.1 and Corollary 8.2.2, $(\mathbf{l}_w^I)_{w \in X_I \cdot C}$ and $(\mathbf{c}_w^L)_{w \in X_I \cdot C}$ can be viewed as A-bases of $\mathcal{H}^L[X_I \cdot C]$.

Since the pair (σ,μ) is left cellular, the A-linear map $\mathcal{H}_I^L[C] \to \mathcal{H}_I^L[\sigma(C)]$, $\mathbf{c}_w \mapsto \mu_w \mathbf{c}_{\sigma(w)}$ is an isomorphism of left $\mathcal{H}_I$-modules. Applying the functor $\mathcal{H} \otimes_{\mathcal{H}_I} -$, we get that the A-linear map

$$\begin{array}{rccc} \theta : & \mathcal{H}^L[X_I \cdot C] & \longrightarrow & \mathcal{H}^L[X_I \cdot \sigma(C)] \\ & \mathbf{l}_w^I & \longmapsto & \mu_{L,w} \mathbf{l}_{\sigma^L(w)}^I \end{array}$$

is an isomorphism of left $\mathcal{H}$-modules.

Now, the left $\mathcal{H}$-modules $\mathcal{H}(X_I \cdot \mathfrak{I})$ and $\mathcal{H}(X_I \cdot \mathfrak{I}^\#)$ are stable under the involution $\overline{}$. So $\mathcal{H}^L[X_I \cdot C]$ inherits an action of the involution $\overline{}$. Similarly, $\mathcal{H}^L[X_I \cdot \sigma(C)]$ inherits an action of the involution $\overline{}$. Moreover, these two A-modules (endowed with $\overline{}$) satisfy the hypotheses (1) and (2) of Chapter 2 (by Theorem 8.2.1).

Also, it follows from the definition that the isomorphism θ commutes with this involution. Therefore, $\overline{\theta(\mathbf{c}_w)} = \theta(\mathbf{c}_w)$ for all $w \in X_I \cdot C$. Moreover, it follows from 8.1.7 that

$$\theta(\mathbf{c}_w) \equiv \mu_{L,w} \mathbf{l}^I_{\sigma^L(w)} \mod \bigoplus_{x \in X_I \cdot \sigma(C)} A_{<0} \mathbf{l}^I_x.$$

But the element $\mathbf{c}_{\sigma^L(w)}$ is stable under the involution $^-$ and, again by Theorem 8.2.1, it satisfies

$$\mathbf{c}_{\sigma^L(w)} \equiv \mathbf{l}^I_{\sigma^L(w)} \mod \bigoplus_{x \in X_I \cdot \sigma(C)} A_{<0} \mathbf{l}^I_x.$$

Therefore, by Corollary 8.2.6,

(8.4.5) $$\theta(\mathbf{c}_w) = \mu_{L,w} \mathbf{c}_{\sigma^L(w)}.$$

• *Second step: partition into left cells.* Now, assume that $w \sim_L w'$. According to Corollary 8.2.3(b), there exists a unique cell C in W_I such that $w, w' \in X_I \cdot C$. By the definition of $\leqslant_L$ and $\sim_L$, there exist four sequences $x_1, \dots, x_m$, $y_1, \dots, y_n$, $w_1, \dots, w_m$, $w'_1, \dots, w'_n$ such that:

$$\begin{cases} w_1 = w, w_m = w', \\ w'_1 = w', w'_n = w, \\ \forall\, i \in \{1, 2, \dots, m-1\},\ h_{x_i, w_i, w_{i+1}} \neq 0, \\ \forall\, j \in \{1, 2, \dots, n-1\},\ h_{y_j, w'_j, w'_{j+1}} \neq 0. \end{cases}$$

Therefore, we have $w' = w_m \leqslant_L \cdots \leqslant_L w_2 \leqslant_L w_1 = w = w'_n \leqslant_L \cdots \leqslant_L w'_2 \leqslant_L w'_1 = w$ and so $w = w_1 \sim_L w_2 \sim_L \cdots \sim_L w_m = w' = w'_1 \sim_L w'_2 \sim_L \cdots \sim_L w'_n = w$. Again by Corollary 8.2.3(b), w_i, $w'_j \in X_I \cdot C$. So it follows from 8.4.5 that $h_{x, \sigma^L(w_i), \sigma^L(w_{i+1})} = \mu_{L,w_i} \mu_{L,w_{i+1}} h_{x, w_i, w_{i+1}}$ and $h_{x, \sigma^L(w'_j), \sigma^L(w'_{j+1})} = \mu_{L,w'_j} \mu_{L,w'_{j+1}} h_{y_j, w'_j, w'_{j+1}}$ for all $x \in W$. Consequently,

$$\begin{cases} \forall\, i \in \{1, 2, \dots, m-1\},\ h_{x_i, \sigma^L(w_i), \sigma^L(w_{i+1})} \neq 0, \\ \forall\, j \in \{1, 2, \dots, n-1\},\ h_{y_j, \sigma^L(w'_j), \sigma^L(w'_{j+1})} \neq 0. \end{cases}$$

It then follows that

$$\begin{aligned} \sigma^L(w') = \sigma^L(w_m) \leqslant_L \cdots \leqslant_L \sigma^L(w_2) &\leqslant_L \sigma^L(w_1) = \sigma^L(w) = \sigma^L(w'_n) \\ &\leqslant_L \cdots \leqslant_L \sigma^L(w'_2) \leqslant_L \sigma^L(w'_1) = \sigma^L(w'), \end{aligned}$$

and so $\sigma^L(w) \sim_L \sigma^L(w')$, as expected. So we have proved that

(∗) *if $w \sim_L w'$, then $\sigma^L(w) \sim_L \sigma^L(w')$.*

Now, let $\sigma_1 : W_I \to W_I$ be the map defined by $\sigma_1(x) = x$ if $x \notin \sigma(C)$ and $\sigma_1(\sigma(x)) = x$ if $x \in C$. Let $\mu_1 : W \to \boldsymbol{\mu}_2$ be defined by $\mu_{1,x} = 1$ if $x \notin \sigma(C)$ and $\mu_{1,\sigma(x)} = \mu_x$ if $x \in C$. Since left cellular maps can be defined "locally" (i.e. left cell by left cell), it is easily checked that (σ_1, μ_1) is left cellular. So, applying $(*)$ to the pair (σ_1, μ_1) with w and w' replaced by $\sigma^L(w)$ and $\sigma^L(w')$, we obtain

(8.4.6) $$w \sim_L w' \textit{ if and only if } \sigma^L(w) \sim_L \sigma^L(w').$$

• *Third step: left cellularity.* Now, let C' be a left cell in W. It follows from 8.4.6 that $\sigma^L(C')$ is also a left cell and it follows from 8.4.5 that the A-linear map $\mathscr{H}^L[C'] \to \mathscr{H}^L[\sigma(C')]$, $\mathbf{c}_w \mapsto \mu_{L,w}\mathbf{c}_{\sigma^L(w)}$ is an isomorphism of left $\mathscr{H}$-modules. In other words, (σ^L, μ_L) is left cellular.

• *Fourth step: strongness.* Assume, moreover, that (σ, μ) is strongly left cellular. Let $w \in W$. Let us write $w = ax$ with $a \in X_I$ and $x \in W_I$. Then $\sigma(x) \sim_R x$ by (LC3) and so $\sigma^L(w) = a\sigma(x) \sim_R ax = w$ by Corollary 8.3.2. ■

Corollary 8.4.7. *Let (σ, μ) be a left cellular pair for (W_I, I, φ_I) and let $x, y \in W$ be such that $\mathrm{pr}_L^I(x) \sim_L \mathrm{pr}_L^I(y)$. Then*

$$p_{x,y}^I = \mu_x \mu_y p_{\sigma^L(x),\sigma^L(y)}^I.$$

Proof. This follows from 8.4.5. ■

As usual, using the involution $*$, we get:

Theorem 8.4.8. *Let I be a subset of S and let (σ, μ) be a right cellular pair for (W_I, I, φ_I). Then (σ^R, μ_R) is a right cellular pair for (W, S, φ). If, moreover, (σ, μ) is strongly right cellular, then (σ^L, μ_L) is strongly right cellular.*

8.4.C. About Strongness

It is probable that left cellular pairs or maps are strongly left cellular (indeed, we will see in Proposition 16.3.1 that, if some conjectures of Lusztig hold, then this is true). Without appealing to these conjectures, we are only able to prove the following related result:

Proposition 8.4.9. *Let (σ, μ) be a left (resp. right) cellular pair and let $w \in W$. Then $\mathscr{L}(\sigma(w)) = \mathscr{L}(w)$ (resp. $\mathscr{R}(\sigma(w)) = \mathscr{R}(w)$).*

Proof. Let C denote the left cell of w and let $s \in S$. Then $s \in \mathscr{L}(w)$ if and only if $\mathbf{C}_s \mathbf{c}_w^L = -(v^{\varphi(s)} + v^{-\varphi(s)})\mathbf{c}_w^L$. So the result follows from the fact that the map $(\sigma, \mu)_C$ is an isomorphism of left $\mathscr{H}$-modules. ■

Notes

All the results of Sections 8.1 and 8.2 about parabolic induction of cells are due to Geck [Gec3]. The results of Section 8.3 about parabolic restriction of cells are due to Lusztig [Lus20, §9].

The first example of a cellular map is given by the $*$-operation (in the equal parameter case) constructed by Kazhdan and Lusztig [KaLu1]. This was extended by Lusztig [Lus20, Proposition 10.7], still in the equal parameter case. The fact that these results can be generalized to the unequal parameters case is due to Geck and the author [BoGe2] where the general framework for Theorem 8.4.4 on induction of left cellular maps is provided. The easy extension to cellular pairs (by adding a sign) is due to the author [Bon5].

Exercises

Exercise 8.1. If $\sigma : W \to W$ is a map, we define $\sigma^{\mathrm{op}} : W \to W$ by $\sigma^{\mathrm{op}}(w) = \sigma(w^{-1})^{-1}$. Show that σ is left cellular if and only if σ^{op} is right cellular.

Exercise 8.2 (Type B_2). As in Examples 3.12.8, 4.4.14, 5.7.10, 6.7.7 and 8.4.3, we assume that $S = \{s, t\}$ with $m_{st} = 4$, so that (W, S) is of type B_2. We assume here that $\varphi(s) = \varphi(t) \neq 0$. Show that the map $\sigma_=$ constructed in Example 8.4.3 is in fact two-sided cellular.

Exercise 8.3 (Type G_2). As in Exercises 3.1, 4.1, 5.1, 6.1, we assume here that $S = \{s, t\}$ with $m_{st} = 6$, so that (W, S) is of type G_2. We set $\varphi(s) = a$ and $\varphi(t) = b$, and assume here that $a, b > 0$. Here is the corresponding Coxeter graph, together with the values of φ:

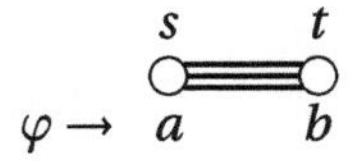

Determine, according to the values of a, b, all the possible left cellular maps and determine which ones are two-sided cellular.

Exercise 8.4. We assume here that $\varphi(s) = 0$ for all $s \in S$. Let $x \in W$. Show that the map $W \to W$, $w \mapsto wx$ (resp. $w \mapsto xw$) is left cellular (resp. right cellular). Show that it is two-sided cellular if and only if $x \in Z(W)$.

Exercise 8.5. Let $\sigma : W \to W$ be a left cellular map and let C be a left cell of W. Show that the map $\sigma : C \xrightarrow{\sim} \sigma(C)$ induces an isomorphism of W-graphs $(C, \mathscr{L}, \lambda) \xrightarrow{\sim} (\sigma(C), \mathscr{L}, \lambda)$ (see Exercise 6.5).

Chapter 9
Descent Sets, Knuth Relations and Vogan Classes

This chapter is devoted to applications of the results of the previous Chapter 8, where we investigated the general problem of relating the (left or right) cells of W and the (left or right) cells of all its standard parabolic subgroups. If $|I|=1$ (the easiest case), then the partition of W_I into cells has been determined in Example 6.1.11: but, in this case, the only interesting result one can obtain is that right (resp. left) descent sets are constant on left (resp. right) cells (and this has been obtained by other methods: see Proposition 6.3.1 and Corollary 6.3.2). If $|I|=2$, then the partition of W_I into cells will be determined in Chapter 21 (the reader can check that these partitions are obtained using only results from Chapters 2 and 6). If $|I| = 3$, then there are many cases where the partition of W_I is not known. Therefore, we focus in this chapter mostly on standard parabolic subgroups W_I where $|I|=2$. We will see later that this is already sufficient for obtaining interesting information about the partition of W into (left or right) cells. The results of this chapter are summarized below:

- Using the case where $|I| = 2$ and Geck's Induction Theorem 8.2.1, we improve slightly Proposition 6.3.1 and Corollary 6.3.2 about descent sets.

- Using the case where $|I|=2$ and Lusztig's Restriction Theorem 8.3.1, we define a partition of W into left *coplactic classes* which is finer than the partition into left cells.

- Using the induction of left cellular maps (see Theorem 8.4.4), we define a partition of W into *left Vogan classes* which is coarser than the partition into left cells.

Although most of these results rely only on the knowledge of cells of dihedral groups, they provide partitions which are first reasonable approximations of the partition into left cells.

C. Bonnafé, *Kazhdan–Lusztig Cells with Unequal Parameters*,
Algebra and Applications 24, https://doi.org/10.1007/978-3-319-70736-5_9

9.1. Enhanced Descent Sets

It has been shown in Corollary 6.3.2 that the map $\mathscr{R}$ is constant on left cells. This can easily be improved by using the knowledge of the partition of dihedral groups into cells. Let us introduce some notation. Set

$$S^{\varphi} = S \cup \{sts \mid s,t \in S \text{ such that } \varphi(s) < \varphi(t)\}.$$

Now, if $w \in W$, we define

$$\mathscr{R}^{\varphi}(w) = \{s \in S^{\varphi} \mid \ell(ws) < \ell(w)\}$$

and
$$\mathscr{L}^{\varphi}(w) = \{s \in S^{\varphi} \mid \ell(sw) < \ell(w)\}.$$

Then $\mathscr{R}^{\varphi}(w)$ (resp. $\mathscr{L}^{\varphi}(w)$) is called the *enhanced (with respect to φ) right* (resp. *left) descent set* of w.

Proposition 9.1.1. *Let x and y be two elements of W. Then:*

(a) *If $x \leqslant_L y$, then $\mathscr{R}^{\varphi}(y) \subseteq \mathscr{R}^{\varphi}(x)$.*
(b) *If $x \sim_L y$, then $\mathscr{R}^{\varphi}(x) = \mathscr{R}^{\varphi}(y)$.*
(a′) *If $x \leqslant_R y$, then $\mathscr{L}^{\varphi}(y) \subseteq \mathscr{L}^{\varphi}(x)$.*
(b′) *If $x \sim_R y$, then $\mathscr{L}^{\varphi}(x) = \mathscr{L}^{\varphi}(y)$.*

Proof. (b) follows immediately from (a), (a′) follows from (a) by applying the anti-automorphism $*$ and (b′) follows immediately from (a′). So it remains to prove (a). Now, assume that $x \leqslant_L y$ and let $s \in \mathscr{R}^{\varphi}(y)$. If $s \in S$, then $s \in \mathscr{R}(y)$ so $s \in \mathscr{R}(x)$ by Corollary 6.3.2(a). If $s \notin S$, then there exists t, $u \in S$ such that $s = tut$ and $\varphi(t) < \varphi(u)$.

Then $\mathrm{pr}_L^{t,u}(x) \leqslant_L \mathrm{pr}_L^{t,u}(y)$ by Corollary 8.2.3. Since $s \in \mathscr{R}(\mathrm{pr}_L^{t,u}(y))$ if and only if $s \in \mathscr{R}^{\varphi}(y)$, we only need to show the result whenever $S = \{t,u\}$. But, then, this follows from the description of cells given in Theorems 24.5.1 (for the case where $m_{tu} = \infty$) and 21.3.1 (for the case where $m_{tu} < \infty$). ■

9.2. Knuth Relations

If s, $t \in S$ and $x \in W$ are such that $\varphi(s) \geqslant \varphi(t)$ and $sx < x < tx < stx$, then we write $x \,\dot{\smile}_L^{\varphi}\, tx$. We denote by $\smile_L^{\varphi}$ the equivalence relation generated by $\dot{\smile}_L^{\varphi}$.

If x, $y \in W$, we write $x \,\dot{\smile}_R^{\varphi}\, y$ (resp. $x \smile_R^{\varphi} y$) if $x^{-1} \,\dot{\smile}_L^{\varphi}\, y^{-1}$ (resp $x^{-1} \smile_L^{\varphi} y^{-1}$). Finally, we denote by $\smile_{LR}^{\varphi}$ the equivalence relation generated by $\smile_L^{\varphi}$ and $\smile_R^{\varphi}$.

Definition 9.2.1. *The relation* $\dot{\smile}_L^{\varphi}$ *(resp.* $\dot{\smile}_R^{\varphi}$*) is called the* **left** *(resp.* **right***)* **Knuth relation***. The relation* $\smile_L^{\varphi}$ *(resp.* $\smile_R^{\varphi}$*) is called the* **coplactic** *(resp.* **plactic***)* **equivalence***.*

The equivalence classes for the coplactic and plactic equivalence are respectively called the **coplactic classes** *and the* **plactic classes***. The equivalence classes for* $\smile_{LR}^{\varphi}$ *are called the* **carpets***.*

If necessary, we talk about φ**-plactic** *or* φ**-coplactic classes***, or about* φ**-carpets***.*

Proposition 9.2.2. *Let* $x, y \in W$ *and let* $? \in \{L, R, LR\}$*. If* $x \smile_?^{\varphi} y$*, then* $x \sim_? y$*.*

Proof. By Lemma 6.1.5, it is sufficient to show that, if $x \dot{\smile}_L^{\varphi} y$, then $x \sim_L^{\varphi} y$. So let s and t in S be such that $\varphi(s) \geqslant \varphi(t)$ and $sx < x < tx = y < stx = sy$. As $\ell(tx) = \ell(t) + \ell(x) = 1 + \ell(x)$, it follows from Example 6.1.2 that $y = tx \leqslant_L x$. Also, as $\varphi(s) \geqslant \varphi(t)$, it follows from the equality (♢) in Example 5.7.3 that $\lambda_{x,y}^s \neq 0$, so $x \leqslant_L y$. Therefore, $x \sim_L y$. ■

If $x \dot{\smile}_L^{\varphi} y$, then it follows from Exercise 3.7 that $\mathscr{R}(x) = \mathscr{R}(y)$: so Proposition 9.2.2 is compatible with Corollary 6.3.2.

***Remark* 9.2.3.** Lusztig's Conjectures 6.5.2 are about properties of the partition of W into cells: it would be natural to ask whether such conjectures hold for the partition into coplactic classes. If it is obvious from the definition that the coplactic classes are left-connected, it is false in general that they always contain an involution, or that $w \smile_{LR}^{\varphi} w^{-1}$ for all $w \in W$. A counter-example can be found, for instance, if W is of type D_4 (see Exercise 9.2). □

9.3. Vogan Classes

Let $\mathbf{LC}(W, S, \varphi)$ be the set of triples (I, σ, μ), where I is a subset of S and (σ, μ) is a left cellular pair on W_I such that σ is bijective. Note that this set depends on φ (as can already be seen when $|S| = 2$: see Theorems 24.5.1 and 21.3.1). It is of course difficult to determine $\mathbf{LC}(W, S, \varphi)$ in general, but we explain in this section how one can use the knowledge of a subset of $\mathbf{LC}(W, S, \varphi)$ to construct a partition of W which is coarser than the partition into cells, but which is already an interesting first approximation. This also opens the possibility of inductive arguments.

So let us fix a subset ∇ of $\mathbf{LC}(W, S, \varphi)$ and a map $\gamma : W \to \mathscr{E}$ (where $\mathscr{E}$ is a set) such that the fibers of γ are unions of left cells (note that this is equivalent to giving an equivalence relation on W which is coarser than $\sim_L$). We denote by $\mathscr{V}_\nabla$ the group of permutations of W generated by the family $(\sigma^L)_{(I,\sigma,\mu) \in \nabla}$. Then, each $w \in W$ defines a map $\tau_w^{\nabla,\gamma} : \mathscr{V}_\nabla \longrightarrow \mathscr{E}$ as follows:

$$\tau_w^{\nabla,\gamma}(\sigma) = \gamma(\sigma(w)) \qquad \text{for all } \sigma \in \mathscr{V}_\nabla.$$

Definition 9.3.1. *Let $x, y \in W$. We say that x and y have the same* ***Vogan $\tau^{\nabla,\gamma}$-invariant*** *if $\tau_x^{\nabla,\gamma} = \tau_y^{\nabla,\gamma}$ (as maps from $\mathscr{V}_\nabla$ to $\mathscr{E}$). The equivalence classes for this relation are called the* ***left Vogan (∇,γ)-classes****.*

It follows from Theorem 8.4.4 that, for all $\sigma \in \mathscr{V}_\nabla$, there exists a sign function $\mu : W \to \mu_2$ such that the pair (σ, μ) is left cellular. In particular,

(9.3.2) $$\textit{if } x \sim_L y, \textit{ then } \sigma(x) \sim_L \sigma(y) \textit{ for all } \sigma \in \mathscr{V}_\nabla.$$

An immediate consequence is the following:

Proposition 9.3.3. *Let $x, y \in W$ be such that $x \sim_L y$. Then x and y have the same Vogan $\tau^{\nabla,\gamma}$-invariant.*

Remark 9.3.4. There is an equivalent formulation of Definition 9.3.1 which is more in the spirit of Vogan's original definition of a *generalized τ-invariant*. We define by induction on n a family of equivalence relations $\approx_n^{\nabla,\gamma}$ on W as follows. Let $x, y \in W$.

- We write $x \approx_0^{\nabla,\gamma} y$ if $\gamma(x) = \gamma(y)$.
- For $n \geq 1$, we write $x \approx_n^{\nabla,\gamma} y$ if $x \approx_{n-1}^{\nabla,\gamma} y$ and $\sigma^L(x) \approx_{n-1}^{\nabla,\gamma} \sigma^L(y)$ for all $(I, \sigma, \mu) \in \nabla$.

Note that the relation $\approx_n^{\nabla,\gamma}$ is finer than $\approx_{n-1}^{\nabla,\gamma}$. It follows from the definition that x, y have the same Vogan $\tau^{\nabla,\gamma}$-invariant if and only if $x \approx_n^{\nabla,\gamma} y$ for all $n \geqslant 0$.

This inductive definition is less easy to write than Definition 9.3.1, but it is more efficient for computational purposes. Indeed, if one finds an n_0 such that the relations $\approx_{n_0}^{\nabla,\gamma}$ and $\approx_{n_0+1}^{\nabla,\gamma}$ coincide, then x and y have the same Vogan $\tau^{\nabla,\gamma}$-invariant precisely when $x \approx_{n_0}^{\nabla,\gamma} y$. Note that such an n_0 always exists if W is finite. Also, even in small Coxeter groups, the group $\mathscr{V}_\nabla$ can become enormous while n_0 is reasonably small and the relation $\approx_{n_0}^{\nabla,\gamma}$ can be computed quickly. □

Commentary 9.3.5. Of course, the bigger ∇ is and the finer the partition of W defined by the fibers of the map γ is, the finer the result of Proposition 9.3.3 is. However, it is interesting to keep some flexibility because, in concrete examples, one only knows a small family ∇ of pairs (I, σ) as above, or one only has in hand the maps $\mathscr{R}$ or $\mathscr{R}^\varphi$ as starting point for γ. □

Example 9.3.6. Let ∇_2 be the set of pairs $(I, \sigma, \mu) \in \mathbf{LC}(W, S, \varphi)$ where $|I| = 2$. This set will be completely determined in Theorems 24.5.1 and 21.3.1. Also, one can take for γ the map $\mathscr{R}$ (or, even better, the map $\mathscr{R}^\varphi$).

Then Proposition 9.3.3 says that the left Vogan $(\nabla_2, \mathscr{R}^\varphi)$-classes are unions of left cells. We will see in Part VII that this is sometimes a very good first approximation. □

Remark 9.3.7. The results of this section, which concern only left cells, can be easily translated into results about right cells using the involution $*$. □

Notes

In 1979, Vogan [Vog1], [Vog3] introduced invariants of left cells which are computable in terms of certain combinatorially defined operators $T_{\alpha\beta}$, $S_{\alpha\beta}$ where α, β are adjacent simple roots of W. In the case where W is the symmetric group $\mathfrak{S}_n$, these invariants completely characterise the left cells [KaLu1, §5], [Vog1, §6] (see Chapter 22). Although Vogan's invariants are not complete invariants in general, they have turned out to be extremely useful in describing left cells.

The extension of these ideas to the unequal parameter case and the use of the general setting of left cellular maps are due to Geck and the author [BoGe2] (as in the previous chapter, the easy extension to left cellular pairs is due to the author [Bon5]).

Exercises

Exercise 9.1. (Type A_3). Assume here that (W,S) is of type A_3, that $\mathscr{A} = \mathbb{Z}$ and that $\varphi(s) = 1$ for all $s \in S$. Show that the coplactic classes coincide with the left Vogan $(\nabla_2, \mathscr{R})$-classes, so that this describes the left cells in W. Recall from Example 9.3.6 that ∇_2 is the set of pairs $(I,\sigma) \in \mathbf{LC}(W,S,\varphi)$ where $|I| = 2$.

Exercise 9.2 (Type D_4). As it is conjectured that any left cell contains an involution (see Lusztig's Conjecture 6.5.2(L1)) and as any left cell is a union of coplactic classes (see Proposition 9.2.2), it is natural to try to check if any coplactic class contains an involution. Show that this is not the case when (W,S) is of type D_4. ***Hint.*** If $S = \{s_1, s_2, s_3, t\}$ with $m_{s_i t} = 3$ and $m_{s_i s_j} = 2$, then check that $\{s_1 s_2 s_3 t s_2, t s_1 s_2 s_3 t s_2\}$ is a coplactic class containing no involution.

Remark. If (W,S) is of type D_4, there are in fact 8 coplactic classes in W which contain no involution (six of cardinality 2, and two of cardinality 6). □

Chapter 10
The Longest Element and Duality in Finite Coxeter Groups

Hypothesis and notation. *In this chapter, and only in this chapter, we assume that W is* ***finite****. Recall that we denote by w_0 the longest element of W (see section 3.7). Let*

$$\begin{array}{rccc} \omega_0 : & W & \longrightarrow & W \\ & w & \longmapsto & w_0 w w_0. \end{array}$$

We show in this chapter that multiplication by w_0 sends a left, right or two-sided cell to a left, right or two-sided cell respectively. During the proof of this important result, we also get some more symmetries (involving multiplication by w_0) for some of the structure constants $h_{x,y,z}$.

10.1. About ω_0

Recall that $w_0^2 = 1$ so that ω_0 is an involutive inner automorphism of W, which is the identity if and only if w_0 is central. The following lemma is easy:

Lemma 10.1.1. *$\omega_0(S) = S$ and $\varphi \circ \omega_0 = \varphi$.*

Proof. Let $s \in S$. Then it follows from Proposition 3.7.1 that $\ell(w_0 s w_0) = \ell(w_0) - \ell(s w_0) = \ell(w_0) - (\ell(w_0) - \ell(s)) = 1$, so $\omega_0(S) \subset S$ and the first statement follows from the fact that S is finite and ω_0 is injective.

Now, $\omega_0(s)$ is conjugate to s by definition, so $\varphi(\omega_0(s)) = \varphi(s)$. ■

C. Bonnafé, *Kazhdan–Lusztig Cells with Unequal Parameters*,
Algebra and Applications 24, https://doi.org/10.1007/978-3-319-70736-5_10

By Lemma 10.1.1, the A-linear map (still denoted by ω_0) defined by

$$\begin{aligned} \omega_0 : \mathscr{H} &\longrightarrow \mathscr{H} \\ T_w &\longmapsto T_{\omega_0(w)} \end{aligned}$$

is an involutive automorphism of $\mathscr{H}$.

Corollary 10.1.2. *If $h \in \mathscr{H}$, then $\omega_0(h) = T_{w_0} h T_{w_0}^{-1} = T_{w_0}^{-1} h T_{w_0}$.*

Proof. If $w \in W$, then it follows from Proposition 3.7.1 that

(10.1.3) $$T_{w_0} = T_{w_0 w^{-1}} T_w = T_w T_{w^{-1} w_0}.$$

Therefore,

$$T_{\omega_0(w)} T_{w_0} = T_{\omega_0(w)} T_{w_0 w^{-1}} T_w = T_{\omega_0(w)} T_{\omega_0(w)^{-1} w_0} T_w = T_{w_0} T_w.$$

This proves the first equality. The second follows as well. ∎

Corollary 10.1.4. *$T_{w_0}^2$ is central in $\mathscr{H}$.*

Now, ω_0 preserves $\mathscr{H}_{<0}$ and commutes with all the involutions *, $^-$ and †. In particular,

(10.1.5) $$\omega_0(\mathbf{C}_w) = \mathbf{C}_{\omega_0(w)}.$$

So

(10.1.6) $$p_{\omega_0(x),\omega_0(y)} = p_{x,y} \qquad \text{and} \qquad P_{\omega_0(x),\omega_0(y)} = P_{x,y}$$

for all $x, y \in W$.

10.2. The Inversion Formula

Since the A-module $\mathscr{H}$ is free of finite rank, it follows from 4.2.5 and 4.2.6 that there exists a unique A-basis $(\mathbf{D}_y)_{y \in W}$ of $\mathscr{H}$ such that

$$\tau(\mathbf{C}_x \mathbf{D}_y) = \delta_{x=y}$$

for all $x, y \in W$. The matrix $(p_{x,y})_{x,y \in W}$ being unipotent triangular (with respect to the Bruhat order) with off-diagonal coefficients in $A_{<0}$, it is invertible and its inverse $(q_{x,y})_{x,y \in W}$ is also triangular with off-diagonal coefficients in $A_{<0}$. Then, by definition,

(10.2.1) $$T_y = \sum_{x \in W} q_{x,y} \mathbf{C}_x.$$

So it follows from 4.2.5 that

(10.2.2) $$\mathbf{D}_y = \sum_{x \in W} q_{y,x} T_x.$$

In fact:

Theorem 10.2.3. *If $y \in W$, then $\mathbf{D}_y = \varepsilon(y w_0)\mathbf{C}^{\dagger}_{y^{-1}w_0} T_{w_0} = \varepsilon(y w_0) T_{w_0} \mathbf{C}^{\dagger}_{w_0 y^{-1}}$.*

Proof. The second equality is immediate from Corollary 10.1.2 and 10.1.5. Let us prove the first one. For this, let

$$\begin{aligned} \tau' : \mathscr{H} &\longrightarrow A \\ h &\longmapsto \tau(h T_{w_0}). \end{aligned}$$

Let $\mathbf{D}'_y = \mathbf{D}_y T_{w_0}^{-1}$. Then it turns out that $(\mathbf{D}'_y)_{y \in W}$ is an A-basis of $\mathscr{H}$ and that

$$\tau'(\mathbf{C}_x \mathbf{D}'_y) = \tau(\mathbf{C}_x \mathbf{D}_y) = \delta_{x=y}.$$

But 4.2.5 shows that $\tau'(T_w) = \delta_{w=w_0}$ for all $w \in W$. By 4.4.3, this implies that

$$\tau'(\overline{h}) = \overline{\tau'(h)}$$

for all $h \in \mathscr{H}$. Therefore,

$$\tau'(\mathbf{C}_x \overline{\mathbf{D}}'_y) = \delta_{x=y}.$$

This shows that $\overline{\mathbf{D}}'_y = \mathbf{D}'_y$. In particular,

$$\mathbf{D}_y T_{w_0}^{-1} = \overline{\mathbf{D}}_y T_{w_0} = \sum_{x \in W} \overline{q}_{y,x} (T_x^{-1} T_{w_0}).$$

So it follows from 10.1.3 that

(10.2.4) $$\mathbf{D}_y T_{w_0}^{-1} = \sum_{x \in W} \overline{q}_{y,x} T_{x^{-1}w_0} \equiv T_{y^{-1}w_0} \mod \mathscr{H}_{>0}.$$

Now, Remark 5.3.4 implies that

$$\mathbf{D}_y T_{w_0}^{-1} = \varepsilon(y w_0) \mathbf{C}^{\dagger}_{y^{-1}w_0},$$

as expected. ■

Corollary 10.2.5. *If $x, y \in W$, then $q_{y,x} = \varepsilon(xy) p_{w_0 x, w_0 y} = \varepsilon(xy) p_{x w_0, y w_0}$.*

Proof. Again, it is sufficient to prove only the first equality. From Theorem 10.2.3 and 10.2.4, we get

$$\sum_{x \in W} \overline{q}_{y,x} T_{x^{-1}w_0} = \varepsilon(y w_0) \mathbf{C}^{\dagger}_{y^{-1}w_0} = \varepsilon(y w_0) \sum_{x \in W} \varepsilon(x) \overline{p}_{x, y^{-1}w_0} T_x.$$

This shows that

$$\overline{q}_{y,x} = \varepsilon(y w_0)\varepsilon(x w_0)\overline{p}_{x^{-1} w_0, y^{-1} w_0} = \varepsilon(xy)\overline{p}_{w_0 x, w_0 y}$$

(see 5.3.2). ■

Corollary 10.2.5 together with 10.2.1 shows that

$$T_y = \sum_{x \in W} \varepsilon(xy) p_{w_0 y, w_0 x} \mathbf{C}_x = \sum_{x \in W} \varepsilon(xy) p_{y w_0, x w_0} \mathbf{C}_x. \tag{10.2.6}$$

10.3. Cells, Structure Constants

The next lemma is an immediate consequence of 10.1.5:

Lemma 10.3.1. *Let $x, y \in W$. Then $x \leqslant_? y$ (resp. $x \sim_? y$) if and only if $\omega_0(x) \leqslant_? \omega_0(y)$ (resp. $\omega_0(x) \sim_? \omega_0(y)$).*

In view of understanding the partition of W into cells, the main result of this chapter is the following:

Theorem 10.3.2. *Let $x, y \in W$. Then $x \leqslant_? y$ (resp. $x \sim_? y$) if and only if $w_0 y \leqslant_? w_0 x$ (resp. $y w_0 \sim_? x w_0$).*

Proof. By Lemma 10.3.1, it is sufficient to prove that $x \leqslant_L y$ if and only if $y w_0 \leqslant_L x w_0$. If $x, y \in W$, we write $y \leqslant'_L x$ if there exists an $h \in \mathcal{H}$ such that $\mathbf{D}_y$ appears with a non-zero coefficient in the decomposition of $\mathbf{D}_x h$ in the basis $(\mathbf{D}_w)_{w \in W}$. On one hand, it follows from Proposition 1.2.2 that $y \leqslant'_L x$ if and only if $x \leqslant_L y$.

On the other hand, by multiplying on the left by $T_{w_0}^{-1}$, we see that $y \leqslant'_L x$ if and only if $\mathbf{C}^\dagger_{w_0 y^{-1}}$ appears with a non-zero coefficient in the decomposition of $\mathbf{C}^\dagger_{w_0 x^{-1}} h$ in the basis $(\mathbf{C}^\dagger_{w_0 w^{-1}})_{w \in W}$. Applying the automorphism †, this shows that $y \leqslant'_L x$ if and only if $w_0 y^{-1} \leqslant_R w_0 x^{-1}$.

Using now Lemma 6.1.5, we get that $x \leqslant_L y$ if and only if $y w_0 \leqslant_L x w_0$. ■

Corollary 10.3.3. *We have:*

(a) *If C is a left (resp. right) cell, then $w_0 C$ and $C w_0$ are left (resp. right) cells.*
(b) *If Γ is a two-sided cell, then $\omega_0(\Gamma) = \Gamma$ and $w_0 \Gamma = \Gamma w_0$ is a two-sided cell.*

Proof. It only remains to prove that $\omega_0(\Gamma)=\Gamma$. Let $w\in\Gamma$. Then it follows from Corollary 10.1.2 and 10.1.5 that

$$\mathbf{C}_{\omega_0(w)} = T_{w_0}\mathbf{C}_w T_{w_0}^{-1}.$$

Therefore, $\omega_0(w)\leqslant_{LR} w$. Similarly, $w\leqslant_{LR}\omega_0(w)$, and the result follows. ∎

If C if a left cell, we define, as in Section 1.2, the right ideals

$$\mathscr{H}^{\vee,\geqslant_L C} = \bigoplus_{w\geqslant_L C} A\mathbf{D}_w \qquad \text{and} \qquad \mathscr{H}^{\vee,>_L C} = \bigoplus_{w>_L C} A\mathbf{D}_w$$

(see Proposition 1.2.2). We can define analogues of these for a right or a two-sided cell.

Corollary 10.3.4. *We have:*

(a) *If C is a left cell, then* $\mathscr{H}^{\vee,\geqslant_L C} = \big(\mathscr{H}^{\leqslant_R C^{-1}w_0}\big)^\dagger$ *and* $\mathscr{H}^{\vee,>_L C} = \big(\mathscr{H}^{<_R C^{-1}w_0}\big)^\dagger$.
(b) *If C is a right cell, then* $\mathscr{H}^{\vee,\geqslant_R C} = {}^\dagger\big(\mathscr{H}^{\leqslant_L w_0 C^{-1}}\big)$ *and* $\mathscr{H}^{\vee,>_R C} = {}^\dagger\big(\mathscr{H}^{<_L w_0 C^{-1}}\big)$.
(c) *If Γ is a two-sided cell, then* $\mathscr{H}^{\vee,\geqslant_{LR}\Gamma} = {}^\dagger\big(\mathscr{H}^{\leqslant_{LR} w_0\Gamma}\big)^\dagger$ *and* $\mathscr{H}^{\vee,>_{LR}\Gamma} = {}^\dagger\big(\mathscr{H}^{<_{LR} w_0\Gamma}\big)^\dagger$.

Proof. We only prove the first statement of (a), since the proofs of all the others are similar. By definition, we have

$$\mathscr{H}^{\vee,\geqslant_L C} = \bigoplus_{w\geqslant_L C} A\mathbf{D}_w.$$

So, by Theorem 10.2.3,

$$\mathscr{H}^{\vee,\geqslant_L C} = \bigoplus_{w\geqslant_L C} A\mathbf{C}^\dagger_{w^{-1}w_0} T_{w_0}.$$

Since $\mathscr{H}^{\vee,\geqslant_L C}$ is a right ideal, we can multiply this equality on the right by $T_{w_0}^{-1}$, and we get

$$\mathscr{H}^{\vee,\geqslant_L C} = \bigoplus_{w\geqslant_L C} A\mathbf{C}^\dagger_{w^{-1}w_0}.$$

Now, using Theorem 10.3.2 (and Lemma 6.1.5), we get

$$\mathscr{H}^{\vee,\geqslant_L C} = \bigoplus_{w\leqslant_L C^{-1}w_0} A\mathbf{C}^\dagger_{w^{-1}w_0} = \big(\mathscr{H}^{\leqslant_R C^{-1}w_0}\big)^\dagger,$$

as expected. ∎

We conclude with a result on structure constants:

Proposition 10.3.5. *Let $s\in S$ and let $x,y\in W$ be such that $sx<x$ and $sy>y$. Then*

$$\lambda^s_{x,y} = -\varepsilon(xy)\lambda^s_{yw_0,xw_0}.$$

Proof. If $h \in \mathcal{H}$ and $y \in W$, let us write $\mathbf{C}_y h = \sum_{x \in W} \alpha_{x,y}(h)\mathbf{C}_x$ and $h\mathbf{D}_y = \sum_{x \in W} \beta_{x,y}(h)\mathbf{D}_x$. Then

$$\alpha_{x,y}(h) = \tau((\mathbf{C}_y h)\mathbf{D}_x) = \tau(\mathbf{C}_y(h\mathbf{D}_x)) = \beta_{y,x}(h).$$

Therefore, if $sx < x$, then

$$\mathbf{C}_s\mathbf{D}_{x^{-1}} = \sum_{y \in W} \rho^s_{x^{-1},y^{-1}}\mathbf{D}_{y^{-1}} = (v^{\varphi(s)} + v^{-\varphi(s)})\mathbf{D}_{x^{-1}} + \sum_{\substack{y \in W \\ sy > y}} \rho^s_{x^{-1},y^{-1}}\mathbf{D}_{y^{-1}}$$

by Corollary 5.5.6. By multiplying on the right by $T_{w_0}^{-1}$ and noticing that $-\mathbf{C}_s^\dagger = \mathbf{C}_s - (v^{\varphi(s)} + v^{-\varphi(s)})$, we get

$$-\varepsilon(xw_0)\mathbf{C}_s^\dagger\mathbf{C}_{xw_0}^\dagger = \sum_{\substack{y \in W \\ sy > y}} \rho^s_{x^{-1},y^{-1}}\varepsilon(yw_0)\mathbf{C}_{yw_0}^\dagger.$$

On the other hand, applying † to equality 5.5.2, we get

$$\mathbf{C}_s^\dagger\mathbf{C}_{xw_0}^\dagger = \sum_{\substack{y \in W \\ sy > y}} \lambda^s_{yw_0,xw_0}\mathbf{C}_{yw_0}^\dagger.$$

So the result follows from 5.5.4. ■

10.4. Cell Modules

If M is a left (resp. right) $\mathcal{H}$-module and ρ is an automorphism of $\mathcal{H}$, we denote by ${}^\rho M$ (resp. M^ρ) the left (resp. right) $\mathcal{H}$-module whose underlying A-module is still M but such that h acts on ${}^\rho M$ (resp. M^ρ) as $\rho^{-1}(h)$ acts on M. We denote by M_* (resp. ${}_*M$) the right (resp. left) module whose underlying A-module is still M but such that h acts on the right (resp. left) on M_* (resp. ${}_*M$) as h^* acts on the left (resp. right) on M. Finally, we set $M^\vee = \mathrm{Hom}_A(M,A)$: it is a right (resp. left) $\mathcal{H}$-module.

Proposition 10.4.1. *We have:*

(a) *If C is a left convex subset of W, then we have an isomorphism of left $\mathcal{H}$-modules $\mathcal{H}^L[Cw_0] \xrightarrow{\sim} {}^\dagger\mathcal{H}^L[C]^\vee_*$.*
(b) *If C is a right convex subset, then we have an isomorphism of right $\mathcal{H}$-modules $\mathcal{H}^R[w_0C] \xrightarrow{\sim} ({}_*\mathcal{H}^R[C]^\vee)^\dagger$.*

Proof. It is sufficient to prove (a). But this follows from Proposition 1.2.2(a) by following the same argument as in the proof of Theorem 10.3.2. ■

10.5. The Element $\mathbf{C}_{w_0}$

The Kazhdan–Lusztig basis element $\mathbf{C}_{w_0}$ is given by the following formula:

(10.5.1) $$\mathbf{C}_{w_0} = \sum_{w \in W} v^{\varphi(w)-\varphi(w_0)} T_w.$$

Proof. This amounts to proving that $p_{w,w_0} = v^{\varphi(w)-\varphi(w_0)}$ for all $w \in W$. But this follows easily from Proposition 5.6.1(a) by a descending induction on $\ell(w)$ (the starting case where $w = w_0$ being trivial). ■

Proposition 10.5.2. *We have:*

(a) *w_0 is alone in its left, right or two-sided cell. Its left (resp. right, respectively two-sided) cell is the unique minimal left (resp. right, resp. two-sided) cell.*
(b) *For all $s \in S$, $\mathbf{C}_s \mathbf{C}_{w_0} = \mathbf{C}_{w_0} \mathbf{C}_s = (v^{\varphi(s)} + v^{-\varphi(s)}) \mathbf{C}_{w_0}$.*
(c) *If $h, h' \in \mathscr{H}$, then $h \mathbf{C}_{w_0} h' = \mathrm{ind}(hh') \mathbf{C}_{w_0}$.*
(d) *$\mathbf{C}_{w_0}^2 = (\sum_{w \in W} v^{2\varphi(w)-\varphi(w_0)}) \mathbf{C}_{w_0}$.*

Proof. If $w \in W$, then it follows from Example 6.1.2 that $w_0 \leqslant_L w$ (resp. $w_0 \leqslant_R w$) because $\ell(w_0) = \ell(w_0 w^{-1}) + \ell(w)$ (resp. $\ell(w_0) = \ell(w^{-1} w_0) + \ell(w)$). On the other hand, if $w \in W$ is such that $w \leqslant_L w_0$ (resp. $w \leqslant_R w_0$), then $S = \mathscr{R}(w_0) \subset \mathscr{R}(w)$ (resp. $S = \mathscr{L}(w_0) \subset \mathscr{L}(w)$) by Corollary 6.3.2, and so $w = w_0$. This shows (a).

(b) follows from Corollary 5.5.6 and (c) follows from (b) and from the fact that $(\mathbf{C}_s)_{s \in S}$ is a generating set for the A-algebra $\mathscr{H}$ (note that $\mathrm{ind}(\mathbf{C}_s) = v^{\varphi(s)} + v^{-\varphi(s)}$).

(d) follows from (c) and 10.5.1. ■

Notes

The results of this section are due to Kazhdan and Lusztig [KaLu1, §3] in the equal parameter case and to Lusztig [Lus20, §11] in the general case. We present here a slightly different proof.

Chapter 11
The Guilhot Induction Process

11.1. Main Result

Definition 11.1.1. *Let $\mathscr{F}$ be a non-empty subset of W and let $\mathscr{F}^\#$ be a non-empty subset of $W \times \mathscr{F}$. We say that the pair $(\mathscr{F}, \mathscr{F}^\#)$* ***satisfies the Guilhot induction process*** *if the following five conditions hold:*

(G1) *If $w \in \mathscr{F}$, then $(1, w) \in \mathscr{F}^\#$.*
(G2) *If $(x, w) \in \mathscr{F}^\#$, then $\ell(xw) = \ell(x) + \ell(w)$.*
(G3) *If (x, w), $(x', w') \in \mathscr{F}^\#$ are such that $w \neq w'$, then $xw \neq x'w'$.*
(G4) *The A-submodule $\mathscr{M} = \sum_{(x,w)\in\mathscr{F}^\#} A\, T_x \mathbf{C}_w$ is a left ideal in $\mathscr{H}$.*
(G5) *If $(x, w) \in \mathscr{F}^\#$, then $T_x \mathbf{C}_w \equiv T_{xw} \mod \mathscr{H}_{<0}$.*

Hypothesis. *We fix in this section, and only in this section, a non-empty subset $\mathscr{F}$ of W and a subset $\mathscr{F}^\#$ of $W \times \mathscr{F}$ such that the pair $(\mathscr{F}, \mathscr{F}^\#)$ satisfies the Guilhot induction process.*

We set $\mathscr{B}^{\mathscr{F}^\#} = (T_x \mathbf{C}_w)_{(x,w)\in\mathscr{F}^\#}$. We define a preorder $\sqsubseteq^{\mathscr{F}^\#}$ on $\mathscr{F}^\#$ as follows: if (x, w) and (x', w') are two elements of $\mathscr{F}^\#$, we write $(x, w) \sqsubseteq^{\mathscr{F}^\#} (x', w')$ if either one of the following two conditions holds:

- $(x, w) = (x', w')$;
- $x < x'$ *and* $xw \leqslant x'w'$.

We write $(x, w) \sqsubset^{\mathscr{F}^\#} (x', w')$ if the second statement is satisfied.

C. Bonnafé, *Kazhdan–Lusztig Cells with Unequal Parameters*,
Algebra and Applications 24, https://doi.org/10.1007/978-3-319-70736-5_11

Note that, by (G2), the statement (G5) is equivalent to:

(G5′) *If* $(x,w)\in\mathscr{F}^\#$ *and* $w'\in W$ *are such that* $w'<w$, *then* $p_{w',w}T_xT_{w'}\in\mathscr{H}_{<0}$.

Also, it follows from (G2) and (G3) that

(11.1.2) $$\mathscr{B}^{\mathscr{F}^\#} \text{ is an } A\text{-basis of } \mathscr{M}.$$

Moreover, if we set $\mathscr{M}_{<0}=\bigoplus_{(x,w)\in\mathscr{F}^\#}A_{<0}\,T_x\mathbf{C}_w$, then

(11.1.3) $$\mathscr{M}_{<0}=\mathscr{M}\cap\mathscr{H}_{<0}$$

by (G5). Now, by (G4), we can define a preorder $\leqslant_L^{\mathscr{F}^\#}$ on $\mathscr{F}^\#$ following a now standard procedure: we say that $(x,w)\overset{\mathscr{F}^\#}{\twoheadleftarrow}(x',w')$ if there exists an $h\in\mathscr{H}$ such that $T_x\mathbf{C}_w$ appears with a non-zero coefficient in the decomposition of $hT_{x'}\mathbf{C}_{w'}$ in the basis $\mathscr{B}^{\mathscr{F}^\#}$ and we denote by $\leqslant_L^{\mathscr{F}^\#}$ the transitive closure of $\overset{\mathscr{F}^\#}{\twoheadleftarrow}$. It is a preorder on $\mathscr{F}^\#$ and we denote by $\sim_L^{\mathscr{F}^\#}$ the associated equivalence relation. Recall that $(1,w)\in\mathscr{F}^\#$ for all $w\in\mathscr{F}$ (by (G1)).

Lemma 11.1.4. *Let* $(x,w)\in\mathscr{F}^\#$. *Then* $(x,w)\sim_L^{\mathscr{F}^\#}(1,w)$.

Proof. It is clear that $(x,w)\leqslant_L^{\mathscr{F}^\#}(1,w)$. The converse follows from the fact that $\mathbf{C}_w=T_x^{-1}(T_x\mathbf{C}_w)$. ■

It follows from Lemma 11.1.4 that we can define a preorder $\leqslant_L^{\mathscr{F}}$ on $\mathscr{F}$ as follows: we write $w\leqslant_L^{\mathscr{F}}w'$ if $(1,w)\leqslant_L^{\mathscr{F}^\#}(1,w')$. We denote by $\sim_L^{\mathscr{F}}$ the associated equivalence relation (the notation is somewhat ambiguous, as $\sim_L^{\mathscr{F}}$ does not depend only on $\mathscr{F}$, but also on $\mathscr{F}^\#$).

We now wish to apply the results of Chapter 2 to the pair $(\mathscr{M},\mathscr{B}^{\mathscr{F}^\#})$.

Proposition 11.1.5. *The left A-module* $\mathscr{M}$ *is stable under the semilinear involution* $\bar{\ }$. *Moreover, the triple* $(\mathscr{M},\mathscr{B}^{\mathscr{F}^\#},\sqsubseteq^{\mathscr{F}^\#})$ *satisfies the hypotheses* (I1) *and* (I2) *of Chapter 2.*

Proof. Let $(x,w)\in\mathscr{F}^\#$. Then $\overline{T_x\mathbf{C}_w}=T_{x^{-1}}^{-1}\mathbf{C}_w\in\mathscr{M}$ because $\mathbf{C}_w\in\mathscr{M}$ by (G1) and $\mathscr{M}$ is a left ideal of $\mathscr{H}$ by (G4). So $\mathscr{M}$ is stable under the semilinear involution $\bar{\ }$.

Also, if $(x,w)\sqsubseteq^{\mathscr{F}^\#}(x',w')$, then $xw\leqslant x'w'$. This proves the property (I1) of Chapter 2, thanks to (G2). It remains to prove (I2). But

$$\overline{T_x\mathbf{C}_w}=\sum_{x'\leqslant x}\bar{r}_{x',x}T_{x'}\mathbf{C}_w$$

and so the result follows from (G4) and 11.1.2. ■

Let $(x,w) \in \mathscr{F}^\#$. Then it follows from Proposition 11.1.5 and Corollary 2.1.3 that there exists a unique $M_{x,w} \in \mathscr{M}$ such that

$$\begin{cases} \overline{M}_{x,w} = M_{x,w} \\ M_{x,w} \equiv T_x \mathbf{C}_w \mod \mathscr{M}_{<0}. \end{cases}$$

But, by (G5) and 11.1.3, $T_x \mathbf{C}_w \equiv T_{xw} \mod \mathscr{H}_{<0}$, so it follows from the characterization of the Kazhdan–Lusztig basis elements that

$$M_{x,w} = \mathbf{C}_{xw}. \tag{11.1.6}$$

As a consequence, we get from Corollary 2.1.6 that

$$\mathscr{M} = \bigoplus_{(x,w)\in\mathscr{F}^\#} A\mathbf{C}_{xw}. \tag{11.1.7}$$

We are now ready to prove the main result of this chapter:

Theorem 11.1.8. *Assume that the pair $(\mathscr{F},\mathscr{F}^\#)$ satisfies the Guilhot induction process. Let $\mathscr{E}$ be a left ideal in $\mathscr{F}$ for the preorder $\leqslant_L^{\mathscr{F}}$. Then*

$$\{xw \mid (x,w) \in \mathscr{F}^\# \cap (W \times \mathscr{E})\}$$

is a left ideal in W (for the preorder $\leqslant_L$).

Proof. Let $\mathscr{E}^\# = \mathscr{F}^\# \cap (W \times \mathscr{E})$. Then it follows immediately from the definitions that the pair $(\mathscr{E},\mathscr{E}^\#)$ satisfies the hypotheses (G1), (G2), (G3), (G4) and (G5) and so 11.1.7 shows that

$$\bigoplus_{(x,w)\in\mathscr{E}^\#} A\mathbf{C}_{xw}$$

is a left ideal of $\mathscr{H}$. This completes the proof of the theorem. ■

Corollary 11.1.9. *Assume that the pair $(\mathscr{F},\mathscr{F}^\#)$ satisfies the Guilhot induction process. Let $\mathscr{E}$ be an equivalence class in $\mathscr{F}$ for the relation $\sim_L^{\mathscr{F}}$. Then*

$$\{xw \mid (x,w) \in \mathscr{F}^\# \cap (W \times \mathscr{E})\}$$

is a union of left cells.

Example 11.1.10. Assume here, and only here, that $\mathscr{F} = W_I$ for some subset I of S and let $\mathscr{F}^\# = X_I \times W_I$. Then (G1), (G2) and (G3) are trivially satisfied. Moreover, $\mathscr{M} = \mathscr{H}$, so that (G4) is also satisfied. Finally, (G5) is exactly 8.1.4. In this case, Theorem 11.1.8 and Corollary 11.1.9 are just Theorem 8.2.1 and Corollary 8.2.2.

Thus the Guilhot induction process is a generalization of Geck's Theorems on parabolic induction. This justifies the name "induction". □

11.2. Application to the Asymptotic Case

We start with a preliminary lemma, which is just a computation in a dihedral group:

Lemma 11.2.1. *Assume here that* $S=\{s,t\}$ *with* m_{st} *even. Let*

$$X=\{x\in X_t s \mid xs>x\}.$$

Then $\sum_{x\in X} T_x \mathbf{C}_s \mathcal{H}_t$ *is a left ideal of* $\mathcal{H}$.

Proof. Recall that $\mathcal{H}_t = A\oplus A\,T_t$. Note also that

$$X=X_s\setminus\{w_0 s\}=\{1,t,st,tst,\dots,(st)^{m-1}\},$$

where $m_{st} = 2m \geqslant 2$ and w_0 denotes the longest element of W. Let $\mathcal{M} = \sum_{x\in X} T_x \mathbf{C}_s \mathcal{H}_t$. It is a right $\mathcal{H}_t$-submodule of $\mathcal{H}$ and $\mathbf{C}_s\mathcal{H}_t \subset \mathcal{M} \subset \mathcal{H}\mathbf{C}_s\mathcal{H}_t$. So it remains to prove that $\mathcal{H}\mathbf{C}_s \subset \mathcal{M}$. We first prove an intermediate result (for all $i\in\{0,1,\dots,m\}$).

$$(\aleph_i) \qquad T_{(st)^i} - T_{(ts)^i} \in \mathcal{M}.$$

The proof is by induction on i. If $i=0$ (or m), then $(\aleph_i)$ trivially holds. Now, let $i\in\{1,2,\dots,m-1\}$ and assume that $(\aleph_{i-1})$ holds. An immediate computation yields

$$T_{(st)^i} - T_{(ts)^i} = T_{(st)^{i-1}}\mathbf{C}_s T_t - T_{(ts)^{i-1}t}\mathbf{C}_s - v^{-\varphi(s)}(T_{(st)^{i-1}} - T_{(ts)^{i-1}})T_t.$$

But $(st)^{i-1}$, $(ts)^{i-1}t \in X$, so $T_{(st)^{i-1}}\mathbf{C}_s T_t - T_{(ts)^{i-1}t}\mathbf{C}_s \in \mathcal{M}$ by construction. On the other hand, $T_{(st)^{i-1}} - T_{(ts)^{i-1}} \in \mathcal{M}$ by the induction hypothesis, and so $-v^{-\varphi(s)}(T_{(st)^{i-1}} - T_{(ts)^{i-1}})T_t \in \mathcal{M}$ because $\mathcal{M}$ is a right $\mathcal{H}_t$-submodule. Thus $(\aleph_i)$ follows.

To complete the proof of the lemma, we only need to show that $T_x\mathbf{C}_s \in \mathcal{M}$ for all $x\in W$. This is done by induction on $\ell(x)$. If $\ell(x)=0$, then $x=1$ and the result is trivial. Assume now that $\ell(x)>0$. Two cases may occur:

• If $xs<x$, then $T_x\mathbf{C}_s = T_{xs}T_s\mathbf{C}_s = v^{\varphi(s)}T_{xs}\mathbf{C}_s$ and the result follows from the induction hypothesis.

• If $xs>x$, then $x\in X_s$. If $x\neq w_0 s$, then $x\in X$ and the result is trivial. So we may assume that $x = w_0 s = (ts)^{m-1}t$ (recall that $w_0 = (st)^m = (ts)^m$ because $m_{st}=2m$). But then

$$T_{(ts)^{m-1}t}\mathbf{C}_s = T_{(st)^{m-1}}\mathbf{C}_s T_t + v^{-\varphi(s)}(T_{(ts)^{m-1}} - T_{(st)^{m-1}})T_t.$$

Since $(st)^{m-1}\in X$ and $T_{(ts)^{m-1}} - T_{(st)^{m-1}} \in \mathcal{M}$ by $(\aleph_{m-1})$, we get that $T_x\mathbf{C}_s \in \mathcal{M}$, as desired. ■

Let us write

$$T_x T_y = \sum_{z \in W} t_{x,y,z} T_z,$$

where $t_{x,y,z} \in A$.

Theorem 11.2.2. *Let $I \in \mathscr{P}(S)$ and let $s \in S \setminus I$ be such that*

$$\varphi(s) > \deg(t_{x,y,z})$$

for all $x, y, z \in W_I$. Then

$$\{xw \mid (x,w) \in X_I \times W_I \text{ and } xs < x\}$$

is a left ideal of W.

Proof. We set $\mathscr{F} = sW_I$ and

$$\mathscr{F}^\# = \{(x,w) \in X_I s \times sW_I \mid xs > x\}.$$

According to Theorem 11.1.8, we only need to prove that $(\mathscr{F}, \mathscr{F}^\#)$ satisfies the Guilhot induction process. Note first the following fact:

(∗) If $t \in I$, then st has even order or infinite order.

Indeed, since $t_{t,t,t} = v^{\varphi(t)} - v^{-\varphi(t)}$, it follows from the hypothesis that $\varphi(s) > \varphi(t)$ and so st cannot have odd order.

Since $s \in X_I$, $1 \in X_I s$ and so $\{1\} \times sW_I \subset \mathscr{F}^\#$. So (G1) follows. Moreover, if $(x,w) \in \mathscr{F}^\#$, then $\ell(xw) = \ell(xssw) = \ell(xs) + \ell(sw)$ because $(xs, sw) \in X_I \times W_I$. But $\ell(xs) = \ell(x) + 1$ and $\ell(sw) = \ell(w) - 1$, so (G2) holds. If (x', w') is another element of $\mathscr{F}^\#$ such that $xw = x'w'$, then $xssw = x'ssw'$ and so $xs = x's$ and $sw = sw'$ because (xs, sw) and $(x's, sw')$ belong to $X_I \times W_I$. So $(x,w) = (x', w')$ and (G3) holds.

(G4) Let $\mathscr{M} = \sum_{(x,w) \in \mathscr{F}^\#} A\, T_x \mathbf{C}_w$. As $\mathscr{F} = sW_I$, it follows from Example 5.7.2 that $\mathbf{C}_w = \mathbf{C}_s \mathbf{C}_{sw}$ for all $w \in \mathscr{F}$. Therefore, $\mathscr{M}$ is a right $\mathscr{H}_I$-submodule of $\mathscr{H}$ and

$$\mathbf{C}_s \mathscr{H}_I \subset \mathscr{M} \subset \mathscr{H} \mathbf{C}_s \mathscr{H}_I.$$

So it remains to show that $\mathscr{H} \mathbf{C}_s \mathscr{H}_I \subset \mathscr{M}$. Since $\mathscr{M}$ is a right $\mathscr{H}_I$-submodule of $\mathscr{H}$, this means that we only need to show that $T_x \mathbf{C}_s \in \mathscr{M}$ for all $x \in W$. We will prove this statement by induction on $\ell(x)$. If $\ell(x) = 0$, then $x = 1$ and the result is obvious.

Now assume that $\ell(x) > 0$. Two cases may occur:

• If $xs < x$, then $T_x = T_{xs} T_s$ and $T_x \mathbf{C}_s = T_{xs} T_s \mathbf{C}_s = v^{\varphi(s)} T_{xs} \mathbf{C}_s$, so $T_x \mathbf{C}_s \in \mathscr{M}$ by the induction hypothesis.

• If $xs > x$, then we may assume that $x \notin X_I s$ (for otherwise, the result is obvious). So $xs \notin X_I$, which means that there exists a $t \in I$ such that $xst < xs$.

So $\{s,t\} \in \mathscr{R}(xs)$, which implies that $W_{s,t}$ is finite (see Lemma 3.7.6) and that st has even order $2m$ by $(*)$, with $m \geq 1$. In particular, $w_{s,t} = (st)^m = (ts)^m$ is central in $W_{s,t}$, and $xs = x_0 w_{s,t}$, with $x_0 \in X_{s,t}$. So $x = x_0(ts)^{m-1}t$ and $\ell(x) = \ell(x_0) + 2m - 1$. In particular, $T_x = T_{x_0} T_{(ts)^{m-1}t}$. Now, following Lemma 11.2.1, we set $X = \{1, t, st, tst, \ldots, (st)^{m-1}\}$ and we get

$$T_{(ts)^{m-1}t}\mathbf{C}_s \in \bigoplus_{y\in X} T_y \mathbf{C}_s \mathscr{H}_t \subset \bigoplus_{y\in X} T_y \mathbf{C}_s \mathscr{H}_I.$$

Consequently,

$$T_x \mathbf{C}_s \in \bigoplus_{y\in X} T_{x_0 y} \mathbf{C}_s \mathscr{H}_I$$

and, by the induction hypothesis, $T_{x_0 y}\mathbf{C}_s \in \mathscr{M}$ for all $y \in X$. So $T_x \mathbf{C}_s \in \mathscr{M}$, as desired.

(G5) Let $(x,w) \in \mathscr{F}^\#$ and let $w' \in W$ be such that $w' < w$. Since $w \in sW_I$, two cases may occur:

- If $w' \in sW_I$, then $\ell(xw') = \ell(x) + \ell(w')$ (by (G2)) and so $p_{w',w} T_x T_{w'} = p_{w',w} T_{xw'} \in \mathscr{H}_{<0}$ because $p_{w',w} \in A_{<0}$. This proves (G5′) in this case.
- If $w' \in W_I$, then $sw' > w'$ and, since $sw < w$, it follows from Proposition 5.6.1 that $p_{w',w} = v^{-\varphi(s)} p_{sw',w}$, and so $\deg(p_{w',w}) \leq -\varphi(s)$. On the other hand, if we write $x = x'w''$ with $x' \in X_I$ and $w'' \in W_I$, then

$$p_{w',w} T_x T_{w'} = \sum_{z\in W_I} p_{w',w} t_{w'',w',z} T_{xz}$$

and so $p_{w',w} T_x T_{w'} \in \mathscr{H}_{<0}$ because $\deg(p_{w',w} t_{w'',w',z}) \leqslant \deg(t_{w'',w',z}) - \varphi(s) < 0$ by hypothesis. This proves (G5′) in this case. ■

Remark 11.2.3. It will be shown in Lemma 14.1.4 that, if $I \in \mathscr{P}_{\mathrm{sph}}(S)$ and x, y, $z \in W_I$, then $\deg(t_{x,y,z}) \leqslant \varphi(w_I)$. □

Corollary 11.2.4. *Let $I \in \mathscr{P}(S)$ be such that*

$$\varphi(s) > \deg(t_{x,y,z})$$

for all $s \in S \setminus I$ and all x, y, $z \in W_I$. Then $W \setminus W_I$ is a two-sided ideal in W and any left, right or two-sided cell of W_I is respectively a left, right or two-sided cell of W.

Proof. If $x \in X_I$ is different from 1, then there exists an $s \in S \setminus I$ such that $xs < x$. Therefore,

$$\bigcup_{s\in S\setminus I} \{x \in X_I \mid xs < x\} = X_I \setminus \{1\}.$$

So it follows from Theorem 11.2.2 that $W \setminus W_I = \{xw \mid (x,w) \in (X_I \setminus \{1\}) \times W_I\}$ is a left ideal of W. Similarly, $W \setminus W_I$ is a right ideal of W, so it is a two-sided ideal of W. The last assertion follows. ■

Remark 11.2.5. Assume in this remark, and only in this remark, that S is finite and that $\Gamma = \mathbb{R}$. Let $\varphi_0 : S \to \Gamma$ be defined by

$$\varphi_0(s) = \begin{cases} \varphi(s) & \text{if } s \notin I, \\ 0 & \text{if } s \in I. \end{cases}$$

Let $\tilde{J} = \{wsw^{-1} \mid w \in W_I \text{ and } s \in S \setminus I\}$, let $\widetilde{W} = \langle \tilde{J} \rangle$ and let $\tilde{\varphi}_0$ denote the restriction to $\tilde{J}$ of the braided map $\varphi_0^\flat : W \to \Gamma$ associated to φ_0 by Matsumoto's Theorem 3.4.7. We are in the set-up of §6.2.C. Since $\tilde{\varphi}_0$ takes only positive values, $\{1\}$ is a left, right or two-sided cell of W, and so W_I is a left, right or two-sided φ_0-cell of W_I.

On the other hand, the Semicontinuity Conjecture 7.2.1 would imply that there exists a constant $\eta \in \mathbb{R}_{>0}$ such that, if $\varphi(t) < \eta\varphi(s)$ for all $s \in S \setminus I$ and $t \in I$, then W_I is a union of two-sided φ-cells (and in particular of left or right φ-cells). We will also see later that Lusztig conjectures that there exists an $m \in \mathbb{R}_{>0}$ such that

$$\deg(t_{x,y,z}) \leqslant m \max_{s \in I} \varphi(s)$$

for all $x, y, z \in W_I$ (see Conjecture 14.1.5). So Corollary 11.2.4 is compatible with the Semicontinuity Conjecture (taking $\eta = 1/m$). □

Notes

The results of this chapter are all due to Guilhot [Gui3], who used them to compute the partition into cells for affine Weyl groups of type $\widetilde{B}_2$ and $\widetilde{G}_2$ for all choices of parameters [Gui4].

Chapter 12
Submaximal Cells (Split Case)

Recall from Example 6.7.2 that $1 \leqslant_L w$ and $1 \leqslant_R w$ for all $w \in W$, so that the left, right or two-sided cell containing 1, which is equal to $\{1\}$ (because $\varphi(s) \neq 0$ for all $s \in S$) is maximal among left, right or two-sided cells respectively. A left (resp. right, resp. two-sided) cell is called *submaximal* if it is maximal in $W \setminus \{1\}$. The aim of this chapter is to describe submaximal cells in the *split case* (see Example 5.7.4). These results will be extended to the general case in Chapter 13, but they mainly rely on the split case.

> **Hypothesis and notation.** *We assume in this chapter, and only in this chapter, that (W,S) is* ***irreducible****, that $\mathscr{A} = \mathbb{Z}$ and that $\varphi(s) = 1$ for all $s \in S$. In other words, we are in the* ***split case****. We denote by Γ^{uni} the set of elements of $W \setminus \{1\}$ which have a unique reduced expression and, if $s \in S$, we set*
>
> $$\Gamma_s^{\mathrm{uni}} = \{w \in \Gamma^{\mathrm{uni}} \mid s \in \mathscr{R}(w)\}.$$

Note that the hypothesis that (W,S) is irreducible is made here just to simplify the statements. The general case may be easily reduced to this case. We prove in this chapter that Γ^{uni} is a two-sided cell and that it is the only submaximal one, while the Γ_s^{uni} are left cells and are the only submaximal ones. We also explain a strange property of the W-graph associated with submaximal left cells, whenever Γ^{uni} is finite.

C. Bonnafé, *Kazhdan–Lusztig Cells with Unequal Parameters*,
Algebra and Applications 24, https://doi.org/10.1007/978-3-319-70736-5_12

12.1. About the Elements of Γ^{uni}

12.1.A. Elementary Properties

First of all, note that

$$(\Gamma^{\mathrm{uni}})^{-1} = \Gamma^{\mathrm{uni}} \tag{12.1.1}$$

and that

$$|\mathscr{L}(x)| = |\mathscr{R}(x)| = 1 \tag{12.1.2}$$

for all $x \in \Gamma^{\mathrm{uni}}$ (by the Exchange Condition 3.1.6). To simplify the notation, we write $\mathscr{L}(x) = s$ (resp. $\mathscr{R}(x) = s$) to say that $\mathscr{L}(x) = \{s\}$ (resp. $\mathscr{R}(x) = \{s\}$) for $x \in \Gamma^{\mathrm{uni}}$. If $(s_1, s_2, \dots, s_r)$ is the reduced expression of an element $x \in \Gamma^{\mathrm{uni}}$, then

$$s_1 = \mathscr{L}(x) \qquad \text{and} \qquad s_r = \mathscr{R}(x). \tag{12.1.3}$$

In particular,

$$\Gamma^{\mathrm{uni}} = \dot{\bigcup_{s \in S}} \Gamma^{\mathrm{uni}}_s \qquad \text{and} \qquad \Gamma^{\mathrm{uni}}_s = \{x \in \Gamma^{\mathrm{uni}} \mid \mathscr{R}(x) = s\}. \tag{12.1.4}$$

Another easy fact is the following: if $x \in \Gamma^{\mathrm{uni}}$ and $(s_1, s_2, \dots, s_r)$ is the reduced expression of x, then

$$s_i s_{i+1} \cdots s_j \in \Gamma^{\mathrm{uni}} \tag{12.1.5}$$

for all $1 \leqslant i \leqslant j \leqslant l$.

12.1.B. Description of Γ^{uni}

The following characterization of elements of Γ^{uni} follows immediately from Example 3.4.8 (which is a reformulation of Matsumoto's Theorem 3.4.7):

Lemma 12.1.6. *Let $w \in W$ and let $(s_1, s_2, \dots, s_r)$ be a reduced expression of w. Then $w \in \Gamma^{\mathrm{uni}}$ if and only if there does not exist $s \neq t \in S$ and $i \in \{0, 1, 2, \dots, r - m_{st}\}$ such that $(s_{i+1}, s_{i+2}, \dots, s_{i+m_{st}}) = \underbrace{(s, t, s, \dots)}_{m_{st} \text{ terms}}$.*

Corollary 12.1.7. *Let $w \in \Gamma^{\mathrm{uni}}$ and let $(s_1, s_2, \dots, s_r)$ be the reduced expression of w. Then $m_{s_i s_{i+1}} \geqslant 3$ for all $i \in \{1, 2, \dots, r-1\}$.*

Corollary 12.1.8. *Let $x \in \Gamma^{\text{uni}}$ and $s \in S$ be such that $sx \notin \Gamma^{\text{uni}}$. Let $t = \mathscr{L}(x)$ and let $(s_1, s_2, \dots, s_r)$ be the unique reduced expression of x. Then $s \neq t$ and*

$$(s_1, s_2, \dots, s_{m_{st}-1}) = \underbrace{(t, s, t, \dots)}_{m_{st}-1 \text{ terms}}.$$

We denote by $\mathscr{U}nbr(W,S)$ the set of non-empty sequences $(s_1, s_2, \dots, s_r)$ of elements of S such that:

(U) There does not exist $s \neq t \in S$ and $i \in \{0, 1, 2, \dots, r - m_{st}\}$ such that $(s_{i+1}, s_{i+2}, \dots, s_{i+m_{st}}) = \underbrace{(s, t, s, \dots)}_{m_{st} \text{ terms}}$.

An element of $\mathscr{U}nbr(W,S)$ is called an *unbraided path* in the Coxeter graph $\mathscr{C}ox(W,S)$. If $(s_1, s_2, \dots, s_r) \in \mathscr{U}nbr(W,S)$, then (U) implies that $m_{s_i s_{i+1}} \geqslant 3$ for all $i \in \{1, 2, \dots, r-1\}$.

Remark 12.1.9 (Paths). A sequence $(s_1, s_2, \dots, s_r)$ of elements of S is called a *path* if $m_{s_i s_{i+1}} \geqslant 3$ for all $i \in \{1, 2, \dots, r-1\}$. Since the Coxeter graph $\mathscr{C}ox(W,S)$ has an edge between s and t if and only if $m_{st} \geqslant 3$, this corresponds to the usual notion of path in a graph. So the set of unbraided paths is a subset of the set of paths in $\mathscr{C}ox(W,S)$.

If $(s_1, s_2, \dots, s_r)$ is a path such that $s_i \neq s_j$ for all $1 \leqslant i < j \leqslant r$, then it is unbraided (by Lemma 12.1.6). □

It follows from Lemma 12.1.6 that the map

(12.1.10)
$$\begin{aligned} \pi : \mathscr{U}nbr(W,S) &\longrightarrow \Gamma^{\text{uni}} \\ (s_1, s_2, \dots, s_r) &\longmapsto s_1 s_2 \cdots s_r \end{aligned}$$

is bijective. If $s \in S$, we denote by $\mathscr{U}nbr(W,S)_s$ the set of unbraided paths $(s_1, s_2, \dots, s_r)$ such that $s_r = s$. Then, 12.1.3 can be rewritten as follows:

(12.1.11) *π restricts to a bijection $\mathscr{U}nbr(W,S)_s \xrightarrow{\sim} \Gamma_s^{\text{uni}}$.*

12.1.C. Further Properties

As we will prove that Γ^{uni} is a two-sided cell and Γ_s^{uni} is a left cell, the properties of these subsets of W gathered in the next proposition are compatible with Conjectures 6.5.2 and 6.5.3.

Proposition 12.1.12. *Assume that (W,S) is irreducible and let $s, t \in S$. Then:*

(a) *Γ_s^{uni} is a left-connected component of Γ^{uni}.*
(b) *$(\Gamma_s^{\text{uni}})^{-1} \cap \Gamma_t^{\text{uni}} \neq \varnothing$.*

Proof. (a) Let L_s denote the left-connected component of s in Γ^{uni}. Let $w \in \Gamma_s^{\mathrm{uni}}$ and let $(s_1, s_2, \dots, s_r)$ be the reduced expression of w. Then $s_r = s$ by 12.1.3 and it follows from 12.1.5 that $s_j s_{j+1} \dots s_r \in \Gamma_s^{\mathrm{uni}}$ for all $j \in \{1, 2, \dots, r\}$, so that $w \in L_s$. Therefore, $\Gamma_s^{\mathrm{uni}} \subset L_s$.

Conversely, it also follows from 12.1.5 that if $w \in \Gamma^{\mathrm{uni}}$ and $s' \in S$ are such that $s'w \in \Gamma^{\mathrm{uni}}$, then $\mathscr{R}(s'w) = \mathscr{R}(w)$ by 12.1.3, so that the map $\mathscr{R}$ is constant on left-connected components of Γ^{uni}. So $L_s \subset \Gamma_s^{\mathrm{uni}}$, and so $L_s = \Gamma_s^{\mathrm{uni}}$.

(b) Since (W, S) is irreducible, there exists a path $(s_1, s_2, \dots, s_r)$ in the graph $\mathscr{C}ox(W, S)$ such that $s = s_1$ and $t = s_r$. We may also assume that $s_i \neq s_j$ whenever $1 \leqslant i < j \leqslant r$. Hence $(s_1, s_2, \dots, s_r)$ is unbraided and so $s_1 s_2 \dots s_r \in (\Gamma_s^{\mathrm{uni}})^{-1} \cap \Gamma_t^{\mathrm{uni}}$. ■

Proposition 12.1.13. *Assume that (W, S) is irreducible. Let $\mathscr{C}$ be a conjugacy class of involutions such that $\mathscr{C} \cap \Gamma^{\mathrm{uni}} \neq \varnothing$. Then all the elements of $\mathscr{C}$ are reflections and $\mathscr{C} \cap \Gamma_s^{\mathrm{uni}} \neq \varnothing$ for all $s \in S$.*

Proof. Let $w \in \mathscr{C} \cap \Gamma^{\mathrm{uni}}$. We prove by induction on $\ell(w)$ that w is a reflection. If $\ell(w) = 1$, then the result is obvious. Assume that $\ell(w) > 1$ and let $(s_1, s_2, \dots, s_r)$ denote the unique reduced expression of w. Since $w = w^{-1}$, we have

$$(*) \qquad (s_1, s_2, \dots, s_r) = (s_r, \dots, s_2, s_1)$$

and so $s_1 = s_r$. Therefore, $s_1 w s_1 = s_2 s_3 \cdots s_{r-1} \in \Gamma^{\mathrm{uni}}$ by 12.1.5. But then $s_1 w s_1 \in \mathscr{C} \cap \Gamma^{\mathrm{uni}}$ and $\ell(s_1 w s_1) < \ell(w)$, so $s_1 w s_1$ is a reflection by the induction hypothesis, and so w is a reflection.

Now, let $t \in \mathscr{C} \cap S$ and let $s \in S$. We want to prove that $\mathscr{C} \cap \Gamma_s^{\mathrm{uni}} \neq \varnothing$. This is done by induction on the distance between s and t in $\mathscr{C}ox(W, S)$. If this distance is 0, then $s = t \in \mathscr{C} \cap \Gamma_s^{\mathrm{uni}}$. Assume that the distance is strictly greater than 0. Let $(s_1, \dots, s_r)$ be a path of minimal length in $\mathscr{C}ox(W, S)$ such that $s_1 = t$ and $s_r = s$, with $r \geqslant 2$. This implies in particular that $m_{s_i s_{i+1}} \geqslant 3$ if $i \in \{1, 2, \dots, r-1\}$ and $s_i \neq s_j$ if $1 \leqslant i < j \leqslant r$. Let $w = s_r \cdots s_2 s_1 s_2 \cdots s_r$. If $m_{s_1 s_2} \geqslant 4$, then $(s_r, \dots, s_2, s_1, s_2, \dots, s_r) \in \mathscr{U}nbr(W, S)$, and so $w \in \mathscr{C} \cap \Gamma_s^{\mathrm{uni}}$ by 12.1.3. If $m_{s_1, s_2} = 3$, then $s_1 s_2 s_1 s_2 s_1 = s_2$, so $s_2 \in \mathscr{C}$ and the distance between s_2 and s is $r - 2$, so $\mathscr{C} \cap \Gamma_s^{\mathrm{uni}} \neq \varnothing$ by the induction hypothesis. ■

12.1.D. Finiteness

Recall that the definitions of a *forest* and a *tree* have been given in Definition 3.12.4. Recall also that S is assumed to be finite.

Proposition 12.1.14. *Assume that (W, S) is irreducible and let $s \in S$. Then the following are equivalent:*

(1) Γ^{uni} *is finite.*
(2) Γ_s^{uni} *is finite.*
(3) $\mathscr{C}ox(W,S)$ *is a finite tree which contains at most one edge whose label is* $\geqslant 4$*, and this label, if it exists, is finite.*

Proof. First, note that, if (1) or (3) holds, then S is finite. Also, by Proposition 12.1.12(b), the statement (2) implies that S is finite. So we may assume here that S is finite. In particular, the equivalence between (1) and (2) follows directly from the next lemma:

Lemma 12.1.15. *Assume that* (W,S) *is irreducible and let* s, $t \in S$. *Then* Γ_s^{uni} *is finite if and only if* Γ_t^{uni} *is finite.*

Proof. By an easy induction argument on the distance between s and t in $\mathscr{C}ox(W,S)$, we may assume that $m_{st} \geqslant 3$. If $w \in \Gamma_s^{\text{uni}}$, we set

$$\psi_{s,t}(w) = \begin{cases} wt & \text{if } wt \in \Gamma^{\text{uni}}, \\ ws & \text{otherwise.} \end{cases}$$

It follows from Corollary 12.1.8 that $\psi_{s,t}(w) \in \Gamma_t^{\text{uni}}$, so that this defines a map $\psi_{s,t} : \Gamma_s^{\text{uni}} \to \Gamma_t^{\text{uni}}$. But it follows from the definition that the fibers of $\psi_{s,t}$ are finite (they have cardinality $\leqslant 2$). So, if Γ_s^{uni} is infinite, then Γ_t^{uni} is infinite. Exchanging the roles of s and t proves the converse. ∎

Let us now prove that (1) implies (3). Assume first that (W,S) contains a loop, that is, a sequence $(s_1, s_2, \ldots, s_r)$ with $r \geqslant 3$, $s_i \neq s_j$ if $i \neq j$ and such that $(s_1, s_2, \ldots, s_r, s_1)$ is a path. We set $w = s_1 s_2 \cdots s_r$. It then follows from Lemma 12.1.6 that $\ell(w^n) = n\ell(w)$ and $w^n \in \Gamma^{\text{uni}}$ for all $n \geqslant 1$. Hence Γ^{uni} is infinite.

Assume now that $\mathscr{C}ox(W,S)$ contains two edges whose labels are $\geqslant 4$. This means that $\mathscr{C}ox(W,S)$ contains a subgraph isomorphic to

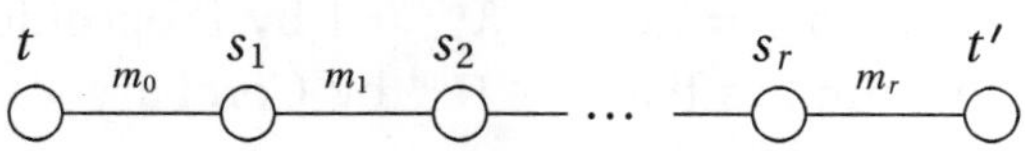

where $r \geqslant 1$, m_0, $m_r \geqslant 4$ and $m_i \geqslant 3$ for $1 \leqslant i \leqslant r-1$. Let $c = s_1 s_2 \cdots s_r$. By Lemma 12.1.6, $w = tct'c^{-1} \in \Gamma^{\text{uni}}$ and, even better, $\ell(w^n) = n\ell(w)$ and $w^n \in \Gamma^{\text{uni}}$ for all $n \geqslant 1$. Hence Γ^{uni} is infinite.

Finally, note that if there exist s, $t \in S$ such that $m_{st} = \infty$, then Γ^{uni} contains $W_{s,t} \setminus \{1\}$, and so is infinite. This completes the proof of the fact that (1) implies (3).

Conversely, let us now prove that (3) implies (1). So assume that $\mathscr{C}ox(W,S)$ is a tree which contains at most one edge whose label is $\geqslant 4$, and that this label, if it exists, is finite. We prove by induction on $|S|$ that $\mathscr{U}nbr(W,S)$ is finite. The case where $|S| = 1$ or 2 is obvious. So assume that $|S| \geqslant 3$. Since S is a tree, it admits at least two *terminal* vertices, namely vertices $s \in S$ such that there exists a unique $t \in S$ such that $m_{st} \geqslant 3$. Since there is at most one edge whose label is $\geqslant 4$, this means that we may assume moreover that $m_{st} = 3$. Now, if s occurs more than once in an unbraided path $(s_1, s_2, \ldots, s_r)$, then necessarily $s_1 = s$, $s_r = s$ and $s_i \neq s$ if $i \in \{2, \ldots, r-1\}$. This shows that $(s_2, \ldots, s_{r-1})$ is an unbraided path in $\mathscr{C}ox(W_{S \setminus \{s\}}, S \setminus \{s\})$. In particular, $|\mathscr{U}nbr(W,S)_s| \leqslant 2|\mathscr{U}nbr(W_{S\setminus\{s\}}, S \setminus \{s\})|$ and so, by the induction hypothesis, $\mathscr{U}nbr(W,S)_s$ is finite. So $\mathscr{U}nbr(W,S)$ is finite (because (1) is equivalent to (2)). This shows that (3) implies (1). ■

12.2. Submaximal Cells

We start with an easy result about the structure constants involving elements of Γ^{uni}:

Lemma 12.2.1. *Assume that we are in the split case. Let $x \in \Gamma^{\mathrm{uni}}$, $y \in W \setminus \{1\}$ and $s \in S$ be such that $\lambda_{x,y}^s \neq 0$. Then:*

(a) $y \in \Gamma^{\mathrm{uni}}$.
(b) *If $x \neq y$, then $y = tx$ for some $t \in S$ and $\lambda_{x,y}^s = 1$.*

Proof. According to Corollary 5.5.6, three cases may occur:

- If $sy < y$, then $x = y \in \Gamma^{\mathrm{uni}}$. This proves (a) in this case and (b) is vacuous.
- If $x = sy > y$, then $y \in \Gamma^{\mathrm{uni}}$ by 12.1.2 and $\lambda_{x,y}^s = 1$. This proves (a) and (b) in this case.
- If $x < y$, then $sx < x < y < sy$. Let $t \in \mathscr{L}(y)$. Then $t \neq s$ and so $tx > x$ because $\mathscr{L}(x) = s$. So $y = tx$ and $\lambda_{x,y}^s = 1$ by Proposition 6.4.1(d). But since $sy > y$, this implies that $y \in \Gamma^{\mathrm{uni}}$ by Corollary 12.1.8. This again proves (a) and (b) in this case. ■

Corollary 12.2.2. *Assume that we are in the split case. Let $x \in \Gamma^{\mathrm{uni}}$, $y \in W \setminus \{1\}$ and $s \in S$ be such that $\rho_{x,y}^s \neq 0$. Then:*

(a) $y \in \Gamma^{\mathrm{uni}}$.
(b) *If $x \neq y$, then $y = xt$ for some $t \in S$ and $\rho_{x,y}^s = 1$.*

The main result of this chapter is the following:

Theorem 12.2.3. *Assume that* (W,S) *is irreducible and that we are in the split case. Then:*

(a) Γ^{uni} *is a two-sided cell and is the unique submaximal one.*
(b) *If* $s \in S$*, then* Γ_s^{uni} *is a left cell and is a submaximal one.*
(c) *If* C *is a submaximal left cell, then there exists a unique* $s \in S$ *such that* $C = \Gamma_s^{\mathrm{uni}}$.

Proof. Let $w \in \Gamma^{\mathrm{uni}}$ and $s \in S$ be such that $sw \in \Gamma^{\mathrm{uni}}$. We want to prove that $sw \sim_L w$. For this, we may assume that $sw > w$. By 12.1.2, if we write $\mathscr{L}(w) = t$, we have $\mathscr{L}(sw) = s$ and so $tw < w < sw < tsw$. This shows that $w \,\dot{\smile}_L^{\varphi}\, sw$, and so $w \sim_L sw$ by Proposition 9.2.2. This implies that any left-connected component of Γ^{uni} is contained in a unique left cell.

If $s \in S$, we denote by $C(s)$ its left cell. We have just proved that $\Gamma_s^{\mathrm{uni}} \subset C(s)$. Since, moreover, $(\Gamma_s^{\mathrm{uni}})^{-1} \cap \Gamma_t^{\mathrm{uni}} \neq \varnothing$ for all s, $t \in S$, this means that Γ^{uni} is contained in a unique two-sided cell (which is denoted by Γ). Now, let x and y be such that $x \xleftarrow{L} y$, $y \neq 1$ and $x \in \Gamma^{\mathrm{uni}}$. Then there exists an $s \in S$ such that $\lambda_{x,y}^s \neq 0$. So $y \in \Gamma^{\mathrm{uni}}$ by Lemma 12.2.1. This shows that $\Gamma = \Gamma^{\mathrm{uni}}$.

This also shows that $C(s) \subset \Gamma^{\mathrm{uni}}$ for all $s \in S$. Since the map $\mathscr{R}$ is constant on left cells, this implies that $C(s) \subset \Gamma_s^{\mathrm{uni}}$. So $C(s) = \Gamma_s^{\mathrm{uni}}$.

Finally, let $w \in W \setminus \{1\}$ and let $s \in \mathscr{L}(w)$ and $t \in \mathscr{R}(w)$. Then $w \leqslant_L t$ and $w \leqslant_R s$ by Example 6.1.2. This proves that Γ^{uni} is a submaximal two-sided cell, that Γ_t^{uni} is a submaximal left cell and that all submaximal cells are obtained in this way. ■

Here are some easy consequences of Theorem 12.2.3 (compare with Conjectures 6.5.2 and 6.5.3):

Corollary 12.2.4. *Assume that* (W,S) *is irreducible and that we are in the split case. Then:*

(a) *Every left cell contained in* Γ^{uni} *contains an involution.*
(b) *If* $w \in \Gamma^{\mathrm{uni}}$*, then* $w \sim_{LR} w^{-1}$.
(c) *Inside* Γ^{uni}*, the equivalence relation* $\sim_{LR}$ *is generated by* $\sim_L$ *and* $\sim_R$.
(d) *If* $x, y \in \Gamma^{\mathrm{uni}}$ *are such that* $x \leqslant_L y$ *(resp.* $x \leqslant_R y$*), then* $x \sim_L y$ *(resp.* $x \sim_R y$*).*
(e) *The left-connected components of* Γ^{uni} *are the left cells contained in* Γ^{uni}.
(f) Γ^{uni} *meets a finite parabolic subgroup.*
(g) *If* $\mathscr{C}$ *is a conjugacy class of involutions such that* $\mathscr{C} \cap \Gamma^{\mathrm{uni}} \neq \varnothing$*, then* $\mathscr{C} \cap \Gamma_s^{\mathrm{uni}} \neq \varnothing$ *for all* $s \in S$.

Proof. Statements (a), (b), (d), (e) and (f) follow immediately from Theorem 12.2.3, Equations 12.1.4 and 12.1.1 and Propositions 8.4.9 and 12.1.12(a). Statement (g) is just a repetition of Proposition 12.1.13. Let us prove (c). So, let x, $y \in \Gamma^{\mathrm{uni}}$, let $s = \mathscr{R}(x)$ and $t = \mathscr{L}(y)$ and let $w \in (\Gamma_s^{\mathrm{uni}})^{-1} \cap \Gamma_t^{\mathrm{uni}}$ (see Proposition 12.1.12(b)). Then $x \sim_L s \sim_R w \sim_L t \sim_R y$ by Theorem 12.2.3(b). ■

Corollary 12.2.5. *The two-sided ideal* $\mathcal{H}^{<_{LR}\Gamma^{\mathrm{uni}}}$ *is generated by* $(\mathbf{C}_{w_I})_{I\in\mathcal{P}_{\mathrm{sph}}(S),|I|=2}$.

Proof. Let $\mathfrak{h}$ denote the two-sided ideal of $\mathcal{H}$ generated by $(\mathbf{C}_{w_I})_{I\in\mathcal{P}_{\mathrm{sph}}(S),|I|=2}$. Then $\mathfrak{h}$ is contained in $\mathcal{H}^{<_{LR}\Gamma^{\mathrm{uni}}}$. Conversely, let $w\in\Gamma\setminus\Gamma^{\mathrm{uni}}$. Then there exist $x, y\in W$ and $I\in\mathcal{P}_{\mathrm{sph}}(S)$, with $|I|=2$, such that $w=xw_Iy$ and $\ell(w)=\ell(x)+\ell(w_I)+\ell(y)$ (see Theorem 3.6.7(a)). We will prove by induction on $\ell(w)$ that $\mathbf{C}_w\in\mathfrak{h}$. By Example 6.1.2,

$$\mathbf{C}_w=\mathbf{C}_x\mathbf{C}_{w_I}\mathbf{C}_y-\sum_{w'<w}\alpha_{w'}\mathbf{C}_{w'}$$

for some $\alpha_{w'}$ in A. But, if $w'<w$ is such that $\alpha_{w'}\neq 0$, then $w'\leqslant_{LR} w$ and so $w'\in W\setminus\Gamma^{\mathrm{uni}}$ by Theorem 12.2.3. In this case, by the induction hypothesis, $\mathbf{C}_{w'}\in\mathfrak{h}$, and so $\mathbf{C}_w\in\mathfrak{h}$. ■

12.3. Cell Modules, W-graphs

12.3.A. Description

Let $s\in S$. We encode the structure of the left $\mathcal{H}$-module $\mathcal{H}^L[\Gamma_s^{\mathrm{uni}}]$ in a vertex-coloured graph as follows (this is a simplification of the notion of W-graph defined in Exercise 6.5). We first define a graph $\mathcal{G}_s^{\mathrm{uni}}$ as follows:

- The vertices of $\mathcal{G}_s^{\mathrm{uni}}$ are the elements of Γ_s^{uni}.
- Two vertices $x, y\in\Gamma_s^{\mathrm{uni}}$ are linked by an edge if and only if there exists a $t\in S$ such that $y=tx$.

Note that two elements x and y of Γ_s^{uni} are linked by an edge in $\mathcal{G}_s^{\mathrm{uni}}$ if and only if $x \smile_L^{\varphi} y$ (Knuth relation). Finally, the vertices are coloured by elements of S through the map $\mathcal{L}:\Gamma_s^{\mathrm{uni}}\to S$. So the pair $(\mathcal{G}_s,\mathcal{L})$ is a vertex-coloured graph. Then:

Proposition 12.3.1. *If* $s, t\in S$ *and* $y\in\Gamma_s^{\mathrm{uni}}$, *then*

$$T_t\mathbf{c}_y=\begin{cases} -v^{-1}\mathbf{c}_y+\displaystyle\sum_{\substack{x\in\Gamma_s^{\mathrm{uni}}\\ \text{such that } \mathcal{L}(x)=t\\ \text{and } x\smile_L^{\varphi}y}}\mathbf{c}_x & \text{if } t\neq\mathcal{L}(y),\\ v\mathbf{c}_y & \text{if } t=\mathcal{L}(y).\end{cases}$$

Proof. This follows from Corollary 5.5.6 and Lemma 12.2.1. ■

Remark 12.3.2. Let $\mathcal{F}_s$ be the set of triples $(t,x,y) \in S \times \Gamma_s^{\mathrm{uni}} \times \Gamma^{\mathrm{uni}}$ such that $t = \mathcal{L}(x)$ and $t \neq \mathcal{L}(y)$. If $(t,x,y) \in \mathcal{F}_s$, then $\lambda_{x,y}^t$ is recovered from the graph $\mathcal{G}_s^{\mathrm{uni}}$ as follows:

$$\lambda_{x,y}^t = \begin{cases} 1 & \text{if } x \mathrel{\dot{\smile}}_L^{\varphi} y, \\ 0 & \text{otherwise.} \end{cases}$$

Then $(\Gamma_s^{\mathrm{uni}}, \mathcal{L}, \lambda)$ is the W-graph associated with Γ_s^{uni} (see Exercise 6.5) and the vertex-coloured graph $(\mathcal{G}_s^{\mathrm{uni}}, \mathcal{L})$ might be seen as its representation. □

Proposition 12.3.3. *If $s \in S$, the graph $\mathcal{G}_s^{\mathrm{uni}}$ contains no loop.*

Proof. Assume that there exist $x_1, x_2, \dots, x_r, x_{r+1} = x_1$ in Γ_s^{uni} (with $r \geqslant 3$) such that $x_i \mathrel{\dot{\smile}}_L x_{i+1}$ and $x_i \neq x_j$ for all $i, j \in \{1,2,\dots,r\}$ such that $i \neq j$. Let s_i denote the element of S such that $x_{i+1} = s_i x_i$. We may assume that $\ell(x_1) = \max\{\ell(x_i) \mid 1 \leqslant i \leqslant r\}$. In particular, $x_r = s_r x_1 < x_1$ and $x_2 = s_1 x_1 < x_1$. Therefore, $s_1, s_r \in \mathcal{L}(x_1)$ and so $s_1 = s_r$ by 12.1.2. Hence, $x_r = x_2$, which is impossible. ■

Now, let t be such that $3 \leqslant m_{st} < \infty$. We denote by $\omega_{s,t}$ the unique automorphism of the Coxeter system $(W_{s,t}, \{s,t\})$ which exchanges s and t and we define a map $\sigma_{s,t} : W_{s,t} \to W_{s,t}$ by

$$\sigma_{s,t}(w) = \begin{cases} w & \text{if } w \in \{1, w_{s,t}\}, \\ \omega_{s,t}(w) w_{s,t} & \text{otherwise.} \end{cases}$$

It is a bijective involution and it will be shown in Chapter 21 that this is a left cellular map on $W_{s,t}$ (in the split case). In particular, $\sigma_{s,t}^L$ is a left cellular map (see Theorem 8.4.4). Since $\sigma_{s,t}(s) = t\, w_{s,t} \in \Gamma^{\mathrm{uni}}$, this implies that $\sigma_{s,t}^L(\Gamma_s^{\mathrm{uni}})$ is a left cell contained in Γ^{uni}. By computing the right descent set of $t\, w_{s,t}$, we get

$$\sigma_{s,t}^L(\Gamma_s^{\mathrm{uni}}) = \begin{cases} \Gamma_t^{\mathrm{uni}} & \text{if } m_{st} \text{ is odd,} \\ \Gamma_s^{\mathrm{uni}} & \text{if } m_{st} \text{ is even.} \end{cases} \tag{12.3.4}$$

In particular, if m_{st} is odd (resp. even), then $\sigma_{s,t}^L$ induces an isomorphism of vertex-coloured graphs $(\mathcal{G}_s^{\mathrm{uni}}, \mathcal{L}) \xrightarrow{\sim} (\mathcal{G}_t^{\mathrm{uni}}, \mathcal{L})$ (resp. a non-trivial involutive automorphism of the vertex-coloured graph $(\mathcal{G}_s^{\mathrm{uni}}, \mathcal{L})$).

Proposition 12.3.5. *Let $s, t \in \Gamma^{\mathrm{uni}}$. Then:*

(a) *If s and t are conjugate in W, then the vertex-coloured graphs $(\mathcal{G}_s^{\mathrm{uni}}, \mathcal{L})$ and $(\mathcal{G}_t^{\mathrm{uni}}, \mathcal{L})$ are isomorphic.*
(b) *If m_{st} is even and $\geqslant 4$, then $\sigma_{s,t}^L$ is a non-trivial involutive automorphism of the vertex-coloured graph $(\mathcal{G}_s^{\mathrm{uni}}, \mathcal{L})$.*

Since $(\mathcal{G}_s^{\mathrm{uni}}, \mathcal{L})$ encodes the structure of the left $\mathcal{H}$-module $\mathcal{H}^L[\Gamma_s^{\mathrm{uni}}]$, Proposition 12.3.5 implies the following facts:

Corollary 12.3.6. *Let $s, t \in \Gamma^{\mathrm{uni}}$. Then:*

(a) *If s and t are conjugate in W, then there exists a bijection $\sigma : \Gamma_s^{\mathrm{uni}} \to \Gamma_t^{\mathrm{uni}}$ such that the A-linear map $\mathscr{H}^L[\Gamma_s^{\mathrm{uni}}] \to \mathscr{H}^L[\Gamma_t^{\mathrm{uni}}]$, $\mathbf{c}_w \mapsto \mathbf{c}_{\sigma(w)}$ is an isomorphism of left $\mathscr{H}$-modules.*

(b) *If m_{st} is even and $\geqslant 4$, then $\sigma_{s,t}^L$ induces a non-trivial involutive automorphism of left $\mathscr{H}$-modules $\mathscr{H}^L[\Gamma_s^{\mathrm{uni}}] \xrightarrow{\sim} \mathscr{H}^L[\Gamma_s^{\mathrm{uni}}]$, $\mathbf{c}_w \mapsto \mathbf{c}_{\sigma_{s,t}^L(w)}$.*

The Coxeter system (W,S) is said *simply-laced* if all the edges of its Coxeter graph have label 3. In this case, there is no sequence of the form (s,t,s) in any admissible path. Also, we may forget about the labels of the edges of the Coxeter graph $\mathscr{C}ox(W,S)$. The following proposition is then immediate:

Proposition 12.3.7. *Assume that (W,S) is irreducible, simply-laced and has no loop in its Coxeter graph. Then:*

(a) *If $s, t \in S$, then $|\Gamma_s^{\mathrm{uni}} \cap (\Gamma_t^{\mathrm{uni}})^{-1}| = 1$.*

(b) *The map $\Gamma_s^{\mathrm{uni}} \longrightarrow S$, $x \mapsto \mathscr{L}(x)$ is bijective and induces an isomorphism of graphs $\mathscr{G}_s^{\mathrm{uni}} \xrightarrow{\sim} \mathscr{C}ox(W,S)$.*

12.3.B. Homomorphisms

We fix $s \in S$. Let us first start with this easy remark:

Lemma 12.3.8. *The left $\mathscr{H}$-module $\mathscr{H}^L[\Gamma_s^{\mathrm{uni}}]$ is generated by $\mathbf{c}_s$.*

Proof. This follows from the fact that $\bigoplus_{ws<w} A\,\mathbf{C}_w$ is generated, as a left ideal, by $\mathbf{C}_s$ (see Theorem 5.5.7). ■

Corollary 12.3.9. *If M is a left $\mathscr{H}$-module, then the map*

$$\begin{array}{ccc} \mathrm{Hom}_{\mathscr{H}}(\mathscr{H}^L[\Gamma_s^{\mathrm{uni}}], M) & \longrightarrow & M \\ f & \longmapsto & f(\mathbf{c}_s) \end{array}$$

is injective.

Now, let $t \in S$ and let $f \in \mathrm{Hom}_{\mathscr{H}}(\mathscr{H}^L[\Gamma_s^{\mathrm{uni}}], \mathscr{H}^L[\Gamma_t^{\mathrm{uni}}])$. Then $\mathbf{C}_s\mathbf{c}_s = (v + v^{-1})\mathbf{c}_s$, so $\mathbf{C}_s f(\mathbf{c}_s) = (v+v^{-1}) f(\mathbf{c}_s)$. Hence, $f(\mathbf{c}_s)$ belongs to $\bigoplus_{w \in (\Gamma_s^{\mathrm{uni}})^{-1} \cap \Gamma_t^{\mathrm{uni}}} A\mathbf{c}_w$. In fact:

Proposition 12.3.10. *Let $s, t \in S$. Then the map*

$$\begin{array}{ccc} \mathrm{Hom}_{\mathscr{H}}(\mathscr{H}^L[\Gamma_s^{\mathrm{uni}}], \mathscr{H}^L[\Gamma_t^{\mathrm{uni}}]) & \longrightarrow & \displaystyle\bigoplus_{w \in (\Gamma_s^{\mathrm{uni}})^{-1} \cap \Gamma_t^{\mathrm{uni}}} A\mathbf{c}_w \\ f & \longmapsto & f(\mathbf{c}_s) \end{array}$$

is an isomorphism of A-modules.

Proof. By Corollary 12.3.9, it is sufficient to prove that the map is surjective. If $h \in \mathcal{H}$, then the right multiplication by h on $\mathcal{H}^{LR}[\Gamma^{\mathrm{uni}}]$ is an endomorphism of left $\mathcal{H}$-module. But, as a left $\mathcal{H}$-module, we have

$$\mathcal{H}^{LR}[\Gamma^{\mathrm{uni}}] = \bigoplus_{u \in S} \mathcal{H}^L[\Gamma_u^{\mathrm{uni}}]$$

(see Corollary 12.2.4(d)). So, if $h \in \bigoplus_{w \in \Gamma_t^{\mathrm{uni}}} A\mathbf{C}_w$, then the image of the right multiplication by h is contained in $\mathcal{H}^L[\Gamma_t^{\mathrm{uni}}]$. In this case, let $\psi_{s,h} : \mathcal{H}^L[\Gamma_s^{\mathrm{uni}}] \to \mathcal{H}^L[\Gamma_t^{\mathrm{uni}}]$ denote the restriction of the right multiplication by h. Now, assume moreover that $h \in \bigoplus_{w \in (\Gamma_s^{\mathrm{uni}})^{-1} \cap \Gamma_t^{\mathrm{uni}}} A\mathbf{C}_w$. Then it follows from Exercise 5.4 that $\psi_{s,h}(m) \in (v+v^{-1})\mathcal{H}^L[\Gamma_t^{\mathrm{uni}}]$ for all $m \in \mathcal{H}^L[\Gamma_s^{\mathrm{uni}}]$. So the map

$$\begin{array}{rccc} \psi'_{s,h} : & \mathcal{H}^L[\Gamma_s^{\mathrm{uni}}] & \longrightarrow & \mathcal{H}^L[\Gamma_t^{\mathrm{uni}}] \\ & m & \longmapsto & \dfrac{1}{v+v^{-1}} m h \end{array}$$

is well-defined, and is a morphism of left $\mathcal{H}$-modules. Moreover, $\psi'_{s,h}(\mathbf{c}_s)$ is the image of h in $\mathcal{H}^L[\Gamma_t^{\mathrm{uni}}]$. This completes the proof of the proposition. ■

12.3.C. Reflection Representation

We fix again $s \in S$. Recall from Section 4.5 that E^φ denotes the $\mathbb{R}(\mathcal{A})$-vector space $\mathbb{R}(\mathcal{A})E$ endowed with its left $\mathbb{R}(\mathcal{A})\mathcal{H}$-module structure described in Theorem 4.5.1 (it was called the *reflection representation* of $\mathcal{H}$). Let

$$\begin{array}{rccc} \pi_s : & \mathbb{R}(\mathcal{A})\mathcal{H} & \longrightarrow & E^\varphi \\ & h & \longmapsto & h e_s^\varphi . \end{array}$$

It is a (surjective) morphism of $\mathbb{R}(\mathcal{A})\mathcal{H}$-modules. Now, $\mathbf{C}_s^\dagger e_s^\varphi = (v^{\varphi(s)} + v^{-\varphi(s)}) e_s^\varphi$, so that π_s restricts to a surjective morphism of left $\mathbb{R}(\mathcal{A})\mathcal{H}$-modules $\pi_s^0 : \mathbb{R}(\mathcal{A})\mathcal{H}\mathbf{C}_s^\dagger \longrightarrow E^\varphi$. By Exercise 5.5, $\pi_s(\mathbf{C}_{w_I}^\dagger) = 0$ for all spherical subsets I of S such that $|I| \geqslant 2$. So π_s^0 factorizes through a surjective morphism of left $\mathbb{R}(\mathcal{A})\mathcal{H}$-modules

$$\pi_s^1 : \mathbb{R}(\mathcal{A})\mathcal{H}\mathbf{C}_s^\dagger / (\mathbb{R}(\mathcal{A})\mathcal{H}\mathbf{C}_s \cap \mathbb{R}(\mathcal{A})\mathcal{H}^{<_{LR}\Gamma^{\mathrm{uni}}})^\dagger \longrightarrow E^\varphi$$

(see Corollary 12.2.5). In other words, we have proved the following result:

Proposition 12.3.11. *There exists a surjective morphism of left $\mathcal{H}$-modules*

$$^\dagger\mathbb{R}(\mathcal{A})\mathcal{H}^L[\Gamma_s^{\mathrm{uni}}] \longrightarrow E^\varphi .$$

12.4. Unfolding Coxeter Graphs

Hypothesis. *We assume in this section, and only in this section, that Γ^{uni} is* ***finite****. According to Proposition 12.1.14, this means that $\mathscr{C}ox(W,S)$ is a tree with no loop, and which admits at most one edge whose label is $\geqslant 4$, and this label is finite. We fix $s \in S$.*

We denote by $(\mathscr{W}_s, \mathscr{S}_s)$ the simply-laced Coxeter system whose Coxeter graph is $\mathscr{G}_s^{\mathrm{uni}}$. In other words, there is a bijection $\Gamma_s^{\mathrm{uni}} \xrightarrow{\sim} \mathscr{S}_s$, $x \mapsto \mathbf{s}_x$ and

$$m_{\mathbf{s}_x \mathbf{s}_y} = \begin{cases} 1 & \text{if } x = y, \\ 3 & \text{if } x \smile_L y, \\ 2 & \text{otherwise.} \end{cases}$$

Note that, if two elements x and y of Γ_s^{uni} are linked by an edge in $\mathscr{G}_s^{\mathrm{uni}}$, then $\mathscr{L}(x) \neq \mathscr{L}(y)$ by construction. In particular, if $t \in S$ and if $x, y \in (\Gamma_t^{\mathrm{uni}})^{-1} \cap \Gamma_s^{\mathrm{uni}}$, then x and y are not linked by an edge in $\mathscr{G}_s$. Therefore, $[\mathbf{s}_x, \mathbf{s}_y] = 1$ in $\mathscr{W}_s$ and so we can define without ambiguity

$$\mathbf{s}(t) = \prod_{x \in (\Gamma_t^{\mathrm{uni}})^{-1} \cap \Gamma_s^{\mathrm{uni}}} \mathbf{s}_x \in \mathscr{W}_s$$

(by using also that $(\Gamma_t^{\mathrm{uni}})^{-1} \cap \Gamma_s^{\mathrm{uni}}$ is finite).

Proposition 12.4.1. *Assume that (W,S) is irreducible and that Γ^{uni} is finite. Then the map $\mathbf{s} : S \longrightarrow \mathscr{W}_s$ extends uniquely to a morphism of groups $\mathbf{s} : W \longrightarrow \mathscr{W}_s$. Moreover:*

(a) $\mathbf{s}$ *is injective.*
(b) *$(\mathscr{W}_s, \mathscr{S}_s)$ is spherical if and only if (W,S) is spherical.*
(c) *If (W,S) is simply-laced and contains no loop, then* $\mathbf{s}$ *is an isomorphism of Coxeter systems.*

Proof. Let us first prove that $\mathbf{s}$ extends to a morphism of groups $\mathbf{s} : W \longrightarrow \mathscr{W}_s$. For this, let $E(s)$ denote the real vector space $\mathbb{R} \otimes_A \mathscr{H}^L[\Gamma_s^{\mathrm{uni}}]$ (through the specialization morphism $A \to \mathbb{R}$, $v^a \mapsto 1$) and view it as a left $\mathbb{R}W$-module. As an $\mathbb{R}W$-module, ${}^\dagger E(s)$ is isomorphic to $E(s) \otimes_{\mathbb{R}} \mathbb{R}_\varepsilon$, where $\mathbb{R}_\varepsilon$ denotes the one-dimensional $\mathbb{R}W$-module $\mathbb{R}$ on which W acts through the sign character $\varepsilon : W \to \{1, -1\}$. This gives a morphism

$$i : W \longrightarrow \mathbf{GL}(E(s) \otimes_{\mathbb{R}} \mathbb{R}_\varepsilon).$$

Now, let $\mathbf{e}'_x$ denote the image of $\mathbf{c}_x$ in $E(s)$ and let $\mathbf{e}_x = \mathbf{e}'_x \otimes_{\mathbb{R}} 1 \in E(s) \otimes_{\mathbb{R}} \mathbb{R}_\varepsilon$. The action of an element $t \in S$ can be expressed in the $\mathbb{R}$-basis $(\mathbf{e}_x)_{x\in\Gamma_s^{\mathrm{uni}}}$ as follows:

$$i(t)(\mathbf{e}_x) = \begin{cases} \mathbf{e}_y - \displaystyle\sum_{\substack{x\in\Gamma_s^{\mathrm{uni}} \\ \mathscr{L}(x)=t \\ x \smile_L y}} \mathbf{e}_x & \text{if } t \neq \mathscr{L}(y), \\ -\mathbf{e}_y & \text{if } t = \mathscr{L}(y). \end{cases}$$

By identifying the underlying space of the reflection representation of $\mathscr{W}_s$ constructed in §3.11.B with $E(s) \otimes_{\mathbb{R}} \mathbb{R}_\varepsilon$ by sending $e_{\mathbf{s}_x}$ to e_x, this gives rise to a morphism of groups

$$i_s : \mathscr{W}_s \longrightarrow \mathbf{GL}(E(s) \otimes_{\mathbb{R}} \mathbb{R}_\varepsilon),$$

where the action of $\mathbf{s}_x \in \mathscr{S}_s$ is given by

$$i_s(\mathbf{s}_x)(\mathbf{e}_y) = \begin{cases} \mathbf{e}_y - \displaystyle\sum_{\substack{x\in\Gamma_s^{\mathrm{uni}} \\ x \smile_L y}} \mathbf{e}_x & \text{if } x \neq y, \\ -\mathbf{e}_x & \text{if } x = y. \end{cases}$$

It then follows that $i(t) = i_s(\mathbf{s}(t))$ and so the morphism $i : W \longrightarrow \mathbf{GL}(E(s) \otimes_{\mathbb{R}} \mathbb{R}_\varepsilon)$ factorizes as $i = i_s \circ \mathbf{s}$, where $\mathbf{s} : W \to \mathscr{W}_s$ is a morphism of groups extending the map $\mathbf{s} : S \to \mathscr{W}_s$.

(a) follows from the fact that i is injective because $E(s) \otimes_{\mathbb{R}} \mathbb{R}_\varepsilon$ admits, as a quotient, the reflection representation E of W (this follows from Proposition 12.3.11 after specialization $v^a \mapsto 1$).

(c) follows immediately from Proposition 12.3.7.

(b) If $\mathscr{W}_s$ is finite, then W is finite by (a). Conversely, assume that W is finite. Let us write $S = \{s_1, s_2, \ldots, s_n\}$ with $n = |S|$ and let $c = s_1 s_2 \cdots s_n$ be a Coxeter element (see Exercise 3.9). Then, by construction, $\mathbf{s}(c)$ is a Coxeter element of $\mathscr{W}_s$. But c has finite order, so $\mathbf{s}(c)$ has finite order too. Hence $\mathscr{W}_s$ is finite by Exercise 3.9(h4). ■

We conclude this chapter by computing the vertex-coloured graph $(\mathscr{G}_s, \mathscr{L})$ (in particular, this describes the Coxeter graph of $(\mathscr{W}_s, \mathscr{S}_s)$) whenever (W,S) is spherical or of affine type (and Γ_s^{uni} is finite: this excludes the case where (W,S) is of type $\tilde{A}_n$ or $\tilde{C}_n$). The result is given in Table 12.a (the simply-laced cases have been omitted, as they are dealt with in Proposition 12.4.1(c)).

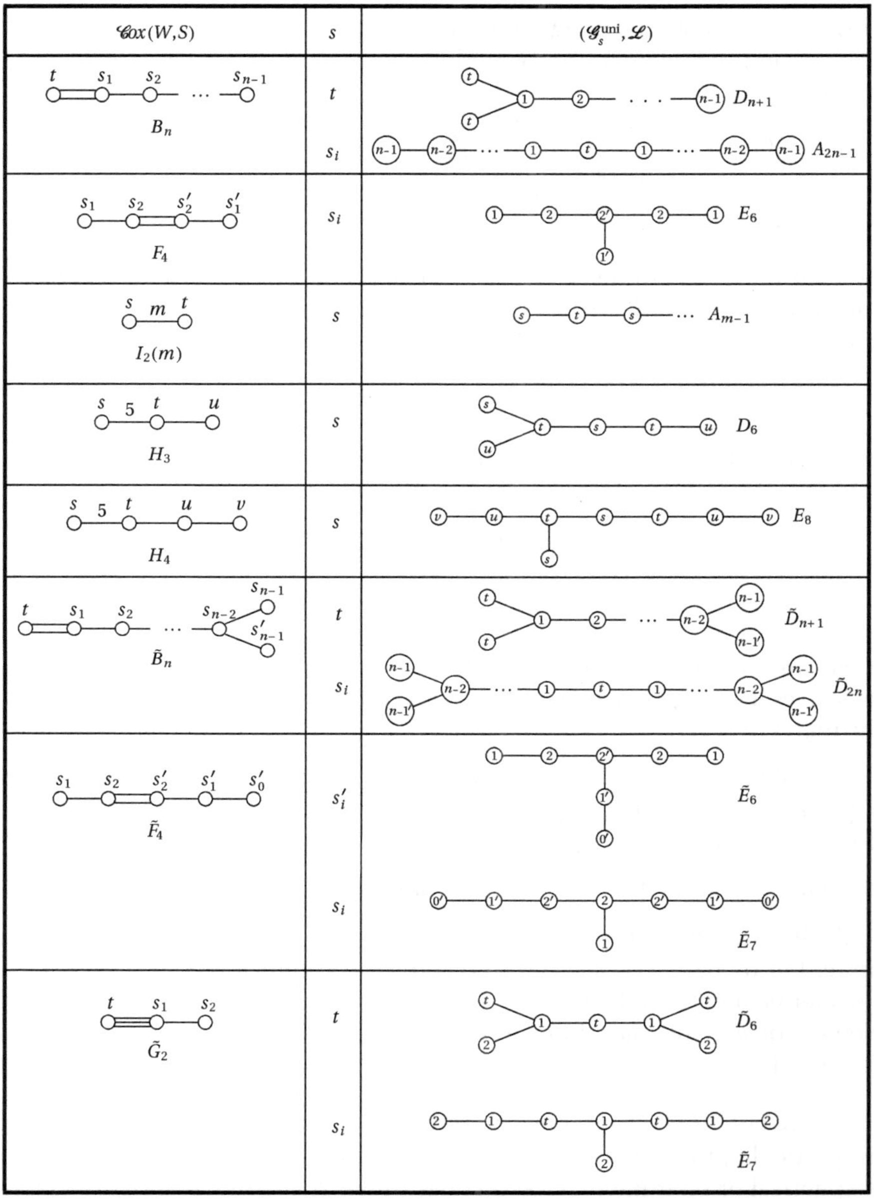

Table 12.a. Some finite vertex-coloured graphs $(\mathscr{G}_s^{\mathrm{uni}}, \mathscr{L})$

Notes

This chapter is inspired by [Lus8, §3]. See also [Dou].

Chapter 13
Submaximal Cells (General Case)

Based on the results of the previous chapter, we describe the submaximal two-sided cells as well as the submaximal left cells which are contained in the submaximal two-sided cells in the general case.

13.1. Submaximal Two-Sided Cells

If $a \in \varphi(S)$, we set $S[a] = \varphi^{-1}(a)$. It is a W-closed subset of S. We denote by $\bowtie_a$ the equivalence relation on $S[a]$ whose equivalence classes correspond to the connected components of $\mathscr{C}ox(W_{S[a]}, S[a])$. We denote by $\bowtie$ the equivalence relation on S defined as follows: $s \bowtie t$ if $\varphi(s) = \varphi(t)$ and $s \bowtie_{\varphi(s)} t$.

We denote by $(S/\bowtie)^{\blacktriangledown}$ the set of $\Omega \in S/\bowtie$ such that $\varphi(s) > \varphi(\Omega)$ for all $s \in S \setminus \Omega$ such that $\mathscr{C}ox(W_{\Omega\cup\{s\}}, \Omega \cup \{s\})$ is connected.

Example 13.1.1. Assume here that $S = \{r, s_1, s_2, s_3, t_1, t_2, t_3, t_4, u_1, u_2, u_3\}$ and that the Coxeter graph of (W,S) and the function $\varphi : S \to \mathbb{Z}$ are given as follows:

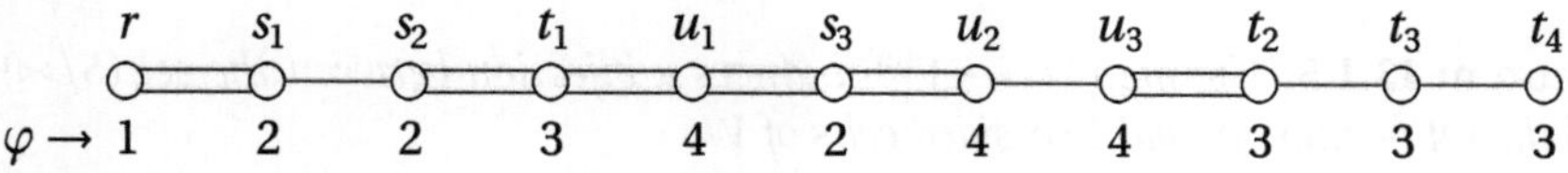

Then

$$S/\bowtie = \{\{r\},\ \{s_1, s_2\},\ \{s_3\},\ \{t_1\},\ \{t_2, t_3, t_4\},\ \{u_1\},\ \{u_2, u_3\}\}$$

and

$$(S/\bowtie)^{\blacktriangledown} = \{\{r\},\ \{s_3\},\ \{t_2, t_3, t_4\}\}.$$

C. Bonnafé, *Kazhdan–Lusztig Cells with Unequal Parameters*,
Algebra and Applications 24, https://doi.org/10.1007/978-3-319-70736-5_13

Indeed (for instance), $\Omega = \{t_2, t_3, t_4\}$ belongs to $(S/\bowtie)^\blacktriangledown$ because u_3 is the unique element s of S such that $\mathscr{C}ox(W_{\Omega\cup\{s\}}, \Omega\cup\{s\})$ is irreducible and $\varphi(u_3) > \varphi(\Omega)$, while $\{t_1\}$ does not belong to $(S/\bowtie)^\blacktriangledown$ because $\mathscr{C}ox(W_{\{t_1\}\cup\{s_2\}}, \{t_1\}\cup\{s_2\})$ is irreducible and $\varphi(s_2) < \varphi(t_1)$. □

If $\Omega \in S/\bowtie$, we set

$$\Gamma_\Omega^{\mathrm{uni}} = \Gamma^{\mathrm{uni}} \cap W_\Omega.$$

Since φ is constant on Ω, Theorem 12.2.3 may apply for the triple $(W_\Omega, \Omega, \varphi_\Omega)$ and we get that

(13.1.2) $\Gamma_\Omega^{\mathrm{uni}}$ *is a submaximal two-sided cell of* W_Ω.

The proof of the next lemma follows essentially the same lines as the proof of Lemma 12.2.1:

Lemma 13.1.3. *Let* $\Omega \in (S/\bowtie)^\blacktriangledown$. *Let* $s \in S$, $y \in \Gamma_\Omega^{\mathrm{uni}}$ *and* $x \in W \setminus \{1\}$ *be such that* $\lambda_{x,y}^s \neq 0$. *Then* $y \in \Gamma_\Omega^{\mathrm{uni}}$.

Proof. According to Corollary 5.5.6, three cases may occur:

- If $sy < y$, then $x = y \in \Gamma_\Omega^{\mathrm{uni}}$ and the lemma follows in this case.

- If $x = sy > y$, then $y \in \Gamma_\Omega^{\mathrm{uni}}$ by 12.1.2, and the lemma follows also in this case.

- If $x < y$, then $sx < x < y < sy$. Since $x \in \Gamma_\Omega^{\mathrm{uni}}$, $s \in \Omega$ and $\mathscr{L}(x) = \{s\}$. Now, let $t \in \mathscr{L}(y)$. Then $t \neq s$ and so $t \notin \mathscr{L}(x)$. Then it follows from Proposition 6.4.1(a) and (b) that $st \neq ts$ and $\varphi(t) \leqslant \varphi(s)$. By definition of $(S/\bowtie)^\blacktriangledown$, this forces $t \in \Omega$ and so $\varphi(t) = \varphi(s)$. So, by Proposition 6.4.1(c), we get that $y = tx \in W_\Omega$. So the fact that $ty < y$ implies that $y \in \Gamma_\Omega^{\mathrm{uni}}$ by Corollary 12.1.8. The proof of the lemma is complete. ■

Corollary 13.1.4. *Let* $\Omega \in (S/\bowtie)^\blacktriangledown$. *Let* $s \in S$, $x \in \Gamma_\Omega^{\mathrm{uni}}$ *and* $y \in W \setminus \{1\}$ *be such that* $\rho_{x,y}^s \neq 0$. *Then* $y \in \Gamma_\Omega^{\mathrm{uni}}$.

Theorem 13.1.5. *The map* $\Omega \longmapsto \Gamma_\Omega^{\mathrm{uni}}$ *induces a bijection between the set* $(S/\bowtie)^\blacktriangledown$ *and the set of submaximal two-sided cells of* W.

Proof. Let $\Omega \in (S/\bowtie)^\blacktriangledown$. We first prove that $\Gamma_\Omega^{\mathrm{uni}}$ is a two-sided cell. First, recall that $\Gamma_\Omega^{\mathrm{uni}}$ is a two-sided cell in W_Ω (see 13.1.2). But Lemma 13.1.3 and Corollary 13.1.4 show that, if $x \leqslant_{LR} y$ with $x \in \Gamma_\Omega^{\mathrm{uni}}$, then $y \in \Gamma_\Omega^{\mathrm{uni}} \cup \{1\}$. So $\Gamma_\Omega^{\mathrm{uni}}$ is a union of two-sided cells of W. Therefore, $\Gamma_\Omega^{\mathrm{uni}}$ is a submaximal two-sided cell of W.

Conversely, let Γ be a submaximal two-sided cell of W. Then Γ contains an element of S (take $x \in \Gamma$ and $s \in \mathscr{R}(x)$, then $x \leqslant_L s$ by Example 6.1.2 and so $s \in \Gamma$ by submaximality). Let $a = \min_{s \in \Gamma\cap S} \varphi(s)$ and let $s \in \Gamma \cap S$ be such that $\varphi(s) = a$. Let Ω be the element of $S/\bowtie$ containing s. By 13.1.2, we just need to prove that $\Omega \in (S/\bowtie)^\blacktriangledown$.

So assume that $\Omega \notin (S/\bowtie)^{\blacktriangledown}$. This means that there exists a $t \in S$ such that $(W_{\Omega\cup\{t\}}, \Omega \cup \{t\})$ is irreducible and $\varphi(t) < \varphi(s)$. So there exists an $s' \in \Omega$ such that $m_{s't} \geq 3$. Since $\varphi(t) \neq \varphi(s')$, this forces $m_{s't}$ to be even, and ≥ 4. Also, $s' \in \Gamma$ because Γ contains $\Gamma_\Omega^{\mathrm{uni}}$ (by 13.1.2). Now, since $\varphi(t) < \varphi(s')$, we get that $ts' \sim_L s'$ by Proposition 9.2.2, and so $ts' \in \Gamma$. But $ts' \leqslant_R t$, so $t \in \Gamma$ because Γ is submaximal. This contradicts the fact that $\varphi(t) < \varphi(s)$. ■

13.2. Submaximal Left Cells

If $s \in S$, we denote by C_s the left cell of s. Note that it depends on φ. Then:

Proposition 13.2.1. *The map $s \longmapsto C_s$ induces a bijection between S and the set of submaximal left cells.*

Proof. Let $s \in S$ and let $w \in W \setminus \{1\}$ be such that $s \leqslant_L w$. Then $\varnothing \neq \mathscr{R}(w) \subset \mathscr{R}(s) = \{s\}$ by Corollary 6.3.2. So $\mathscr{R}(w) = \{s\}$. Then $\ell(w) = \ell(ws) + \ell(s)$, so $w \leqslant_L s$ by Example 6.1.2. Therefore $w \sim_L s$. This shows that C_s is a submaximal left cell of W.

Now, let C be a submaximal left cell. Let $w \in C$. Then $\mathscr{R}(w) \neq \varnothing$. Let $s \in \mathscr{R}(w)$. Then $\ell(w) = \ell(ws) + \ell(s)$, so $w \leqslant_L s$ again by Example 6.1.2. Since C is submaximal, this forces $s \in C$, and so $C = C_s$. This shows that the map is surjective. It is clearly injective, since the right descent set is constant on a left cell by Corollary 6.3.2. ■

Proposition 13.2.2. *Let $\Omega \in (S/\bowtie)^{\blacktriangledown}$ and let $s \in \Omega$. Then $C_s = \Gamma_s^{\mathrm{uni}} \cap W_\Omega$.*

Proof. By Theorem 12.2.3 applied to W_Ω, we have $\Gamma_s^{\mathrm{uni}} \cap W_\Omega \subset C_s$ and, by Theorem 13.1.5, $C_s \subset \Gamma_\Omega^{\mathrm{uni}}$. Since $\mathscr{R}(C_s) = \{s\}$, we get the result. ■

***Remark* 13.2.3.** We do not provide in this chapter a description of C_s in general, even if we assume that Lusztig's Conjectures hold. One of the reasons is that it might not be contained in Γ^{uni}: for instance, if W is an affine Weyl group of type $\widetilde{B}_2$, with Coxeter graph given as follows

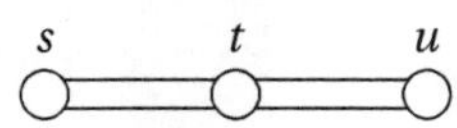

and if $\varphi(s) = a$, $\varphi(t) = b$ and $\varphi(u) = c$, then one can check (see Exercise 13.1 below) that $sut \sim_L t$ whenever $b \geq a + c$ and $a, c > 0$ (but $sut = ust \notin \Gamma^{\mathrm{uni}}$). □

Exercises

Exercise 13.1. We aim to prove here the last statements of Remark 13.2.3. Assume here that W is an affine Weyl group of type $\widetilde{B}_2$, with Coxeter graph given as follows

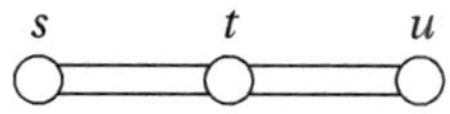

We set $\varphi(s)=a$, $\varphi(t)=b$ and $\varphi(u)=c$, and we assume that $b \geqslant a+c$ and a, b, $c>0$. Show that $\lambda_{t,sut}^{t} \neq 0$. Deduce that $t \sim_L sut = ust$.

Part V
Lusztig's a-Function

Lusztig's **a**-function is the most subtle invariant of cells that has been introduced so far. Despite the fact that it is very difficult to compute, its behaviour seems to be governed by a set of conjectures [Lus20, §14, Conjectures P1, P2,..., P15] which allow one to deduce (in the cases where they are proved) very deep properties of cells. For this purpose, we work under the following hypothesis:

> **Hypothesis.** *From now on, and until the end of this book, we assume that S is* ***finite*** *and recall that $\varphi > 0$.*

Lusztig's **a**-function is the map $\mathbf{a} : W \to \mathscr{A}$ defined by

$$\mathbf{a}(z) = \max_{x,y \in W} \big(\deg(h_{x,y,z})\big)$$

for all $z \in W$. The aim of this part is to study the properties of this map **a**, together with its links to the partition of W into left, right or two-sided cells. Of course, as is readily seen from the definition, Lusztig's **a**-function depends heavily on φ. It must also be said that it is still a conjecture that the above definition makes sense for any Coxeter group (of course, it does make sense if W is finite): this is discussed in Chapter 14.

Chapter 14
Lusztig's Conjectures

14.1. Bounds

Let us write

$$T_x T_y = \sum_{z \in W} t_{x,y,z} T_z,$$

where $t_{x,y,z}$ belongs to A. Note that the structure constants $t_{x,y,z}$ can be computed in the generic Hecke algebra $\mathcal{H}_{\mathbb{Z}}(W,S)$. By property 4.2.5 of the symmetrizing form $\tau : \mathcal{H} \to A$, we have

$$t_{x,y,z} = \tau(T_x T_y T_{z^{-1}}). \tag{14.1.1}$$

Also, by successively applying the involutions $\bar{\ }$ and †, we get

$$\overline{t}_{x,y,z} = \varepsilon(xyz) t_{x,y,z}. \tag{14.1.2}$$

In particular,

$$\deg(\overline{t}_{x,y,z}) = \deg(t_{x,y,z}). \tag{14.1.3}$$

Lemma 14.1.4. *Let x, y, $z \in W$. Then* $\deg(t_{x,y,z}) \leqslant \min\{\varphi(x), \varphi(y), \varphi(z)\}$.

Proof. Since $\tau(T_x T_y T_z) = \tau(T_y T_z T_x) = \tau(T_z T_x T_y)$, it is sufficient to prove that $\tau(T_x T_y T_z) \leqslant \varphi(x)$. We argue by induction on $\ell(x)$. If $\ell(x) = 0$, then $x = 1$ and $\tau(T_x T_y T_z) = \delta_{yz=1}$ by 4.2.5, so the result holds in this case. Now, assume that $\ell(x) \geqslant 1$ and that $\tau(T_{x'} T_{y'} T_{z'}) \leqslant \varphi(x')$ for all x', y', $z' \in W$ such that $\ell(x') < \ell(x)$. Let $s \in \mathscr{R}(x)$ and write $x = x's$. Then $\ell(x') = \ell(x) - 1$, so $\varphi(x) = \varphi(x') + \varphi(s) > \varphi(x')$ and

$$T_x T_y T_z = T_{x'} T_s T_y T_z = T_{x'} T_{sy} T_z + \delta_{sy<y}(v^{\varphi(s)} - v^{-\varphi(s)}) T_{x'} T_y T_z.$$

So the result follows from the induction hypothesis. ∎

C. Bonnafé, *Kazhdan–Lusztig Cells with Unequal Parameters*,
Algebra and Applications 24, https://doi.org/10.1007/978-3-319-70736-5_14

Recall from Definition 3.7.3 that $\mathscr{P}_{\mathrm{sph}}(S)$ denotes the set of spherical subsets of S and that, if I is a spherical subset of S (that is, if W_I is finite), we denote by w_I the longest element of W_I. We can now state the first conjecture of this chapter: it does not involve the Kazhdan–Lusztig basis.

Conjecture 14.1.5 (Lusztig). *For all $x, y, z \in W$,*

$$\deg(t_{x,y,z}) \leqslant \max_{I \in \mathscr{P}_{sph}(S)} \big(\varphi(w_I)\big).$$

Remark 14.1.6. Of course, Conjecture 14.1.5 holds whenever W is finite (by Lemma 14.1.4). It also holds if W is an affine Weyl group [Lus12, Theorem 7.2], or for instance if the Coxeter graph is complete [ShYa2] or if $|S| = 3$ (see [Zho]). □

Lemma 14.1.7. *Let $a \in \mathscr{A}$ be such that $\deg(t_{x,y,z}) \leqslant a$ for all x, y and z in W. Then*

$$\deg(h_{x,y,z}) \leqslant a$$

for all x, y and z in W.

Proof. Write $T_y = \sum_{x \in W} q_{x,y} \mathbf{C}_x$, where $q_{x,y} \in A$. By Exercise 5.3, we have that $q_{x,y} \in A_{\leqslant 0}$. Now, it follows from the definitions that, if $x, y, z \in W$, then

$$h_{x,y,z} = \sum_{x', y', z' \in W} p_{x',x} p_{y',y} q_{z,z'} t_{x',y',z'}.$$

By assumption, we have $\deg(t_{x',y',z'}) \leqslant a$ for all $x', y', z' \in W$. Now, $p_{x',x}$, $p_{y',y}$, $q_{z,z'}$ belong to $A_{\leqslant 0}$, so $\deg(h_{x,y,z}) \leqslant a$, as desired. ■

14.2. Lusztig's a-Function and Cells

14.2.A. Definition

Let us start with a second conjecture:

Conjecture 14.2.1 (Lusztig). *If $z \in W$, then the subset*

$$\{\deg(h_{x,y,z}) \mid x, y \in W\}$$

of $\mathscr{A}$ admits a maximum.

At the time this book is written, we only know that Conjecture 14.2.1 holds in the following cases:

- Whenever W is finite.

• Of course, if $\mathscr{A} = \mathbb{Z}$ and Conjecture 14.1.5 holds, then Conjecture 14.2.1 holds: this shows that it holds for affine Weyl groups and $\mathscr{A} = \mathbb{Z}$ (see Remark 14.1.6 for some other cases).

• Whenever φ is constant: this follows from the deep results of Elias–Williamson [ElWi] on *Soergel bimodules* [ElWi]. This is explained in Chapter 15 (see Corollary 15.2.4).

• Whenever (W, S, φ) is in the so-called *quasi-split case*: see Lusztig [Lus20, §16]. Again, this relies on Elias–Williamson's work [ElWi]. The definition, as well as a proof, is given in Chapter 15 (see Proposition 15.3.2).

Hypothesis. *From now on, and until the end of this chapter, we assume that Conjecture 14.2.1 holds for all triples (W_I, I, φ_I), where I runs over the subsets of S.*

This allows us to define *Lusztig's* **a***-function* as follows:

$$\mathbf{a}(z) = \max_{x,y \in W} \deg(h_{x,y,z}),$$

for $z \in W$. Note that it depends on φ. If $I \subset S$, we denote by $\mathbf{a}_I : W_I \longrightarrow \mathscr{A}$ Lusztig's **a**-function for the group W_I. If $x, y, z \in W$, we set

$$\gamma_{x,y,z^{-1}} = \mathrm{coef}_{\mathbf{a}(z)}(h_{x,y,z}) \in \mathbb{Z}.$$

We also define another map $\mathbf{\Delta} : W \longrightarrow \mathscr{A}$ by

$$\mathbf{\Delta}(z) = -\deg(p_{1,z})$$

and we set

$$n_z = \mathrm{coef}_{-\mathbf{\Delta}(z)}(p_{1,z}) \in \mathbb{Z}.$$

In other words,

$$v^{\mathbf{\Delta}(z)} p_{1,z} \equiv n_z \mod A_{<0}.$$

By 5.3.1 and Proposition 5.6.4(b), we have

$$(14.2.2) \qquad 0 \leqslant \mathbf{\Delta}(z) \leqslant \varphi(z)$$

for all $z \in W$. Moreover,

$$(14.2.3) \qquad \mathbf{\Delta}(1) = 0 \quad \text{and} \quad \mathbf{\Delta}(z) > 0$$

if $z \neq 1$. Also, it follows from 5.3.2 and 5.5.1 that

$$(14.2.4) \qquad \mathbf{\Delta}(z^{-1}) = \mathbf{\Delta}(z), \qquad n_{z^{-1}} = n_z \qquad \text{and} \qquad \mathbf{a}(z^{-1}) = \mathbf{a}(z).$$

It also follows from 5.5.1 that, if $x, y \in W$, then

(14.2.5) $$\gamma_{x,y,z} = \gamma_{y^{-1},x^{-1},z^{-1}}.$$

We set

$$\mathscr{D} = \{z \in W \mid \mathbf{\Delta}(z) = \mathbf{a}(z)\}.$$

Then 14.2.4 implies that

(14.2.6) $$z \in \mathscr{D} \textit{ if and only if } z^{-1} \in \mathscr{D}.$$

Remark 14.2.7. Let $z \in W$ be such that $\mathbf{a}(z) \geqslant \mathbf{a}(z')$ for all $z' \in W$. Let $x, y \in W$. Then $\deg(t_{x,y,z}) \leqslant \mathbf{a}(z)$ and

$$\gamma_{x,y,z^{-1}} = \mathrm{coef}_{\mathbf{a}(z)}(t_{x,y,z}),$$

as follows immediately from the equalities in the proof of Lemma 14.1.7. □

Example 14.2.8. If $h_{x,y,1} \neq 0$, then $1 \leqslant_L y$ and $1 \leqslant_R x$, so $x = y = 1$ by Example 6.7.2. Since $h_{1,1,1} = 1$, we get $\mathbf{a}(1) = 0 = \mathbf{\Delta}(1)$, and so $1 \in \mathscr{D}$. □

Example 14.2.9 (Type A_1). Assume here, and only here, that $S = \{s\}$. Since $\mathbf{C}_1^2 = \mathbf{C}_1$, $\mathbf{C}_1\mathbf{C}_s = \mathbf{C}_s\mathbf{C}_1 = \mathbf{C}_s$ and $\mathbf{C}_s^2 = (v^{\varphi(s)} + v^{-\varphi(s)})\mathbf{C}_s$, we have $\mathbf{a}(s) = \varphi(s)$. Since $p_{1,s} = v^{-\varphi(s)}$, we also have $\mathbf{\Delta}(s) = \varphi(s)$ and so $s \in \mathscr{D}$. Hence, $\mathscr{D} = W$ in this case. □

Example 14.2.10. More generally, assume here, and only here, that W is finite. It follows from Proposition 10.5.2(d) that $\mathbf{a}(w_0) \geqslant \varphi(w_0)$. But, by Lemma 14.1.7 and Remark 14.1.6, $\mathbf{a}(w_0) \leqslant \varphi(w_0)$. So $\mathbf{a}(w_0) = \varphi(w_0)$. Since $p_{1,w_0} = v^{-\varphi(w_0)}$ by 10.5.1, we also have $\mathbf{\Delta}(w_0) = \varphi(w_0)$ and so $w_0 \in \mathscr{D}$. □

Note also that the following useful congruence holds for any $x, y \in W$:

(14.2.11) $$\sum_{z \in W} h_{x,y,z}\, p_{1,z} \equiv \delta_{xy=1} \mod A_{<0}.$$

Proof. Applying the symmetrizing form τ to the equality $\mathbf{C}_x\mathbf{C}_y = \sum_{z \in W} h_{x,y,z}\mathbf{C}_z$ leads to

$$\sum_{z \in W} h_{x,y,z}\, p_{1,z} = \sum_{x',y' \in W} p_{x',x}\, p_{y',y}\, \tau(T_{x'} T_{y'}).$$

Since $\tau(T_{x'} T_{y'}) = \delta_{x'y'=1}$ (see 4.2.5), this gives

$$\sum_{z \in W} h_{x,y,z}\, p_{1,z} = \sum_{x',y' \in W} p_{x',x}\, p_{y',y}\, \delta_{x'y'=1},$$

and the result follows from the usual properties of the elements $p_{w',w}$ (see 5.3.1). ■

14.2.B. Base Change

Let $(B_w)_{w\in W}$ and $(B'_w)_{w\in W}$ be two A-bases of $\mathscr{H}$ such that

$$B_w \equiv B'_w \equiv T_w \mod \mathscr{H}_{<0}$$

for all $w \in W$. Let us write

$$B_x B'_y = \sum_{z\in W} h^\flat_{x,y,z} \mathbf{C}_z,$$

and let us denote by $\gamma^\flat_{x,y,z^{-1}}$ the coefficient of $v^{\mathbf{a}(z)}$ in $h^\flat_{x,y,z}$. Then

(14.2.12) $$h^\flat_{x,y,z} \equiv \gamma_{x,y,z^{-1}} v^{\mathbf{a}(z)} \mod A_{<\mathbf{a}(z)}.$$

Proof. Indeed, let us write

$$B_w = \sum_{w'\in W} \beta_{w',w} \mathbf{C}_{w'} \quad \text{and} \quad B'_w = \sum_{w'\in W} \beta'_{w',w} \mathbf{C}_{w'}.$$

Then it results from the hypothesis that

$$\beta_{w,w} \equiv \beta'_{w,w} \equiv 1 \mod \mathscr{H}_{<0} \qquad \text{and} \qquad \beta_{w',w}, \beta'_{w',w} \in A_{<0}$$

if $w' \neq w$. Since

$$h^\flat_{x,y,z} = \sum_{x',y'\in W} \beta_{x',x} \beta'_{y',y} h_{x',y',z}$$

for all $x, y, z \in W$, the congruence 14.2.12 follows. ■

We then deduce that

(14.2.13) $$\mathbf{a}(z) = \max_{x,y\in W} \deg(h^\flat_{x,y,z}) \qquad \text{and} \qquad \gamma^\flat_{x,y,z} = \gamma_{x,y,z}.$$

Example 14.2.14. The results 14.2.12 and 14.2.13 apply in particular if $B_w = B'_w = T_w$ or if $B_w = B'_w = \mathbf{L}^I_w$ for some $I \subset S$ (see the definition of the Kazhdan–Lusztig–Geck basis in §8.1). □

14.2.C. Conjectures

The following important conjectures were proposed by Lusztig [Lus20, §14].

Lusztig's Conjectures 14.2.15. *The following properties hold:*

(P1) *If $z \in W$, then $\mathbf{a}(z) \leqslant \mathbf{\Delta}(z)$.*
(P2) *If $d \in \mathscr{D}$ and $x, y \in W$ satisfy $\gamma_{x,y,d} \neq 0$, then $x = y^{-1}$.*

(P3) *If* $y \in W$, *then there exists a unique* $d \in \mathcal{D}$ *such that* $\gamma_{y^{-1},y,d} \neq 0$.
(P4) *If* $x \leqslant_{LR} y$, *then* $\mathbf{a}(x) \geqslant \mathbf{a}(y)$. *Hence, if* $x \sim_{LR} y$, *then* $\mathbf{a}(x) = \mathbf{a}(y)$.
(P5) *If* $d \in \mathcal{D}$ *and* $y \in W$ *are such that* $\gamma_{y^{-1},y,d} \neq 0$, *then* $\gamma_{y^{-1},y,d} = n_d = \pm 1$.
(P6) *If* $d \in \mathcal{D}$, *then* $d^2 = 1$.
(P7) *If* $x, y, z \in W$, *then* $\gamma_{x,y,z} = \gamma_{z,x,y}$.
(P8) *Let* $x, y, z \in W$ *be such that* $\gamma_{x,y,z} \neq 0$. *Then* $x \sim_L y^{-1}$, $y \sim_L z^{-1}$ *and* $z \sim_L x^{-1}$.
(P9) *If* $x \leqslant_L y$ *and* $\mathbf{a}(x) = \mathbf{a}(y)$, *then* $x \sim_L y$.
(P10) *If* $x \leqslant_R y$ *and* $\mathbf{a}(x) = \mathbf{a}(y)$, *then* $x \sim_R y$.
(P11) *If* $x \leqslant_{LR} y$ *and* $\mathbf{a}(x) = \mathbf{a}(y)$, *then* $x \sim_{LR} y$.
(P12) *If* $I \subseteq S$ *and* $z \in W_I$, *then* $\mathbf{a}(z) = \mathbf{a}_I(z)$.
(P13) *Any left cell* C *contains a unique element* $d \in \mathcal{D}$. *In this case,* $\gamma_{y^{-1},y,d} \neq 0$ *for all* $y \in C$.
(P14) *If* $z \in W$, *then* $z \sim_{LR} z^{-1}$.
(P15) *If* x, x', y, $y' \in W$ *are such that* $\mathbf{a}(y) = \mathbf{a}(y')$, *then we have the following equality*

$$\sum_{z \in W} h_{x,z,y} \otimes_{\mathbb{Z}} h_{y',x',z} = \sum_{z \in W} h_{x,y',z} \otimes_{\mathbb{Z}} h_{z,x',y}$$

in $A \otimes_{\mathbb{Z}} A$.

Definition 14.2.16. *If Lusztig's Conjecture* (P6) *holds, then the elements of* $\mathcal{D}$ *are called the* ***Duflo involutions*** *of* W.

In Chapter 6, we have provided several conjectures about the partition of W into cells (see Lusztig's Conjectures 6.5.2, numbered (L1), (L2), ..., (L8)). Some of them are consequences of the above Lusztig's Conjectures:

Proposition 14.2.17. *We have:*

(a) (P3) *and* (P6) *imply* (L1).
(b) (P14) *is exactly* (L2).
(c) (P4) *and* (P9) *(or* (P10)*) imply* (L3) *and* (L4).
(d) *If* (P4) *and* (P9) *hold, and if* Γ *is a two-sided cell, then any left cell contained in* Γ *is a union of left-connected components of* Γ *(compare with* (L6)*).*

Proof. (a) and (b) are obvious.

(c) Assume that (P4) and (P9) hold. First, let $x,y \in W$ be such that $x \sim_{LR} y$. Then there exist two sequences $x = x_0, x_1, \ldots, x_m = y$ and $y = y_0, y_1, \ldots, y_n = x$ such that $x_{i-1} \leqslant_L x_i$ or $x_{i-1} \leqslant_R x_i$, and $y_{j-1} \leqslant_L y_j$ or $y_{j-1} \leqslant_R y_j$ for all $i \in \{1,2,\ldots,m\}$ and $j \in \{1,2,\ldots,n\}$. It then follows from (P4) that

$$\mathbf{a}(x) = \mathbf{a}(x_0) \geqslant \mathbf{a}(x_1) \geqslant \cdots \geqslant \mathbf{a}(x_m) = \mathbf{a}(y) = \mathbf{a}(y_0) \geqslant \mathbf{a}(y_1) \geqslant \cdots \geqslant \mathbf{a}(y_n) = \mathbf{a}(x).$$

Therefore,

$$\mathbf{a}(x) = \mathbf{a}(x_0) = \mathbf{a}(x_1) = \cdots = \mathbf{a}(x_m) = \mathbf{a}(y) = \mathbf{a}(y_0) = \mathbf{a}(y_1) = \cdots = \mathbf{a}(y_n).$$

Now, since (P9) and (P10) hold (indeed, (P9) and (P10) are equivalent thanks to Lemma 6.1.5 and 14.2.4), it follows that $x_{i-1} \sim_L x_i$ or $x_{i-1} \sim_R x_i$ for all $i \in \{1,2,\dots,m\}$. This proves (L3).

Now, assume moreover that $x \leqslant_L y$. Then $\mathbf{a}(x) = \mathbf{a}(y)$ by (P4) and $x \sim_L y$ by (P9). This proves (L4).

(d) Assume that (P4) and (P9) hold and let Γ be a two-sided cell. Let x, y be two elements belonging to the same left-connected component of Γ. We only need to show that $x \sim_L y$. By an easy induction argument, we may assume that there exists an $s \in S$ such that $y = sx$. By exchanging the roles of x and y if necessary, we may assume that $y = sx > x$. Then $y \leqslant_L x$ by Example 6.1.2. But $\mathbf{a}(x) = \mathbf{a}(y)$ by (P4), and so $x \sim_L y$ by (P9), as desired. ■

Remark 14.2.18. Let $a \in \mathscr{A}$. We set

$$W(\geqslant a) = \{w \in W \mid \mathbf{a}(w) \geqslant a\}, \qquad W(> a) = \{w \in W \mid \mathbf{a}(w) > a\}$$

and
$$W(a) = \{w \in W \mid \mathbf{a}(w) = a\}.$$

If (P4) holds, then $W(\geqslant a)$ and $W(> a)$ are two-sided ideals of W, while $W(a)$ is a two-sided convex subset of W. □

Commentary 14.2.19. Some of the results of this chapter rely on the hypothesis that $\mathbf{a}$ takes only finitely many values: this would be a consequence of Conjectures 6.5.2(L8) and (P4). Indeed, Conjecture 6.5.2(L8) states that the number of two-sided cells is finite. □

14.3. Reductions

The proof of Lusztig's Conjectures seems out of reach in full generality at the current time. It is known that they all hold in the following cases:

- Whenever φ is constant and $\mathbf{a}$ is bounded [Lus20, §15]: this is based on the results of Elias–Williamson [ElWi] (see Theorem 15.1.4). This will be explained in Chapter 15 (see Theorem 15.2.5).
- Whenever (W, S, φ) is in the so-called *quasi-split case* and $\mathbf{a}$ is bounded [Lus20, §16] (see also Elias [Eli]): this is also based on the results of Elias–Williamson [ElWi]. We will also treat this case in Chapter 15 (see Theorem 15.3.4).
- Whenever $|S| = 2$ (see Lusztig [Lus20, §17] for the case where W is infinite and Geck [Gec8, Proposition 5.1] whenever W is finite): the finite case will be treated in Chapter 21 while the infinite case will be seen as a special case of free (i.e. universal) Coxeter groups (see below).

- Whenever (W,S) is a free Coxeter system, i.e. if $m_{st}=\infty$ for all $s,\ t\in S$ such that $s\neq t$ (see Shi–Yang [ShYa1]). This case will be studied in full detail in Chapter 24.
- Whenever (W,S) is of type F_4 (Geck [Gec4]).
- Whenever (W,S) is of type B_n, with $n\leqslant 5$ (Geck).
- Whenever (W,S) is of type B_n (with $S=\{t,s_1,s_2,\dots,s_{n-1}\}$, $m_{ts_1}=4$, $m_{ts_i}=2$ for all $i\geqslant 2$) and $\varphi(t)>(n-1)\varphi(s_1)$ (see Bonnafé–Iancu [BoIa], Bonnafé [Bon1] and Geck–Iancu [GeIa1]).
- Whenever (W,S) is of type $\widetilde{G}_2$ (see Guilhot-Parkinson [GuPa]).

One of the aims of the present part is to give an account of the proof of Lusztig's Conjectures in the first two cases. The fifth and sixth cases follow from (clever) computer calculations. First of all, we explain in this section some preliminary reductions.

14.3.A. First Consequences of the Conjectures

The next lemma contains some useful consequences of Lusztig's Conjectures: it will be used as an intermediate step towards the proof of Proposition 14.3.3, which aims to provide some logical links between all these conjectures.

If $w,\ x,\ y,\ z\in W$, we denote by $h_{x,y,z,w}$ the coefficient of $\mathbf{C}_w$ in the expansion of $\mathbf{C}_x\mathbf{C}_y\mathbf{C}_z$ in the Kazhdan–Lusztig basis. We also denote by $\gamma_{x,y,z,w}$ the coefficient of $v^{2\mathbf{a}(w)}$ in $h_{x,y,z,w}$. By computing $\mathbf{C}_x\mathbf{C}_y\mathbf{C}_z$ in two ways, we get

$$h_{x,y,z,w}=\sum_{u\in W}h_{x,y,u}\,h_{u,z,w}=\sum_{u\in W}h_{x,u,w}\,h_{y,z,u}. \tag{14.3.1}$$

Lemma 14.3.2. *We have:*

(a) *If* (P1) *holds and* $x,\ y\in W$, *then* $\sum_{d\in\mathscr{D}}\gamma_{x,y,d}\,n_d=\delta_{xy=1}$.

(b) *If* (P4) *holds and* $w,\ x,\ y,\ z\in W$, *then* $\deg(h_{x,y,z,w})\leqslant 2\mathbf{a}(w)$ *and*

$$\gamma_{x,y,z,w}=\sum_{\substack{u\in W\\ \mathbf{a}(u)=\mathbf{a}(w)}}\gamma_{x,y,u^{-1}}\gamma_{u,z,w^{-1}}=\sum_{\substack{u\in W\\ \mathbf{a}(u)=\mathbf{a}(w)}}\gamma_{x,u,w^{-1}}\gamma_{y,z,u^{-1}}.$$

(c) *If* (P2) *and* (P4) *hold, if* $x,\ y,\ z\in W$ *and* $d\in\mathscr{D}$, *then*

$$\gamma_{x,y,z,d}=\begin{cases}\gamma_{x,y,z}\gamma_{z^{-1},z,d^{-1}} & \textit{if } \mathbf{a}(z)=\mathbf{a}(d),\\ 0 & \textit{if } \mathbf{a}(z)<\mathbf{a}(d),\end{cases}$$

and

$$\gamma_{x,y,z,d}=\begin{cases}\gamma_{y,z,x}\gamma_{x,x^{-1},d^{-1}} & \textit{if } \mathbf{a}(x)=\mathbf{a}(d),\\ 0 & \textit{if } \mathbf{a}(x)<\mathbf{a}(d).\end{cases}$$

(d) *Assume that* (P2), (P3), (P4), (P5) *hold and that* **a** *takes only finitely many values. If* $z \in W$ *and* $d \in \mathcal{D}$ *are such that* $\gamma_{z^{-1},z,d} \neq 0$, *then* $\mathbf{a}(z) = \mathbf{a}(d)$.

Proof. Statement (a) follows from 14.2.11, 14.2.4 and 14.2.6.

(b) Assume that (P4) holds and let $w, x, y, z \in W$. If $h_{u,z,w} \neq 0$ (resp. $h_{x,u,w} \neq 0$), then $w \leqslant_L u$ (resp. $w \leqslant_R u$) and so it follows from (P4) that $\mathbf{a}(u) \leqslant \mathbf{a}(w)$. Consequently, $\deg(h_{x,y,u}) \leqslant \mathbf{a}(w)$, $\deg(h_{y,z,u}) \leqslant \mathbf{a}(w)$ and so

$$\deg(h_{x,y,z,w}) \leqslant 2\mathbf{a}(w)$$

by 14.3.1. The two equalities now follow from these observations and again from 14.3.1.

Statement (c) follows from (b) and (P2).

(d) Assume that (P2), (P3), (P4), (P5) hold and that **a** takes only finitely many values $a_1 > a_2 > \cdots > a_n$. Let $z \in W$ and $d \in \mathcal{D}$ be such that $\gamma_{z^{-1},z,d} \neq 0$. First, note that $\mathbf{a}(z) \leqslant \mathbf{a}(d)$ by (P4). We prove (d) by descending induction on $\mathbf{a}(z)$. Of course, if $\mathbf{a}(z) = a_1$, then $\mathbf{a}(d) = a_1 = \mathbf{a}(z)$ necessarily.

Now, assume that $\mathbf{a}(z) = a_m$ (with $m > 1$) and that (d) holds for $z' \in W$ and $d \in \mathcal{D}$ if $\mathbf{a}(z') > a_m$. Assume that $\mathbf{a}(z) \neq \mathbf{a}(d)$. Then $\mathbf{a}(d) > \mathbf{a}(z)$. By (P3), there exists an $e \in \mathcal{D}$ such that $\gamma_{d^{-1},d,e^{-1}} \neq 0$. It follows from the induction hypothesis that $\mathbf{a}(e) = \mathbf{a}(d)$. But, by (c), we get

$$\gamma_{z^{-1},z,d,e} = \gamma_{z^{-1},z,d}\gamma_{d^{-1},d,e^{-1}} \neq 0,$$

and it follows again from (c) that $\gamma_{z,d,z^{-1}} \neq 0$. So $z \leqslant_R d$ and so $\mathbf{a}(z) \geqslant \mathbf{a}(d)$ by (P4), which contradicts the fact that $\mathbf{a}(d) > \mathbf{a}(z)$. So $\mathbf{a}(z) = \mathbf{a}(d)$. ∎

14.3.B. Some Implications between Lusztig's Conjectures

Following Lusztig [Lus20, §14.3], we introduce an auxiliary statement:

(P^+) Let $x, y, z, z' \in W$ be such that $\gamma_{x,y,z^{-1}} \neq 0$ and $z' \overset{L}{\longleftarrow} z$. Then there exists an $x' \in W$ such that $\mathrm{coef}_{\mathbf{a}(z)}(h_{x',y,z'}) \neq 0$.

This statement does not hold in full generality (see Exercise 14.1), but it happens to be very useful whenever it holds (this happens, for instance, in the split case), as is shown by Proposition 14.3.3 below.

Proposition 14.3.3. *We have the following implications between Lusztig's Conjectures and statement* (P^+)*:*

(a) (P^+) *implies* (P4).
(b) (P1) *and* (P3) *imply* (P5).
(c) (P2) *and* (P3) *imply* (P6).

(d) *Assume that* **a** *takes only finitely many values. Then* (P2), (P3), (P4) *and* (P5) *imply* (P7).
(e) (P7) *implies* (P8).
(f) (P^+) *and* (P8) *imply* (P9).
(g) (P9) *and* (P10) *are equivalent.*
(h) (P4) *and* (P9) *(or* (P10)*) imply* (P11).
(i) (P3), (P4) *and* (P8) *for* W *and* W_I *imply* (P12).
(j) (P^+), (P2), (P3) *and* (P7) *imply* (P13).
(k) (P6) *and* (P13) *imply* (P14).

Proof. (a) Assume that (P^+) holds and let x and y be elements of W such that $x \leqslant_{LR} y$. We want to show that $\mathbf{a}(x) \geqslant \mathbf{a}(y)$. By transitivity, we may assume that $x \xleftarrow{L} y$ or that $x \xleftarrow{R} y$. By 14.2.4, we may assume that $x \xleftarrow{L} y$. Let u, $v \in W$ be such that $\deg(h_{u,v,y}) = \mathbf{a}(y)$. Then it follows from ($P^+$) that there exists a $u' \in W$ such that $\deg(h_{u',v,x}) \geqslant \mathbf{a}(y)$, so the result follows.

(b) Assume that (P1) and (P3) hold and let $d \in \mathcal{D}$ and $y \in W$ be such that $\gamma_{y^{-1},y,d} \neq 0$. It follows from Lemma 14.3.2(a), applied to the case where $x = y^{-1}$, that

$$\sum_{z \in \mathcal{D}} \gamma_{y^{-1},y,z} n_z = 1.$$

But it follows from (P3) that, if $z \in \mathcal{D} \setminus \{d\}$, then $\gamma_{y^{-1},y,z} = 0$. So $\gamma_{y^{-1},y,d} n_d = 1$, and (P5) follows.

(c) Assume that (P2) and (P3) hold and let $d \in \mathcal{D}$. By the very definition of $\mathbf{a}(d)$, there exist x, $y \in W$ such that $\gamma_{x,y,d} \neq 0$. By (P2), $x = y^{-1}$. So it follows from 14.2.5 that $\gamma_{y^{-1},y,d^{-1}} \neq 0$ and so $d = d^{-1}$ by the uniqueness statement in (P3).

(d) Assume that **a** takes only finitely many values and that (P2), (P3), (P4) and (P5) hold. Let x, y, $z \in W$. By (P3), there exists a $d \in \mathcal{D}$ such that $\gamma_{z^{-1},z,d} \neq 0$. Note that (P6) holds by (c). By Lemma 14.3.2(d), we get $\mathbf{a}(z) = \mathbf{a}(d)$ and so it follows from the first equality in Lemma 14.3.2(c) that

$$\gamma_{x,y,z,d} = \gamma_{x,y,z} \gamma_{z^{-1},z,d}.$$

Hence, if $\gamma_{x,y,z} \neq 0$, it follows from the second equality in Lemma 14.3.2(c) that $\gamma_{x,x^{-1},d} \neq 0$, and so $\gamma_{z^{-1},z,d} = \gamma_{x,x^{-1},d} = n_d = \pm 1$ by (P5). Using again Lemma 14.3.2(c), this shows that, if $\gamma_{x,y,z} \neq 0$, then $\gamma_{x,y,z} = \gamma_{y,z,x}$. Now, if $\gamma_{x,y,z} = 0$, then necessarily $\gamma_{y,z,x} = 0$ (indeed, if $\gamma_{y,z,x} \neq 0$, then it follows from the previous case that $\gamma_{y,z,x} = \gamma_{z,x,y} = \gamma_{x,y,z} \neq 0$). This completes the proof of (P7).

(e) Assume that (P7) holds and let x, y, $z \in W$ be such that $\gamma_{x,y,z} \neq 0$. Then $z^{-1} \leqslant_L y$ and $z^{-1} \leqslant_R x$. But, by (P7), $\gamma_{y,z,x} \neq 0$ and so $x^{-1} \leqslant_L z$ and $x^{-1} \leqslant_R y$. Similarly, $y^{-1} \leqslant_L x$ and $y^{-1} \leqslant_R z$. So the result follows from Lemma 6.1.5(b).

(f) Assume that (P$^+$) and (P8) hold and let x, $y \in W$ be such that $x \leqslant_L y$ and $\mathbf{a}(x) = \mathbf{a}(y)$. Since (P$^+$) implies (P4) (see (a)), we may assume that $x \overset{L}{\longleftarrow} y$. Now, let u, $v \in W$ be such that $\gamma_{u,v,y^{-1}} \neq 0$. Then it follows from (P$^+$) that there exists a $u' \in W$ such that $\mathrm{coef}_{\mathbf{a}(y)}(h_{u',v,x}) \neq 0$. Since $\mathbf{a}(x) = \mathbf{a}(y)$, this forces $\gamma_{u',v,x^{-1}} \neq 0$. So it follows from (P8) that $x \sim_L v$ and $y \sim_L v$, so $x \sim_L y$, as expected.

Statement (g) is clear from Lemma 6.1.5 and 14.2.4.

Statement (h) is clear.

(i) Assume that (P3), (P4) and (P8) hold for W and W_I, and let $z \in W_I$. It is clear from the definition that $\mathbf{a}_I(z) \leqslant \mathbf{a}(z)$. By (P3), there exists a $d \in \mathscr{D}$ such $\gamma_{z^{-1},z,d} \neq 0$. In particular, $h_{z^{-1},z,d} \neq 0$ and, since $\mathbf{C}_{z^{-1}}\mathbf{C}_z \in \mathscr{H}_I$, this forces $d \in W_I$. So this shows that $\mathbf{a}_I(d) \geqslant \deg(h_{z^{-1},z,d}) = \mathbf{a}(d)$ and so $\mathbf{a}_I(d) = \mathbf{a}(d)$. Now, by (P8), we get that $z \sim_L d$ and so $z \sim_L^I d$ by Corollary 8.2.4. By (P4) for W and W_I, we get $\mathbf{a}(z) = \mathbf{a}(d) = \mathbf{a}_I(d) = \mathbf{a}_I(z)$.

(j) Assume that (P$^+$), (P2), (P3) and (P7) hold and let C be a left cell. Recall from (a), (c) and (e) that this implies that (P4), (P6) and (P8) hold. If $x \in C$, we denote by ∂_x the unique element of $\mathscr{D}$ such that $\gamma_{x^{-1},x,\partial_x} \neq 0$ (this is well-defined by (P3)). It then follows from (P8) that $\partial_x^{-1} \in C$. Since $\partial_x = \partial_x^{-1}$ by (P6), this shows that $\partial_x \in C$. So $\mathscr{D} \cap C \neq \varnothing$.

Now, let x, $y \in C$. We will show that $\partial_x = \partial_y$. By definition, there exist $x = x_0, x_1, \ldots, x_l = y = y_0, y_1, \ldots, y_m = x$ such that $x_i \overset{L}{\longleftarrow} x_{i+1}$ and $y_j \overset{L}{\longleftarrow} y_{j+1}$ for all i, j. Then all the x_i and all the y_j belong to C and this shows that we may assume that $x \overset{L}{\longleftarrow} y$. Now, by (P7), we have $\gamma_{y,\partial_y,y^{-1}} \neq 0$. Since $x \overset{L}{\longleftarrow} y$ and $\mathbf{a}(x) = \mathbf{a}(y)$ by (P4), it follows from (P$^+$) that there exists an $x' \in W$ such that $\mathrm{coef}_{\mathbf{a}(x)}(h_{x',\partial_y,x}) \neq 0$. But $\deg(h_{x',\partial_y,x}) \leqslant \mathbf{a}(x)$ by definition, so $\gamma_{x',\partial_y,x^{-1}} \neq 0$. So, from (P7), we get that $\gamma_{x^{-1},x',\partial_y} \neq 0$. But since $\partial_y \in \mathscr{D}$, it follows from (P2) that $x' = x$, and from the uniqueness statement in (P3), that $\partial_x = \partial_y$.

To prove (P13), it just remains to show that, if $d \in \mathscr{D} \cap C$, then $\partial_d = d$. But, by (P7), $\gamma_{d,d,\partial_d} = \gamma_{\partial_d,d,d}$ and so it follows from (P2) and (P6) that $\partial_d = d^{-1} = d$.

(k) Assume that (P6) and (P13) hold and let $z \in W$. Let C denote the left cell of z. By (P13), there exists a $d \in \mathscr{D} \cap C$. So $z \sim_L d$. Therefore, by Corollary 6.1.6, $z^{-1} \sim_R d^{-1}$. Since $d = d^{-1}$ by (P6), we get that $z \sim_{LR} z^{-1}$. ■

14.3.C. About Conjecture (P15)

If Γ is a two-sided convex subset of W, we denote by $\mathscr{H}_{\#}^{LR}[\Gamma]$ the sub-$\mathbb{Z}$-module of $\mathscr{H}^{LR}[\Gamma]$ defined by

$$\mathscr{H}_{\#}^{LR}[\Gamma] = \bigoplus_{w \in \Gamma} \mathbb{Z}\mathbf{c}_w.$$

Next, we set

$$\hat{\mathcal{H}}^{LR}[\Gamma] = A \otimes_{\mathbb{Z}} \mathcal{H}_{\#}^{LR}[\Gamma] \otimes_{\mathbb{Z}} A$$

and, for $w \in \Gamma$, we set $\hat{\mathbf{c}}_w = 1 \otimes_{\mathbb{Z}} \mathbf{c}_w \otimes_{\mathbb{Z}} 1 \in \hat{\mathcal{H}}^{LR}[\Gamma]$. Since $\mathcal{H}^{LR}[\Gamma] \simeq A \otimes_{\mathbb{Z}} \mathcal{H}_{\#}^{LR}[\Gamma] \simeq \mathcal{H}_{\#}^{LR}[\Gamma] \otimes_{\mathbb{Z}} A$, we get two isomorphisms

$$i_L^{\Gamma} : \hat{\mathcal{H}}^{LR}[\Gamma] \xrightarrow{\sim} A \otimes_{\mathbb{Z}} \mathcal{H}^{LR}[\Gamma] \qquad \text{and} \qquad i_R^{\Gamma} : \hat{\mathcal{H}}^{LR}[\Gamma] \xrightarrow{\sim} \mathcal{H}^{LR}[\Gamma] \otimes_{\mathbb{Z}} A$$

which are defined by $i_L^{\Gamma}(\hat{\mathbf{c}}_w) = 1 \otimes_{\mathbb{Z}} \mathbf{c}_w$ and $i_R^{\Gamma}(\hat{\mathbf{c}}_w) = \mathbf{c}_w \otimes_{\mathbb{Z}} 1$. This allows us to endow $\hat{\mathcal{H}}^{LR}[\Gamma]$ with the structure of a left $\mathcal{H}$-module and of a right $\mathcal{H}$-module.

As might already be seen in the case where $|S| = 1$ and $\Gamma = W$, these two module structures do not commute in general. However, the next proposition is immediate (recall from Remark 14.2.18 that $W(a) = \{w \in W \mid \mathbf{a}(w) = a\}$ is a two-sided convex subset of W if (P4) holds):

Proposition 14.3.4. *Assume that* (P4) *holds. Then the following statements are equivalent:*

(1) (P15) *holds.*
(2) *For any $a \in \mathcal{A}$, the left $\mathcal{H}$-module and the right $\mathcal{H}$-module structures on $\hat{\mathcal{H}}^{LR}[W(a)]$ commute.*

Note the following partial result:

Lemma 14.3.5. *Assume that* (P4) *holds. Let $a \in \mathcal{A}$, let $s, s' \in S$ and let $y \in W(a)$. If $sy < y$ or if $ys' < y$, then*

$$\mathbf{C}_s(\hat{\mathbf{c}}_y \mathbf{C}_{s'}) = (\mathbf{C}_s \hat{\mathbf{c}}_y)\mathbf{C}_{s'} \qquad \textit{in } \hat{\mathcal{H}}^{LR}[W(a)].$$

Proof. By symmetry, we may assume that $sy < y$. Then Corollary 5.5.6 implies that

$$(\mathbf{C}_s \hat{\mathbf{c}}_y)\mathbf{C}_{s'} = (v^{\varphi(s)} + v^{-\varphi(s)}) \otimes_{\mathbb{Z}} \hat{\mathbf{c}}_y \mathbf{C}_{s'},$$

so it remains to show that, if $\mathbf{C}_{y'}$ (with $y' \in W(a)$) occurs with a non-zero coefficient in $\mathbf{C}_y \mathbf{C}_{s'}$, then $sy' < y'$. But, if $\mathbf{C}_{y'}$ occurs with a non-zero coefficient in $\mathbf{C}_y \mathbf{C}_{s'}$, then $y' \leqslant_R y$ and so $\mathcal{L}(y) \subset \mathcal{L}(y')$ by Corollary 6.3.2, so $sy' < y'$, as desired. ■

Corollary 14.3.6. *Assume that* (P4) *and* (P9) *hold. Then the following are equivalent:*

(1) (P15) *holds.*
(2) *If $s, s' \in W$ and $y, y' \in W$ are such that $\mathbf{a}(y) = \mathbf{a}(y')$, $sy > y$, $ys' > y$, $sy' < y'$ and $y's' < y'$, then we have the following equality*

$$\sum_{\substack{z \in W \\ sz < z < zs'}} \lambda_{z,y}^{s} \otimes_{\mathbb{Z}} \rho_{y',z}^{s'} = \sum_{\substack{z \in W \\ zs' < z < sz}} \lambda_{y',z}^{s} \otimes_{\mathbb{Z}} \rho_{z,y}^{s'}$$

in $A \otimes_{\mathbb{Z}} A$.

Proof. By Proposition 14.3.4, (P15) holds if and only if, for all $a \in \mathscr{A}$, we have

$$(*) \qquad \mathbf{C}_s(\hat{\mathbf{c}}_y \mathbf{C}_{s'}) = (\mathbf{C}_s \hat{\mathbf{c}}_y)\mathbf{C}_{s'}$$

for all $y \in W(a)$ and $s, s' \in S$. But, by Lemma 14.3.5, this is automatically satisfied if $sy < y$ or $ys' < y$. This means that it is sufficient to prove $(*)$ whenever $sy > y$ and $ys' > y$.

Note also that, since we assume that (P9) holds, two elements x and x' in $W(a)$ such that $x \leqslant_L x'$ satisfy $x \sim_L x'$, and so $\mathscr{R}(x) = \mathscr{R}(x')$ by Corollary 6.3.2.

By the same argument as in the proof of Lemma 14.3.5, we only need to compute the coefficient of $\hat{\mathbf{c}}_{y'}$ on both sides of $(*)$ whenever $sy' < y'$ and $y's' < y'$. So (P15) is equivalent to

$$\sum_{\substack{z \in W \\ sz < z < zs' \\ \mathbf{a}(z) = \mathbf{a}(y)}} \lambda_{z,y}^s \otimes_{\mathbb{Z}} \rho_{y',z}^{s'} = \sum_{\substack{z \in W \\ zs' < z < sz \\ \mathbf{a}(z) = \mathbf{a}(y)}} \lambda_{y',z}^s \otimes_{\mathbb{Z}} \rho_{z,y}^{s'}$$

for all $s, s' \in W$ and $y, y' \in W$ such that $\mathbf{a}(y) = \mathbf{a}(y')$, $sy > y$, $ys' > y$, $sy' < y'$ and $y's' < y'$. But if $\lambda_{z,y}^s \otimes_{\mathbb{Z}} \rho_{y',z}^{s'} \neq 0$, then $y' \leqslant_R z$ and $z \leqslant_L y$, so $\mathbf{a}(y') \geqslant \mathbf{a}(z) \geqslant \mathbf{a}(y)$ by (P4), so $\mathbf{a}(z) = \mathbf{a}(y)$. This proves that the condition "$\mathbf{a}(z) = \mathbf{a}(y)$" can be removed from the left-hand side and a similar argument shows that it can be removed from the right-hand side too. We retrieve the formula given in (2), which proves the equivalence of (1) and (2). ■

Remark 14.3.7. Let

$$\begin{array}{rccc} \pi_\Gamma : & \hat{\mathscr{H}}^{LR}[\Gamma] & \longrightarrow & \mathscr{H}^{LR}[\Gamma] \\ & \alpha \otimes_{\mathbb{Z}} h \otimes_{\mathbb{Z}} \beta & \longmapsto & \alpha\beta h. \end{array}$$

Then π_Γ is a surjective morphism of (A,A)-bimodules. Also, it is a morphism of left $\mathscr{H}$-modules as well as a morphism of right $\mathscr{H}$-modules. But, on $\mathscr{H}^{LR}[\Gamma]$, the left and the right actions of $\mathscr{H}$ commute. So the same argument as in the proof of Corollary 14.3.6 shows that, if $s, s' \in W$ and $y, y' \in W$ are such that $\mathbf{a}(y) = \mathbf{a}(y')$, $sy > y$, $ys' > y$, $sy' < y'$ and $y's' < y'$, then

$$\sum_{\substack{z \in W \\ sz < z < zs'}} \lambda_{z,y}^s \rho_{y',z}^{s'} = \sum_{\substack{z \in W \\ zs' < z < sz}} \lambda_{y',z}^s \rho_{z,y}^{s'}$$

in A. □

Exercises

Exercise 14.1. Assume that W is of type A_2, B_2 or G_2. Compute Lusztig's $\mathbf{a}$-function and prove that Lusztig's Conjectures hold in all these cases.

Use these computations to show for instance that, if W is of type B_2, if $\mathscr{A}=\mathbb{Z}$, if $\varphi(s)=1$ and $\varphi(t)=10$, if $x=y=z=tst$ and $z'=w_0$, then (P^+) does not hold.

Exercise 14.2. Let $z \in W$. Show that

$$\mathbf{a}(z) \geqslant \max_{s \in \mathscr{L}(z) \cup \mathscr{R}(z)} \varphi(s).$$

In particular, $\mathbf{a}(z) > 0$ if $z \neq 1$.

Exercise 14.3. Find an example of a finite Coxeter group W and some $d \in \mathscr{D}$ such that $w_0 d \notin \mathscr{D}$.

Exercise 14.4. Let $z \in W$. Show that

$$\mathbf{a}(z) \geqslant \mathbf{a}_I(\mathrm{pr}_L^I(z))$$

for all $I \subset S$. Deduce that

$$\mathbf{a}(z) \geqslant \max\{\varphi(w_{\mathscr{L}(z)}), \varphi(w_{\mathscr{R}(z)})\}.$$

Find an example where this inequality is strict.

Chapter 15
Split and Quasi-Split Cases

In the first section of this chapter, we recall briefly how the category of Soergel bimodules categorifies the Hecke algebra (whenever φ is the length function) and that, through this categorification, indecomposable Soergel bimodules correspond to the Kazhdan–Lusztig basis (Elias–Williamson Theorem [ElWi, Theorem 1.1]). In the same section, we explain how this can be extended to the *quasi-split* case (see the definition below), following Lusztig [Lus20, §16] (see also the work of Elias [Eli]). We do not provide proofs of the deep results of Elias–Williamson that are stated in this section.

We then show how these results on the category of Soergel bimodules allow one to show that Lusztig's Conjectures (P1), (P2),..., (P15) hold in the split and the quasi-split case. Note that we do not assume that Conjecture 14.2.1 holds (i.e. that Lusztig's **a**-function is well-defined): it will be a consequence of the results of the first section that it holds both in the split and the quasi-split case.

Hypothesis. *We assume in this chapter, and only in this chapter, that $\mathscr{A} = \mathbb{Z}$ and $\varphi : S \to \mathbb{Z}$, $s \mapsto 1$. Note that this implies that $A = \mathbb{Z}[v, v^{-1}]$. We also fix an automorphism σ of W stabilizing S.*

We denote by n the order of σ and we fix a primitive n-th root of unity ξ in $\mathbb{C}^\times$. Let $\mathcal{O} = \mathbb{Z}[\xi]$.

Following Theorem 3.10.6, we denote by $(S/\sigma)_{\mathrm{sph}}$ the set of σ-orbits I in S such that W_I is finite, we set

$$S_\sigma = \{w_I \mid I \in (S/\sigma)_{\mathrm{sph}}\}$$

and we recall that (W^σ, S_σ) is a Coxeter system. We denote by φ_σ the restriction of the length function ℓ to S_σ. It is a weight function on (W^σ, S_σ) by Corollary 3.10.7.

C. Bonnafé, *Kazhdan–Lusztig Cells with Unequal Parameters*,
Algebra and Applications 24, https://doi.org/10.1007/978-3-319-70736-5_15

Let $\mathscr{H}_\sigma$ denote the Hecke algebra $\mathscr{H}(W^\sigma, S_\sigma, \varphi_\sigma)$. If $w \in W^\sigma$, we denote by T_w^σ the corresponding standard basis element of $\mathscr{H}_\sigma$. Let $(\mathbf{C}_w^\sigma)_{w \in W^\sigma}$ denote the Kazhdan–Lusztig basis of $\mathscr{H}_\sigma$ and let $\leqslant_?^\sigma$ and $\sim_?^\sigma$ be the analogues for W^σ of $\leqslant_?$ and $\sim_?$, respectively. If x, y, $z \in W^\sigma$, we denote by $p_{x,y}^\sigma$ and $h_{x,y,z}^\sigma$ the analogues of $p_{x,y}$ and $h_{x,y,z}$, respectively. Finally, we set

$$\mathbf{\Delta}_\sigma(z) = -\deg(p_{1,z}^\sigma)$$

so that $\mathbf{\Delta}_\sigma : W^\sigma \to \mathbb{Z}$ is the analogue of $\mathbf{\Delta}$ for W^σ, and we denote by n_z^σ the coefficient of $v^{\mathbf{\Delta}_\sigma(z)}$ in $p_{1,z}^\sigma$.

Terminology. By analogy with the representation theory of finite and p-adic groups, the triple (W, S, ℓ) is said to be in the *split* case (as in Example 5.7.4), while the triple $(W^\sigma, S_\sigma, \varphi_\sigma)$ is said to be in the *quasi-split* case: the split case is a particular quasi-split case, for which $\sigma = \mathrm{Id}_W$.

15.1. Recollection on Soergel Bimodules

A *reflection-faithful* representation of (W, S) over $\mathbb{C}$ is a finite-dimensional representation $\tilde{E}$ of W such that:

(1) The representation is faithful;
(2) If $w \in W$ is such that $\mathrm{codim}_{\mathbb{C}}(\tilde{E}^w) = 1$, then $w \in T$ (i.e. w is conjugate to an element of S).

If W is finite, it is straightforward to see that the (extension to $\mathbb{C}$ of the) reflection representation E over $\mathbb{R}$ (see §3.11.B) is reflection-faithful [Hum, Proposition 5.3], because the bilinear form B_W is positive definite. But it might happen that, for infinite W, $\mathbb{C} \otimes_{\mathbb{R}} E$ is not reflection-faithful. However [Soe2, Proposition 2.1 and its proof], there always exists a reflection-faithful representation of (W, S) admitting an extension to $W \rtimes \langle\sigma\rangle$.

Notation. *We fix once and for all a reflection-faithful representation $\tilde{E}$ of (W, S) over $\mathbb{C}$ admitting an extension to $W \rtimes \langle\sigma\rangle$.*

15.1.A. Polynomial Functions

We denote by $\mathbf{S}$ the algebra of polynomial functions on $\tilde{E}$, which is endowed with the unique grading such that the linear forms $\tilde{E} \to \mathbb{C}$ are of degree 2. We denote by $\mathbf{S}^{>0}$ the unique maximal homogeneous ideal of $\mathbf{S}$:

$$\mathbf{S}^{>0} = \{f \in \mathbf{S} \mid f(0) = 0\}.$$

We denote by $\mathbb{C}_0$ the $\mathbf{S}$-module (left or right) defined by $\mathbb{C}_0 = \mathbf{S}/\mathbf{S}^{>0}$. The group W acts on $\mathbf{S}$ and we denote by $\mathbf{S} \rtimes W$ the algebra which is the semidirect product of $\mathbf{S}$ by W: as a left $\mathbf{S}$-module, we have $\mathbf{S} \rtimes W = \bigoplus_{w \in W} \mathbf{S}w$.

By construction, the automorphism σ of W acts on $\tilde{E}$ and so it induces an automorphism of $\mathbf{S}$ (and an automorphism of $\mathbf{S} \rtimes W$) still denoted by σ.

If M is a $\mathbb{Z}$-graded $\mathbb{C}$-vector space, we denote by M_i its component of degree i. If $d \in \mathbb{Z}$, we denote by $M[d]$ the *shift* of M by d: in other words, $M[d] = M$ as a $\mathbb{C}$-vector space and $M[d]_i = M_{d+i}$. If M is finite-dimensional, we set

$$\mathrm{grdim}(M) = \sum_{i \in \mathbb{Z}} \dim(M_i)\, v^i \quad \in \mathbb{N}[v, v^{-1}].$$

If, moreover, θ is a graded endomorphism of M, we set

$$\mathrm{grTr}(\theta, M) = \sum_{i \in \mathbb{Z}} \mathrm{Tr}(\theta, M_i)\, v^i \quad \in \mathbb{C}[v, v^{-1}].$$

If θ is an automorphism of order dividing n, then

$$\mathrm{grTr}(\theta, M) \in \mathcal{O}[v, v^{-1}].$$

15.1.B. Graded Bimodules

Let $\mathcal{B}im$ denote the category of $\mathbb{Z}$-graded $(\mathbf{S}, \mathbf{S})$-bimodules. If $B \in \mathcal{B}im$ and $d \in \mathbb{Z}$, then the shift $B[d]$ of B (as a vector space) is also naturally an object in $\mathcal{B}im$, also denoted by $B[d]$. If B and B' are objects in $\mathcal{B}im$, then $B \otimes_{\mathbf{S}} B'$ is also an object in $\mathcal{B}im$, which is denoted by BB'. If $f \in \mathrm{Hom}_{\mathcal{B}im}(B, B'[i])$, $s \in \mathbf{S}_j$ and $s' \in \mathbf{S}_k$, then the map $sfs' : B \to B'$, $b \mapsto sf(b)s'$ belongs to $\mathrm{Hom}_{\mathcal{B}im}(B, B'[i+j+k])$. So, if we set

$$\mathbf{Hom}^{\mathbf{sh}}(B, B') = \bigoplus_{i \in \mathbb{Z}} \mathrm{Hom}_{\mathcal{B}im}(B, B'[i]),$$

then $\mathbf{Hom}^{\mathbf{sh}}(B, B')$ becomes an object in $\mathcal{B}im$. Finally, if $w \in W$, note that the vector subspace $\mathbf{S}w$ of $\mathbf{S} \rtimes W$ is also an object in $\mathcal{B}im$.

15.1.C. Soergel Bimodules

If $s \in S$, we set $\mathbf{S}^s = \{f \in \mathbf{S} \mid {}^s f = f\}$ and

$$\mathbf{B}_s = \mathbf{S} \otimes_{\mathbf{S}^s} \mathbf{S}[1].$$

It is an indecomposable object in the category $\mathcal{B}\!\mathit{im}$ (see [Soe2]). Now, if $w \in W$, then [Soe2, 6.16] there exists a unique indecomposable $\mathbb{Z}$-graded $(\mathbf{S},\mathbf{S})$-bimodule $\mathbf{B}_w$ such that, for any reduced expression $(s_1, s_2, \dots, s_r)$ of w, $\mathbf{B}_w$ is a direct summand of $\mathbf{B}_{s_1}\mathbf{B}_{s_2}\cdots\mathbf{B}_{s_r}$ and $\mathbf{B}_w$ is not a direct summand of some $\mathbf{B}_{t_1}\mathbf{B}_{t_2}\cdots\mathbf{B}_{t_m}$ with $t_j \in S$ and $m < r$. Note that $\mathbf{B}_1 = \mathbf{S}$.

Let $\mathcal{S}\!\mathit{oe}$ (resp. $\mathcal{S}\!\mathit{oe}_\heartsuit$) denote the full subcategory in $\mathcal{B}\!\mathit{im}$ whose objects are isomorphic to finite direct sums of objects of the form $\mathbf{B}_w[n]$, where $w \in W$ and $n \in \mathbb{Z}$ (resp. of the form $\mathbf{B}_w$, where $w \in W$). The objects of $\mathcal{S}\!\mathit{oe}$ are called *Soergel bimodules* while the objects of $\mathcal{S}\!\mathit{oe}_\heartsuit$ are called *perverse Soergel bimodules*. If B, $B' \in \mathcal{S}\!\mathit{oe}$, then $BB' \in \mathcal{S}\!\mathit{oe}$: this property does not hold in general for the subcategory $\mathcal{S}\!\mathit{oe}_\heartsuit$.

If $w \in W$ and $B \in \mathcal{S}\!\mathit{oe}$, then $\mathbf{Hom}^{\mathbf{sh}}(B, \mathbf{S}w)$ is a finitely generated free *right* $\mathbf{S}$-module [Soe2, 6.15], so $\mathbf{Hom}^{\mathbf{sh}}(B, \mathbf{S}w) \otimes_{\mathbf{S}} \mathbb{C}_0$ is a finite-dimensional graded vector space. We can then define

$$\operatorname{char}(B) = \sum_{w \in W} \overline{\operatorname{grdim}(\mathbf{Hom}^{\mathbf{sh}}(B, \mathbf{S}w) \otimes_{\mathbf{S}} \mathbb{C}_0)}\, v^{\ell(w)} T_w \in \mathcal{H}.$$

Recall from [Soe2, 1.10, 5.3] that

(15.1.1) $$\operatorname{char}(BB') = \operatorname{char}(B)\operatorname{char}(B')$$

for all B, B' in $\mathcal{S}\!\mathit{oe}$. If $d \in \mathbb{Z}$, we denote by $\mathrm{H}^d : \mathcal{S}\!\mathit{oe} \longrightarrow \mathcal{S}\!\mathit{oe}_\heartsuit$ the functor constructed by Elias and Williamson [ElWi, §6.2]. Then we have a non-canonical isomorphism

(15.1.2) $$B \simeq \bigoplus_{d \in \mathbb{Z}} \mathrm{H}^d(B)[-d].$$

If x, $y \in W$, then [ElWi]

(15.1.3) $$\operatorname{Hom}_{\mathcal{B}\!\mathit{im}}(\mathbf{B}_x, \mathbf{B}_y) = \begin{cases} \mathbb{C} & \text{if } x = y, \\ 0 & \text{if } x \neq y. \end{cases}$$

Theorem 15.1.4 (Elias–Williamson). *Let x, y, $z \in W$. Then*

$$\operatorname{char}(\mathbf{B}_z) = \mathbf{C}_z$$

and, consequently,

$$p_{y,z} = \overline{\operatorname{grdim}(\mathbf{Hom}^{\mathbf{sh}}(\mathbf{B}_z, \mathbf{S}y) \otimes_{\mathbf{S}} \mathbb{C}_0)}$$

and
$$h_{x,y,z} = \sum_{n \in \mathbb{Z}} \dim(\operatorname{Hom}_{\mathcal{B}\!\mathit{im}}(\mathbf{B}_z, \mathrm{H}^n(\mathbf{B}_x\mathbf{B}_y))\, v^n.$$

15.1.D. Action of σ

If M is an object in $\mathcal{B}im$, we denote by $\sigma^*(M)$ the object in $\mathcal{B}im$ whose underlying space is M but whose left (resp. right) action by $s \in \mathbf{S}$ is given by left (resp. right) multiplication by $\sigma(s)$. This induces an auto-equivalence of categories

$$\sigma^* : \mathcal{B}im \xrightarrow{\sim} \mathcal{B}im.$$

Of course,

(15.1.5) $$(\sigma^*)^n \textit{ is the identity functor}.$$

For instance, the map σ extends to an automorphism of $\mathbf{S} \rtimes W$ and

(15.1.6) $$\sigma^*(\mathbf{S}w) \simeq \mathbf{S}\sigma^{-1}(w),$$

canonically (the isomorphism is given by the map σ^{-1}). It is also clear that $\sigma^*(\mathbf{B}_s) \simeq \mathbf{B}_{\sigma^{-1}(s)}$ for all $s \in S$ so it follows from the characterization of the Soergel bimodule $\mathbf{B}_w$ that

(15.1.7) $$\sigma^*(\mathbf{B}_w) \simeq \mathbf{B}_{\sigma^{-1}(w)}.$$

However, it must be noticed that this isomorphism is *a priori* not canonical in general. But this shows that σ^* restricts to auto-equivalences of categories

$$\sigma^* : \mathcal{S}oe \xrightarrow{\sim} \mathcal{S}oe \qquad \text{and} \qquad \sigma^* : \mathcal{S}oe_\heartsuit \xrightarrow{\sim} \mathcal{S}oe_\heartsuit.$$

Following Lusztig (see [Lus18, §11] and [Lus20, §16.1]), we denote by $\mathcal{B}im^{(\sigma)}$, $\mathcal{S}oe^{(\sigma)}$ and $\mathcal{S}oe_\heartsuit^{(\sigma)}$ the categories whose objects are pairs (B,ϕ) where B is an object in $\mathcal{B}im$, $\mathcal{S}oe$ or $\mathcal{S}oe_\heartsuit$, respectively, and ϕ is an isomorphism $B \xrightarrow{\sim} \sigma^*(B)$ such that $(\sigma^*)^{n-1}(\phi)\cdots\sigma^*(\phi)\phi = \mathrm{Id}_B$. If (B,ϕ) and (B',ϕ') are two objects in $\mathcal{B}im$, then

$$\mathrm{Hom}_{\mathcal{B}im^{(\sigma)}}((B,\phi),(B',\phi')) = \{f \in \mathrm{Hom}_{\mathcal{B}im}(B,B') \mid \sigma^*(f) = \phi' f \phi^{-1}\}.$$

We set

$$(B,\phi)\oplus(B',\phi') = (B\oplus B',\phi\oplus\phi') \quad \text{and} \quad (B,\phi)\otimes(B',\phi') = (BB',\phi\otimes_{\mathbf{S}}\phi').$$

Note that these two operations preserve $\mathcal{S}oe^{(\sigma)}$. Of course, $(B,\xi\phi)$ is another object of the corresponding category. Also, the functor H^n commutes with σ^*.

Example 15.1.8. Let $w \in W^\sigma$. We denote by σ_w the restriction of the automorphism σ of $\mathbf{S} \rtimes W$ to $\mathbf{S}w$. Then it follows from 15.1.6 that $(\mathbf{S}w, \sigma_w^{-1})$ is an object in $\mathcal{B}im^{(\sigma)}$. □

We denote by $\boldsymbol{\sigma}_{\phi,\phi'}$ the automorphism of the graded vector space $\mathbf{Hom}^{\mathbf{sh}}(B,B')$ given by $f \mapsto \phi'^{-1}\sigma^*(f)\phi$. Note that

$$\text{(15.1.9)} \qquad \operatorname{Hom}_{\mathscr{B}im^{(\sigma)}}((B,\phi),(B',\phi')) = \mathbf{Hom}^{\mathbf{sh}}(B,B')^{\boldsymbol{\sigma}_{\phi,\phi'}}.$$

If $i \in \mathbb{Z}$, we denote by $\boldsymbol{\sigma}_{\phi,\phi',i}$ the degree i component of $\boldsymbol{\sigma}_{\phi,\phi'}$.

Now, let $x \in W^\sigma$. Since $\sigma^*(\mathbf{B}_x) \simeq \mathbf{B}_x$ by 15.1.7, it follows from 15.1.3 that we can choose an isomorphism $\phi : \mathbf{B}_x \xrightarrow{\sim} \sigma^*(\mathbf{B}_x)$ so that $(\mathbf{B}_x,\phi) \in \mathscr{S}\boldsymbol{oe}^{(\sigma)}$. Recall from Example 15.1.8 that $(\mathbf{S}x,\sigma_x^{-1})$ is an object in $\mathscr{B}\boldsymbol{im}^{(\sigma)}$. Then the automorphism $\boldsymbol{\sigma}_{\phi,\sigma_x^{-1},\ell(x)}$ has order n and the homogeneous component of $\mathbf{Hom}^{\mathbf{sh}}(\mathbf{B}_x,\mathbf{S}x) \otimes_{\mathbf{S}} \mathbb{C}_0$ of degree $\ell(x)$ is one-dimensional [Soe2]. By multiplication by a power of ξ, we can choose ϕ so that $\boldsymbol{\sigma}_{\phi,\sigma_x^{-1},\ell(x)}$ is the identity on this one-dimensional vector space: we denote by ϕ_x this normalized isomorphism $\phi_x : \mathbf{B}_x \xrightarrow{\sim} \sigma^*(\mathbf{B}_x)$. Since $\operatorname{Hom}_{\mathscr{B}im}(\mathbf{B}_x,\mathbf{B}_x) = \mathbb{C}$ by 15.1.3, this determines ϕ_x uniquely and

$$\text{(15.1.10)} \qquad (\mathbf{B}_x,\phi_x) \in \mathscr{S}\boldsymbol{oe}_{\heartsuit}^{(\sigma)}.$$

If $(B,\phi) \in \mathscr{S}\boldsymbol{oe}^{(\sigma)}$, we set

$$\operatorname{char}^\sigma(B,\phi) = \sum_{w \in W^\sigma} \overline{\operatorname{grTr}(\boldsymbol{\sigma}_{\phi,\sigma_w^{-1}}, \mathbf{Hom}^{\mathbf{sh}}(B,\mathbf{S}w) \otimes_{\mathbf{S}} \mathbb{C}_0)}\, v^{\ell(w)} T_w^\sigma \in \mathscr{H}_\sigma.$$

Then [Lus20, §16.7]

$$\text{(15.1.11)} \qquad \operatorname{char}^\sigma((B,\phi)\otimes(B',\phi')) = \operatorname{char}^\sigma(B,\phi)\operatorname{char}^\sigma(B',\phi').$$

Now, let x, y, $z \in W^\sigma$ and $d \in \mathbb{Z}$. We denote by $\phi_{x,y,z}^{(d)}$ the automorphism of the graded vector space $\operatorname{Hom}_{\mathscr{B}im}(\mathbf{B}_z, \mathrm{H}^d(\mathbf{B}_x\mathbf{B}_y))$ such that

$$\phi_{x,y,z}^{(d)}(f) = \mathrm{H}^d(\phi_x \otimes_{\mathbf{S}} \phi_y)^{-1}\sigma^*(f)\phi_z$$

for all $f \in \operatorname{Hom}_{\mathscr{B}im}(\mathbf{B}_z, \mathrm{H}^d(\mathbf{B}_x\mathbf{B}_y))$. Then:

Theorem 15.1.12 (Lusztig). *If x, y, $z \in W^\sigma$, then*

$$\operatorname{char}^\sigma(B_z,\phi_z) = \mathbf{C}_z^\sigma$$

and, consequently,

$$p_{y,z}^\sigma = \overline{\operatorname{grTr}(\boldsymbol{\sigma}_{\phi_z,\sigma_y^{-1}}, \mathbf{Hom}^{\mathbf{sh}}(\mathbf{B}_z,\mathbf{S}y) \otimes_{\mathbf{S}} \mathbb{C}_0)}$$

and
$$h_{x,y,z}^\sigma = \sum_{n\in\mathbb{Z}} \operatorname{grTr}(\phi_{x,y,z}^{(n)}, \operatorname{Hom}_{\mathscr{B}im}(\mathbf{B}_z, \mathrm{H}^n(\mathbf{B}_x\mathbf{B}_y)))v^n.$$

***Remark* 15.1.13.** One might think that Elias–Williamson's Theorem 15.1.4 is a particular case of Lusztig's Theorem 15.1.12 (take $\sigma = \mathrm{Id}_W$). But it must be said that the proof of Lusztig's Theorem 15.1.12 relies on Elias–Williamson's Theorem 15.1.4. □

15.2. Lusztig's Conjectures: the Split Case

We aim to prove that Lusztig's Conjectures (Pk) holds for the triple (W,S,ℓ) and for all $1 \leqslant k \leqslant 15$. For this, we almost only use the (crucial!) fact that Theorem 15.1.4 implies that

$$p_{x,y} \in \mathbb{N}[v^{-1}] \qquad \text{and} \qquad h_{x,y,z} \in \mathbb{N}[v,v^{-1}] \tag{15.2.1}$$

for all $x, y, z \in W$. This positivity property together with Proposition 5.6.4(c) implies that

$$h_{s,x,y} = \lambda^s_{y,x} \in \mathbb{N} \qquad (\text{resp. } h_{x,s,y} = \rho^s_{y,x} \in \mathbb{N}) \tag{15.2.2}$$

if $s \in S$ is such that $sx > x$ (resp. $xs > x$). First of all, we need to show that Lusztig's $\mathbf{a}$-function is well-defined.

Proposition 15.2.3. *If $x, y, z \in W$, then* $\deg(h_{x,y,z}) \leqslant \boldsymbol{\Delta}(z)$.

Proof. Let $x, y \in W$. Then, by 14.2.11,

$$\sum_{z \in W} h_{x,y,z}\, p_{1,z} \equiv \delta_{xy=1} \mod A_{<0}.$$

But, by 15.2.1, $h_{x,y,z}$ and $p_{y,z}$ have coefficients in $\mathbb{N}$. This forces the inequality $\deg(h_{x,y,z}\, p_{1,z}) \leqslant 0$ for all $z \in W$. The result follows. ■

Corollary 15.2.4. *Conjecture 14.2.1 holds for (W,S,ℓ) (so Lusztig's $\mathbf{a}$-function is well-defined) and $\mathbf{a}(z) \leqslant \boldsymbol{\Delta}(z)$ for all $z \in W$.*

In other words, this proves (P1) for the triple (W,S,ℓ). Thus, the set $\mathscr{D} = \{z \in W \mid \mathbf{a}(z) = \boldsymbol{\Delta}(z)\}$ is well-defined and we are now ready to prove the following theorem:

Theorem 15.2.5. *If $1 \leqslant k \leqslant 6$, then Lusztig's Conjecture* (Pk) *holds for (W,S,ℓ). If $7 \leqslant k \leqslant 15$ and if, moreover, $\mathbf{a}$ is bounded, then Lusztig's Conjecture* (Pk) *holds for (W,S,ℓ).*

Proof. Since (P1) holds by Corollary 15.2.4, it follows from Lemma 14.3.2(a) that

$$\sum_{d \in \mathscr{D}} \gamma_{x,y,d}\, n_d = \delta_{xy=1} \tag{$*$}$$

for all $x, y \in W$. Since $\gamma_{x,y,z} \geqslant 0$ and $n_z > 0$ by 15.2.1, we deduce from $(*)$ that $\gamma_{x,y,d} = 0$ if $d \in \mathscr{D}$ and $xy \neq 1$. This proves (P2). On the other hand, $(*)$ applied with $x = y^{-1}$ yields that there exists a unique $d \in \mathscr{D}$ such that $\gamma_{y^{-1},y,d} \neq 0$ and, in this case, $\gamma_{y^{-1},y,d}\, n_d = 1$. This shows (P3) (and (P5)).

Lemma 15.2.6. *The statement* (P$^+$) *holds for* (W, S, ℓ).

Proof. Let $x, y, z, z' \in W$ be such that $\gamma_{x,y,z^{-1}} \neq 0$ and $z' \overset{L}{\longleftarrow} z$. Then there exists an $s \in S$ such that $\lambda^s_{z',z} \neq 0$. If $sz < z$, this forces $z' = z$ by Corollary 5.5.6(a) and there is nothing to prove. So we may assume that $sz > z$. Note that $sz' < z$ by Corollary 5.5.6(b). Also, since $h_{x,y,z} \neq 0$, we have $z \leqslant_R x$ and so $\mathscr{L}(x) \subset \mathscr{L}(z)$ by Corollary 6.3.2(a′). In particular, $sx > x$. Hence, if we compute in two ways the coefficient of $\mathbf{C}_{z'}$ in $\mathbf{C}_s \mathbf{C}_x \mathbf{C}_y$, we get

$$\sum_{x' \in W} h_{s,x,x'} h_{x',y,z'} = \sum_{x' \in W} h_{s,x',z'} h_{x,y,x'}.$$

Therefore,

$$\sum_{x' \in W} h_{s,x,x'} h_{x',y,z'} = h_{s,z,z'} h_{x,y,z} + \sum_{\substack{x' \in W \\ x' \neq z}} h_{s,x',z'} h_{x,y,x'}.$$

Since $h_{s,z,z'} \in \mathbb{N}^*$ by 15.2.2, this shows that $\mathrm{coef}_{\mathbf{a}(z)}(h_{s,z,z'} h_{x,y,z}) > 0$. By 15.2.1, this implies that there exists an $x' \in W$ such that

$$\mathrm{coef}_{\mathbf{a}(z)}(h_{s,x,x'} h_{x',y,z'}) > 0.$$

But again $h_{s,x,x'} \in \mathbb{Z}$ by 15.2.2. So $\mathrm{coef}_{\mathbf{a}(z)}(h_{x',y,z'}) \neq 0$, as desired. ∎

Let us come back to the proof of Theorem 15.2.5. Since (P$^+$), (P1), (P2) and (P3) hold for (W, S, ℓ), this shows that (Pk) holds for $1 \leqslant k \leqslant 6$ (see Proposition 14.3.3(a), (b) and (c)). Also, note that, since $\mathscr{A} = \mathbb{Z}$, the fact that $\mathbf{a}$ is bounded is equivalent to the fact that $\mathbf{a}$ takes only finitely many values. So, assume now that $\mathbf{a}$ is bounded. Then $\mathbf{a}_I$ is also bounded for all $I \subset S$, because *a priori* $\mathbf{a}_I(z) \leqslant \mathbf{a}(z)$ for all $z \in W_I$. It then follows from Proposition 14.3.3(d), (e),…, (k) that (Pk) holds for $7 \leqslant k \leqslant 14$.

It remains to prove that (P15) holds. By Proposition 14.3.4, this is equivalent to proving that, if $s, s' \in S$ and $y, y' \in W$ are such that $\mathbf{a}(y) = \mathbf{a}(y')$, $sy > y$, $ys' > y$, $sy' < y'$ and $y's' < y'$, then we have the following equality

$$\sum_{\substack{z \in W \\ sz < z < zs'}} \lambda^s_{z,y} \otimes_{\mathbb{Z}} \rho^{s'}_{y',z} = \sum_{\substack{z \in W \\ zs' < z < sz}} \lambda^s_{y',z} \otimes_{\mathbb{Z}} \rho^{s'}_{z,y}$$

in $A \otimes_{\mathbb{Z}} A$. By 15.2.2, all the terms occurring in this equality belong to $\mathbb{Z}$, so this amounts to proving that

$$\sum_{\substack{z\in W\\ sz<z<zs'}} \lambda^s_{z,y}\rho^{s'}_{y',z} = \sum_{\substack{z\in W\\ zs'<z<sz}} \lambda^s_{y',z}\rho^{s'}_{z,y}.$$

But this follows from Remark 14.3.7. ∎

15.3. Lusztig's Conjectures: the Quasi-Split Case

We aim to prove here that Lusztig's Conjectures (Pk) holds for the triple $(W^\sigma, S_\sigma, \varphi_\sigma)$ and for all $1 \leqslant k \leqslant 14$[1], where we recall that φ_σ is the restriction to W^σ of the length function $\ell : W \to \mathbb{Z}$. This relies on the fact that these conjectures hold for the triple (W,S,ℓ) and the following lemma, which follows from Theorems 15.1.4 and 15.1.12.

Lemma 15.3.1. *Let $x, y, z \in W^\sigma$ and let $n \in \mathbb{Z}$.*

(a) *If* $\mathrm{coef}_n(p^\sigma_{x,y}) \neq 0$ *(resp.* $\mathrm{coef}_n(h^\sigma_{x,y,z}) \neq 0$*), then* $\mathrm{coef}_n(p_{x,y}) \neq 0$ *(resp.* $\mathrm{coef}_n(h_{x,y,z}) \neq 0$*).*

(b) *If* $\mathrm{coef}_n(p_{x,y}) = 1$ *(resp.* $\mathrm{coef}_n(h_{x,y,z}) = 1$*), then* $\mathrm{coef}_n(p^\sigma_{x,y}) = \pm 1$ *(resp.* $\mathrm{coef}_n(h^\sigma_{x,y,z}) \pm 1$*).*

Proof. Theorems 15.1.4 and 15.1.12 imply that $\mathrm{coef}_n(p^\sigma_{x,y})$ (resp. $\mathrm{coef}_n(h^\sigma_{x,y,z})$) is obtained as the trace of an automorphism of finite order of a vector space of dimension $\mathrm{coef}_n(p_{x,y})$ (resp. $\mathrm{coef}_n(h_{x,y,z})$). So (a) follows immediately. If, moreover, $\mathrm{coef}_n(p_{x,y}) = 1$ (resp. $\mathrm{coef}_n(h_{x,y,z}) = 1$), then $\mathrm{coef}_n(p^\sigma_{x,y})$ (resp. $\mathrm{coef}_n(h^\sigma_{x,y,z})$) is a root of unity, which moreover belongs to $\mathbb{Z}$. So (b) follows. ∎

Again, we first need to show that Lusztig's **a**-function for the triple $(W^\sigma, S_\sigma, \varphi_\sigma)$ is well-defined:

Proposition 15.3.2. *If $x, y, z \in W^\sigma$, then* $\deg(h^\sigma_{x,y,z}) \leqslant \mathbf{\Delta}(z)$.

Proof. This follows from Lemma 15.3.1(a) and Proposition 15.2.3. ∎

As a consequence, we can define Lusztig's **a**-function $\mathbf{a}_\sigma : W^\sigma \to \mathbb{Z}$ by

$$\mathbf{a}_\sigma(z) = \max_{x,y\in W^\sigma} \deg(h^\sigma_{x,y,z})$$

for all $z \in W^\sigma$. This also allows us to define the set

$$\mathscr{D}_\sigma = \{z \in W^\sigma \mid \mathbf{a}_\sigma(z) = \mathbf{\Delta}_\sigma(z)\}.$$

We first compare the different "Kazhdan–Lusztig objects" associated with W and W^σ (cells, **a**-function, Duflo involutions,...).

[1] (P15) is stated without proof in [Lus20, §16.24].

Proposition 15.3.3. *Let $x, y, z \in W^\sigma$. Then:*

(a) $\mathbf{a}_\sigma(z) = \mathbf{a}(z)$ *and* $\boldsymbol{\Delta}(z) \leqslant \boldsymbol{\Delta}_\sigma(z)$. *Hence,* $\mathbf{a}_\sigma(z) \leqslant \boldsymbol{\Delta}_\sigma(z)$.
(b) $\gamma_{x,y,z} \neq 0$ *if and only if* $\gamma^\sigma_{x,y,z} \neq 0$.
(c) *If* $\gamma_{x,y,z} = 1$, *then* $\gamma^\sigma_{x,y,z} = \pm 1$.
(d) $\mathscr{D}_\sigma = \mathscr{D} \cap W^\sigma$, *i.e.* $\mathscr{D}_\sigma = \mathscr{D}^\sigma$.
(e) *If* $x \leqslant^\sigma_? y$, *then* $x \leqslant_? y$ *(for* $? \in \{L, R, LR\}$*).*
(f) *If* $x \sim^\sigma_{LR} y$, *then* $x \sim_{LR} y$.
(g) $x \sim^\sigma_? y$ *if and only if* $x \sim_? y$ *(for* $? \in \{L, R\}$*).*

Proof. (a) It follows from Lemma 15.3.1(a) that $\deg(h^\sigma_{x,y,z}) \leqslant \deg(h_{x,y,z})$ and $\deg(p^\sigma_{1,z}) \leqslant \deg(p_{1,z})$. This shows that $\mathbf{a}_\sigma(z) \leqslant \mathbf{a}(z) \leqslant \boldsymbol{\Delta}(z) \leqslant \boldsymbol{\Delta}_\sigma(z)$ (the second inequality follows from the property (P1) for (W, S, ℓ)).

Now, by (P3) and (P5) for (W, S, ℓ), there exists a unique $d \in \mathscr{D}$ such that $\gamma_{z^{-1},z,d} = 1$ (see also 15.2.1). By the uniqueness of d, we have that $d \in W^\sigma$. Now, by (P7) for (W, S, ℓ), we get that $\gamma_{z,d,z^{-1}} = 1$ and it then follows from Lemma 15.3.1(b) that $\mathrm{coef}_{\mathbf{a}(z)}(h^\sigma_{z,d,z}) = \pm 1 \neq 0$. So $\mathbf{a}_\sigma(z) \geqslant \mathbf{a}(z)$ and (a) follows.

Statements (b) and (c) follow from (a) and Lemma 15.3.1.

(d) If $d \in \mathscr{D}_\sigma$, it follows from (a) that necessarily $\mathbf{a}(d) = \boldsymbol{\Delta}(d)$, so $d \in \mathscr{D} \cap W^\sigma$. Conversely, let $d \in \mathscr{D} \cap W^\sigma$. Then $n_d = 1$ and so $\mathrm{coef}_{-\boldsymbol{\Delta}(d)}(p^\sigma_{1,d}) = \pm 1$ by Lemma 15.3.1(b). This shows that $\boldsymbol{\Delta}_\sigma(d) \leqslant \boldsymbol{\Delta}(d)$, and so it follows from (a) that $\boldsymbol{\Delta}_\sigma(d) = \boldsymbol{\Delta}(d) = \mathbf{a}(d) = \mathbf{a}_\sigma(d)$, so $d \in \mathscr{D}_\sigma$.

Statement (e) follows immediately from Lemma 15.3.1(a).

Statement (f) follows from (e).

(g) We only need to prove the result in the case where $? = L$. It follows from (e) that if $x \sim^\sigma_L y$, then $x \sim_L y$. Conversely, assume that $x \sim_L y$. Let C denote the left cell of x and y (it is σ-stable). By (P5) and (P13) for (W, S, ℓ), there exists a unique $d \in C \cap \mathscr{D}$ and it satisfies $\gamma_{x^{-1},x,d} = \gamma_{y^{-1},y,d} = 1$ (see also 15.2.1). By the uniqueness property, $d \in \mathscr{D} \cap W^\sigma = \mathscr{D}_\sigma$. By (P7) for (W, S, ℓ), we get that $\gamma_{y,d,y^{-1}} = \gamma_{x,d,x^{-1}} = 1$ and so this implies that

$$\gamma^\sigma_{x^{-1},x,d}, \gamma^\sigma_{y^{-1},y,d}, \gamma^\sigma_{y,d,y^{-1}}, \gamma^\sigma_{x,d,x^{-1}} \in \{1, -1\}$$

by Lemma 15.3.1(b). Since $d = d^{-1}$ (by (P6) for (W, S, ℓ)), we get that $d \leqslant^\sigma_L x \leqslant^\sigma_L d$ and $d \leqslant^\sigma_L y \leqslant^\sigma_L d$, so $x \sim^\sigma_L d \sim^\sigma_L y$, as desired. ■

We are now ready to prove the following theorem:

Theorem 15.3.4. *If* $1 \leqslant k \leqslant 6$, *then Lusztig's Conjecture* (Pk) *holds for* $(W^\sigma, S_\sigma, \varphi_\sigma)$. *If* $7 \leqslant k \leqslant 14$ *and if, moreover,* $\mathbf{a}$ *is bounded, then Lusztig's Conjecture* (Pk) *holds for* $(W^\sigma, S_\sigma, \varphi_\sigma)$.

Proof. (P1) has been proved in Proposition 15.3.3(a).

(P2) Let x, $y \in W^\sigma$ and $d \in \mathcal{D}_\sigma$ such that $\gamma^\sigma_{x,y,d} \neq 0$. Then $\gamma_{x,y,d} \neq 0$ by Lemma 15.3.1(a) and $d \in \mathcal{D}$ by Proposition 15.3.3(d). So $x = y^{-1}$ since (P2) holds for (W,S,ℓ).

(P3) Let $y \in W^\sigma$. Then, by (P3) and (P5) for (W,S,ℓ), there exists a unique $d \in \mathcal{D}$ such that $\gamma_{y^{-1},y,d} = 1$. By the uniqueness of d, we have that $\sigma(d) = d$, so $d \in \mathcal{D}_\sigma$ by Proposition 15.3.3(d). Then it follows from Lemma 15.3.1(b) that $\gamma^\sigma_{y^{-1},y,d} = \pm 1$. This shows the existence statement in (P3). The uniqueness statement follows from Lemma 15.3.1(a), Proposition 15.3.3(d) and the uniqueness statement for (P3) for (W,S,ℓ).

(P4) Let x, $y \in W^\sigma$ be such that $x \leqslant^\sigma_{LR} y$. Then $x \leqslant_{LR} y$ by Proposition 15.3.3(e) and so $\mathbf{a}(x) \geqslant \mathbf{a}(y)$ by (P4) for (W,S,ℓ). Since $\mathbf{a}(x) = \mathbf{a}_\sigma(x)$ and $\mathbf{a}(y) = \mathbf{a}_\sigma(y)$ by Proposition 15.3.3(a), (P4) follows.

(P5) and (P6) follow from Proposition 14.3.3(c).

Assume now, moreover, that $\mathbf{a}$ is bounded. Then $\mathbf{a}_\sigma$ is also bounded by Proposition 15.3.3(a), so it takes only finitely many values because $\mathscr{A} = \mathbb{Z}$. Then (P7) follows from Proposition 14.3.3(d).

(P8) follows from Proposition 14.3.3(e).

(P9) follows from (P9) for (W,S,ℓ) and from Proposition 15.3.3(e) and (g).

(P10), (P11) and (P12) now follow from Proposition 14.3.3(g), (h) and (i), respectively.

(P13) follows from (P13) for (W,S,ℓ) and Proposition 15.3.3(c), (d) and (g).

(P14) follows from Proposition 14.3.3(k). ■

Notes

The theory of Soergel bimodules is due to Soergel (see [Soe1] and [Soe2]) and was intended to provide a strategy for proving the positivity conjecture of Kazhdan–Lusztig polynomials and structure constants. This aim was achieved in 2014 by Elias and Williamson [ElWi]. This allows us in particular to deduce Lusztig's Conjectures (Pk) in the split case following Lusztig's argument given in 2002 in [Lus20, §15]. Soon after the publication of [ElWi], Lusztig posted an updated version of [Lus20], where he used the Elias–Williamson categorification result to obtain a proof of Lusztig's Conjectures (Pk) in the quasi-split case (except for (P15) which is stated without proof).

Part VI
Applications of Lusztig's Conjectures

The goal of this part is to explain some of the consequences of Lusztig's Conjectures on the theory of Kazhdan–Lusztig cells: most of the results are valid under the condition that most of Lusztig's Conjectures hold. In particular, this implies that they hold in the split and the quasi-split case (see Chapter 15), so that their scope of validity is not empty.

Recall from the hypotheses of Part V that S is finite and that $\varphi > 0$. We denote by $(\mathscr{P})$ the following property:

$(\mathscr{P})$ for all $I \in \mathscr{P}(S)$, Conjecture 14.2.1 holds for the triple (W_I, I, φ_I), so that Lusztig's **a**-function $\mathbf{a}_I : W_I \to \mathscr{A}$, the numbers $\gamma^I_{x,y,z} \in \mathbb{Z}$ (which we define as the analogues of $\gamma^I_{x,y,z}$ but for W_I) and the set $\mathscr{D}_I = \{z \in W_I \mid \mathbf{\Delta}(z) = \mathbf{a}_I(z)\}$ are well-defined. Moreover, Lusztig's Conjecture (Pk) holds for the triple (W_I, I, φ_I) for all $1 \leqslant k \leqslant 14$.

By Proposition 14.3.3, note that the last assertion is equivalent to saying that, for all $I \in \mathscr{P}(S)$, Lusztig's Conjectures (P1), (P2), (P3), (P4), (P9) and (P13) hold for (W_I, I, φ_I). We will also use two variants of $(\mathscr{P})$. The first one, denoted by $(\mathscr{P}_{\mathbf{sph}})$, is weaker:

$(\mathscr{P}_{\mathbf{sph}})$ For all $I \in \mathscr{P}_{\mathrm{sph}}(S)$, we assume that Lusztig's Conjecture (Pk) holds for the triple (W_I, I, φ_I) for all $1 \leqslant k \leqslant 14$.

The second one, involving the action of a group G, is much stronger. It is used in Chapter 20 and is explained there.

> **Hypothesis and notation.** *Recall that S is finite and that $\varphi > 0$. Throughout this part (except in Chapter 18), we assume that $(\mathscr{P})$ holds. In Chapter 18, we only assume that $(\mathscr{P}_{\mathbf{sph}})$ holds.*

Note that we do not assume that Lusztig's Conjecture (P15) holds (it is often the most difficult to obtain): it is needed for only one (important) result in Chapter 19 (see Theorem 19.2.5). Under these hypotheses, we treat in this part the following themes:

• Action of the left (or right) multiplication by T_{w_0} inside a two-sided cell (see Chapter 17): this is then applied to define an action of the cactus group of W which permutes the cells (see Chapter 18).

• Construction and some properties of the *asymptotic algebra*: see Chapter 19. This is the only chapter where we need Lusztig's Conjecture (P15).

• Comparison of Kazhdan–Lusztig cells of W and W^G, where G is a finite solvable group of automorphisms of W stabilizing S: see Chapter 20.

⚠ *Warning*. As most, but not all, of the results of this part are valid only if Lusztig's Conjectures (Pk) hold for $1 \leqslant k \leqslant 14$, we use a way to emphasize which results depend on these Conjectures. We will use the symbol $(\mathscr{P})$

(after the number of the statement) to say that this statement is valid if the above hypothesis ($\mathscr{P}$) holds. For instance, we prove in Chapter 16 the following result:

Proposition 16.2.3 ($\mathscr{P}$). *Let I be a subset of S. Then:*

(a) $\mathscr{D}_I = \mathscr{D} \cap W_I$.
(b) *If* $w \in W_I$, *then* $\boldsymbol{\partial}_w^L = \boldsymbol{\partial}_{I,w}^L$ *and* $\boldsymbol{\partial}_w^R = \boldsymbol{\partial}_{R,w}^L$.

We use the same convention for exercises. Similarly, we use the symbol ($\mathscr{P}_{\mathbf{sph}}$) (after the number of a statement), to indicate that we do not need that ($\mathscr{P}$) holds but we need that ($\mathscr{P}_{\mathbf{sph}}$) holds. For instance, we prove the following result in Chapter 18:

Theorem 18.1.1 ($\mathscr{P}_{\mathrm{sph}}$). *If $I \in \mathscr{P}_{\mathbf{sph}}(S)$, then the pair (λ_I^L, η_L^I) (resp. (ρ_I^L, η_L^I)) is strongly left (resp. right) cellular.*

Chapter 16
Miscellanea

If $\mathscr{E}$ is a subset of a two-sided cell Γ, we denote by $\mathbf{a}(\mathscr{E})$ the value $\mathbf{a}(x)$, where x is some (or any) element of $\mathscr{E}$ (see (P4)).

16.1. About Cells

Lemma 16.1.1 ($\mathscr{P}$). *We have:*

(a) *If $x, y \in W$ are such that $x \sim_{LR} y$ and $x \leqslant_L y$ (resp. $x \leqslant_R y$), then $x \sim_L y$ (resp. $x \sim_R y$).*
(b) *The relation $\sim_{LR}$ is generated by $\sim_L$ and $\sim_R$.*

Proof. (a) Let x, $y \in W$ be such that $x \sim_{LR} y$ and $x \leqslant_L y$. Then $\mathbf{a}(x) = \mathbf{a}(y)$ by (P4) and so $x \sim_L y$ by (P9).

(b) Let x and y be two elements of W such that $x \sim_{LR} y$. In particular, $x \leqslant_{LR} y$ so there exist $x = x_0, x_1, \dots, x_n = y$ in W such that $x_{i-1} \leqslant_L x_i$ or $x_{i-1} \leqslant_R x_i$ for all $i \in \{1,2,\dots,n\}$. It follows from (P4) that $\mathbf{a}(x) = \mathbf{a}(x_0) \geqslant \mathbf{a}(x_1) \geqslant \cdots \geqslant \mathbf{a}(x_n) = \mathbf{a}(y)$. Since $x \sim_{LR} y$, we have $\mathbf{a}(x) = \mathbf{a}(y)$ by (P4). So $\mathbf{a}(x) = \mathbf{a}(x_0) = \mathbf{a}(x_1) = \cdots = \mathbf{a}(x_n) = \mathbf{a}(y)$. Hence it follows from (P9) and (P10) that $x_{i-1} \sim_L x_i$ or $x_{i-1} \sim_R x_i$. This shows (b). ■

Lemma 16.1.2 ($\mathscr{P}$). *Let I be a subset of S and let C_I be a left cell of W_I. Let C be a left cell of W contained in $X_I \cdot C_I$ and we denote by C_I^W the left cell of W containing C_I. Then:*

(a) $C \leqslant_L C_I^W$.

C. Bonnafé, *Kazhdan–Lusztig Cells with Unequal Parameters*,
Algebra and Applications 24, https://doi.org/10.1007/978-3-319-70736-5_16

(b) $C_I^W = \{w \in X_I \cdot C_I \mid \mathbf{a}(w) = \mathbf{a}_I(C_I)\}$.
(c) *If* $C \neq C_I^W$, *then* $\mathbf{a}(C) > \mathbf{a}(C_I^W) = \mathbf{a}_I(C_I)$.

Proof. (a) Let $w \in X_I \cdot C_I$. Write $w = xw'$, with $x \in X_I$ and $w' \in C_I$. Since $\ell(w) = \ell(x) + \ell(w')$, it follows from Example 6.1.2 that $w \leqslant_L w'$.

(b) and (c) follow from (a) and (P9). ■

The following result gives a positive answer to Question 6.1.8, if one assumes Lusztig's Conjectures:

Proposition 16.1.3 ($\mathscr{P}$). *Assume that* $\mathbf{a}$ *takes only finitely many values. Let* C *be a left (resp. right, resp. two-sided) convex subset of* W. *Then* C *contains a maximal and a minimal left (resp. right, resp. two-sided) cell.*

Proof. We just prove the result when C is a left convex subset, the other cases being similar. Let $\mathscr{E} = \mathbf{a}(C) \subset \mathscr{A}$. By hypothesis, $\mathscr{E}$ is finite. If $a \in \mathscr{E}$, we set

$$C[a] = \{w \in C \mid \mathbf{a}(w) = a\}.$$

Let a_m (resp. a_M) be the minimum (resp. maximum) of $\mathscr{E}$ and let C_m (resp. C_M) be a left cell contained in $C[a_m]$ (resp. $C[a_M]$). Then C_m is a maximal left cell and C_M is a minimal left cell by (P4) and (P9). ■

Remark 16.1.4 **($\mathscr{P}$).** This Proposition 16.1.3 applies, for instance, to the case where $C = W$: it says that there exists at least one minimal left, right or two-sided cell. □

16.2. About Duflo Involutions

By (P13), we can define a map $\partial^L : W \longrightarrow \mathscr{D}$, $x \mapsto \partial_x^L$ which is characterized by the property that

$$\partial_x^L \sim_L x$$

for all $x \in W$. Using Corollary 6.1.6, we can similarly define a map $\partial^R : W \longrightarrow \mathscr{D}$, $x \mapsto \partial_x^R$ which is characterized by the property that

$$\partial_x^R \sim_R x$$

for all $x \in W$.

Proposition 16.2.1 ($\mathscr{P}$). *The fibers of* ∂^L *(resp.* ∂^R*) are the left (resp. right) cells of* W. *Moreover:*

(a) *If* $x \in W$, *then* $\partial_x^R = \partial_{x^{-1}}^L$.
(b) *If* $d \in \mathscr{D}$, *then* $\partial_d^L = \partial_d^R = d$ *and* $\gamma_{d,d,d} = n_d = \pm 1$.

Proof. The first statement follows from (P13). (a) follows from (P6). The first assertion of (b) is clear and the second one follows from (P5) and (P13). ■

If I is a subset of S, we denote by $\mathscr{D}_I$ the set of Duflo involutions of W_I and by $\partial_I^L : W_I \to \mathscr{D}_I$ and $\partial_I^R : W_I \to \mathscr{D}_I$ the respective analogues of ∂^L and ∂^R for W_I. Since $\mathbf{a}_I(z) = \mathbf{a}(z)$ by (P12), we have:

Lemma 16.2.2 ($\mathscr{P}$). *If $x, y, z \in W_I$, then $\gamma^I_{x,y,z} = \gamma_{x,y,z}$.*

Proposition 16.2.3 ($\mathscr{P}$). *Let I be a subset of S. Then:*

(a) $\mathscr{D}_I = \mathscr{D} \cap W_I$.
(b) *If $w \in W_I$, then $\partial_w^L = \partial_{I,w}^L$ and $\partial_w^R = \partial_{I,w}^R$.*

Proof. (a) follows from the fact that $\mathbf{a}_I$ is the restriction of $\mathbf{a}$ to W_I by (P12). (b) follows from (a) and (P13). ■

We conclude this section with a result that will be used in Chapter 18. Let Γ be a two-sided cell of W. We denote by

$$\theta_\Gamma : \mathscr{H}^{\leqslant_{LR}\Gamma} \longrightarrow A$$

the unique A-linear map such that

$$\theta_\Gamma(\mathbf{C}_x) = \delta_{x \in \mathscr{D} \cap \Gamma}$$

for all $x \leqslant_{LR} \Gamma$. Since θ_Γ vanishes on $\mathscr{H}^{<_{LR}\Gamma}$, θ_Γ may be viewed as an A-linear map

$$\theta_\Gamma : \mathscr{H}^{LR}[\Gamma] \longrightarrow A.$$

Lemma 16.2.4 ($\mathscr{P}$). *Let $x, y \in \Gamma$. Then*

$$v^{-\mathbf{a}(\Gamma)} \theta_\Gamma(\mathbf{C}_{x^{-1}} \mathbf{C}_y) \equiv n_{\partial_x^L} \delta_{x=y} \mod A_{<0}.$$

Proof. By definition,

$$v^{-\mathbf{a}(\Gamma)} \theta_\Gamma(\mathbf{C}_{x^{-1}} \mathbf{C}_y) = v^{-\mathbf{a}(\Gamma)} \sum_{d \in \mathscr{D} \cap \Gamma} h_{x^{-1},y,d} = \sum_{d \in \mathscr{D} \cap \Gamma} v^{-\mathbf{a}(d)} h_{x^{-1},y,d}.$$

So the result follows from (P2) and (P13). ■

16.3. Strongness of Cellular Pairs

Proposition 16.3.1 ($\mathscr{P}$). *A left (resp. right) cellular pair is strongly left (resp. right) cellular.*

Proof. Let (σ,μ) be a left cellular pair for W. Let $x \in W$ and let $d = \partial_x^R$. Then d satisfies $\gamma_{x,x^{-1},d} \neq 0$ and $d \sim_R x$ by (P13). So $\gamma_{d,x,x^{-1}} \neq 0$ by (P7). This proves that $h_{d,x,x} \neq 0$. Now, since (σ,μ) is left cellular, we have $h_{d,x,x} = h_{d,\sigma(x),\sigma(x)}$, so $h_{d,\sigma(x),\sigma(x)} \neq 0$. This proves that $\sigma(x) \leqslant_R d$ and so $\sigma(x) \leqslant_R x$.

Similarly, let $d' = \partial_{\sigma(x)}^R$. Then d' satisfies $\gamma_{\sigma(x),\sigma(x)^{-1},d'} \neq 0$ and $d' \sim_R x$ by (P13). So $\gamma_{d',\sigma(x),\sigma(x)^{-1}} \neq 0$ by (P7). This proves that $h_{d',\sigma(x),\sigma(x)} \neq 0$ and so $h_{d',x,x} \neq 0$ because (σ,μ) is left cellular. So $x \leqslant_R d' \sim_R \sigma(x)$, and $x \sim_R \sigma(x)$, as desired. ∎

Corollary 16.3.2 ($\mathscr{P}$). *If (σ,μ) is a left (resp. right) cellular pair, then $\sigma(\Gamma) \subset \Gamma$ for all two-sided cells Γ.*

Exercises

Exercise 16.1 ($\mathscr{P}$). Let I be a subset of S such that W_I is finite and $\varphi(w_I) = \max_{J \in \mathscr{P}_{\mathrm{sph}}(S)} \varphi(w_J)$. Assume here that Conjecture 14.1.5 holds. Show that $X_I w_I$ is a (minimal) left cell.

Exercise 16.2 ($\mathscr{P}$). Let $I \in \mathscr{P}_{\mathrm{sph}}(S)$. Show that $\mathbf{a}(w_I) = \varphi(w_I)$ and that $w_I \in \mathscr{D}$.

Chapter 17

Multiplication by T_{w_0}

> **Hypothesis.** *We assume in this chapter, and only in this chapter, that W is **finite**.*

17.1. The Mathas–Lusztig Theorem

If $y \in W$ and $n \in \mathbb{Z}$, we set

$$T_{w_0}^n \mathbf{C}_y = \sum_{x \in W} \lambda_{x,y}^{(n)} \mathbf{C}_x.$$

Note that $\lambda_{x,y}^{(n)} = 0$ if $x \not\leqslant_L y$. We are interested here in the properties of the $\lambda_{x,y}^{(n)}$. For this, we set

$$\boldsymbol{c}(z) = \mathbf{a}(z) - \mathbf{a}(w_0 z)$$

for all $z \in W$. By (P4) and Corollary 10.3.3(b), $\boldsymbol{c}$ is constant on two-sided cells. Note that

(17.1.1) $$\sum_{z \in W} \boldsymbol{c}(z) = 0.$$

Proposition 17.1.2 ($\mathscr{P}$). *Let $n \in \mathbb{Z}$ and let x and y be two elements of W such that $x \leqslant_L y$. Then:*

(a) *If $n \geqslant 0$, then $\deg \lambda_{x,y}^{(n)} \leqslant n\boldsymbol{c}(x)$. If, moreover, $x <_L y$, then $\deg \lambda_{x,y}^{(n)} < n\boldsymbol{c}(x)$.*
(b) *If $n \leqslant 0$, then $\deg \lambda_{x,y}^{(n)} \leqslant n\boldsymbol{c}(y)$. If, moreover, $x <_L y$, then $\deg \lambda_{x,y}^{(n)} < n\boldsymbol{c}(y)$.*
(c) *If $x \sim_L y$, then $v^{-n\boldsymbol{c}(x)} \lambda_{x,y}^{(n)} \in \mathbb{Z}$.*

C. Bonnafé, *Kazhdan–Lusztig Cells with Unequal Parameters*,
Algebra and Applications 24, https://doi.org/10.1007/978-3-319-70736-5_17

Proof. If $n=0$, then (a), (b) and (c) are easily checked. Let us now prove (a) and (b). By 10.2.6,

$$T_{w_0}=\sum_{u\in W}\varepsilon(w_0u)p_{1,w_0u}\mathbf{C}_u.$$

Consequently,

$$\lambda_{x,y}^{(1)}=\sum_{\substack{u\in W\\ x\leqslant_R u}}\varepsilon(w_0u)p_{1,w_0u}h_{u,y,x}.$$

But, by (P1), we have $\deg p_{1,w_0u}\leqslant -\mathbf{a}(w_0u)$. If, moreover, $x\leqslant_R u$, then $w_0u\leqslant_R w_0x$ and so $-\mathbf{a}(w_0u)\leqslant -\mathbf{a}(w_0x)$ by (P4). Therefore

$$\deg\lambda_{x,y}^{(1)}\leqslant \boldsymbol{c}(x).$$

On the other hand, if $\deg\lambda_{x,y}^{(1)}=\boldsymbol{c}(x)$, then there exists a $u\in W$ such that $x\leqslant_R u$ and $\deg h_{u,y,x}=\mathbf{a}(x)$. So, by (P8), we get that $x\sim_L y$. This shows (a) for $n=1$.

Now, let $\nu:\mathscr{H}\to\mathscr{H}$ denote the A-linear map such that $\nu(\mathbf{C}_w)=v^{-\boldsymbol{c}(w)}\mathbf{C}_w$ for all $w\in W$ and let $\mu:\mathscr{H}\to\mathscr{H}$, $h\mapsto T_{w_0}h$. Then, if $w\in W$, we have

$$\nu\mu(\mathbf{C}_y)=\sum_{u\leqslant_L y}v^{-\boldsymbol{c}(u)}\lambda_{u,y}^{(1)}\mathbf{C}_u.$$

So, by the previous discussion, we have $v^{-\boldsymbol{c}(u)}\lambda_{u,y}^{(1)}\in A_{\leqslant 0}$. Moreover, if $u<_L y$, then $v^{-\boldsymbol{c}(u)}\lambda_{u,y}^{(1)}\in A_{<0}$. On the other hand, $\det\mu=\pm 1$ (see Exercise 4.7) and $\det\nu=1$ (by 17.1.1). Therefore, if we write

$$\mu^{-1}\nu^{-1}(\mathbf{C}_y)=\sum_{u\leqslant_L y}\beta_{u,y}\mathbf{C}_u,$$

then $\beta_{u,y}\in A_{\leqslant 0}$ and, if $u<_L y$, then $\beta_{u,y}\in A_{<0}$. Finally,

$$\begin{aligned}T_{w_0}^{-1}\mathbf{C}_y&=\mu^{-1}(\mathbf{C}_y)\\&=\mu^{-1}\nu^{-1}\nu(\mathbf{C}_y)\\&=v^{-\boldsymbol{c}(y)}\mu^{-1}\nu^{-1}(\mathbf{C}_y)\\&=\sum_{u\leqslant_L y}v^{-\boldsymbol{c}(y)}\beta_{u,y}\mathbf{C}_u.\end{aligned}$$

In other words, $\lambda_{x,y}^{(-1)}=v^{-\boldsymbol{c}(y)}\beta_{x,y}$. This shows that (b) holds if $n=-1$. An elementary induction argument using (P4) shows (a) and (b) in full generality.

Let us now prove (c). Applying the involution $\bar{\ }$ to the equality defining the structure constants $\lambda_{x,y}^{(n)}$, we get

$$\lambda_{x,y}^{(-n)}=\overline{\lambda_{x,y}^{(n)}}.$$

This proves that we may assume that $n \geqslant 0$. Now, assume that $x \sim_L y$ and let $\alpha = v^{-n\boldsymbol{c}(x)}\lambda_{x,y}^{(n)}$. By (a), we have $\deg(\alpha) \leqslant 0$. On the other hand, since $x \sim_L y$, we have $\boldsymbol{c}(x) = \boldsymbol{c}(y)$ by (P4), so it follows from Proposition 17.1.2(b) that

$$\deg(\overline{\lambda_{x,y}^{(n)}}) \leqslant -n\boldsymbol{c}(x).$$

In other words, $\operatorname{val}(\lambda_{x,y}^{(n)}) \geqslant n\boldsymbol{c}(x)$. Namely, $\operatorname{val}(\alpha) \geqslant 0 \geqslant \deg(\alpha)$. So $\alpha \in \mathbb{Z}$. ■

Corollary 17.1.3 ($\mathscr{P}$). *Let Γ be a two-sided cell and let $w \in \Gamma$. Then*

$$v^{-2\boldsymbol{c}(\Gamma)} T_{w_0}^2 \mathbf{c}_w = \mathbf{c}_w$$

in $\mathscr{H}^{LR}[\Gamma]$.

Proof. Recall from Lemma 16.1.1 that if $w' <_L w$, then $w' <_{LR} w$. So, by Proposition 17.1.2(c), the matrix of the multiplication by $v^{-2\boldsymbol{c}(\Gamma)} T_{w_0}^2$ in the basis $(\mathbf{c}_w)_{w \in \Gamma}$ has coefficients in $\mathbb{Z}$. So these coefficients can be computed after specialization $A \to \mathbb{Z}$, $v^a \mapsto 1$. But then $\mathscr{H}$ specializes to the group algebra $\mathbb{Z}W$ and $T_{w_0}^2$ specializes to $w_0^2 = 1$, so the result follows. ■

Corollary 17.1.4 ($\mathscr{P}$). *Let $w \in W$ and let $n \in \mathbb{N}$. Then*

$$\deg \tau(T_{w_0}^{-n}\mathbf{C}_w) \leqslant -\mathbf{a}(w) - n\boldsymbol{c}(w).$$

Moreover, $\deg \tau(T_{w_0}^{-n}\mathbf{C}_w) = -\mathbf{a}(w) - n\boldsymbol{c}(w)$ if and only if $w_0^n w^{-1} \in \mathscr{D}$.

Proof. Assume first that n is even. In particular, $w_0^n = 1$. By Proposition 17.1.2, we have

$$\tau(T_{w_0}^{-n}\mathbf{C}_w) = v^{-n\boldsymbol{c}(w)}\tau(\mathbf{C}_w) + \sum_{x <_L w} \lambda_{x,w}^{(-n)} \tau(\mathbf{C}_x).$$

But, if $x <_L w$, then $\deg \tau(\mathbf{C}_x) = -\boldsymbol{\Delta}(x) \leqslant -\mathbf{a}(x) \leqslant -\mathbf{a}(w)$ by (P1) and (P4). So, by Proposition 17.1.2(b), we have that $\deg \lambda_{x,w}^{(-n)} \tau(\mathbf{C}_x) < -\mathbf{a}(w) - n\boldsymbol{c}(w)$. Moreover, again by (P1), we have $\deg \tau(\mathbf{C}_w) = -\boldsymbol{\Delta}(w) \leqslant -\mathbf{a}(w)$. This shows that $\deg \tau(T_{w_0}^{-n}\mathbf{C}_w) \leqslant -\mathbf{a}(w) - n\boldsymbol{c}(w)$ and that equality holds if and only if $\boldsymbol{\Delta}(w) = \mathbf{a}(w)$, that is, if and only if $w \in \mathscr{D}$, as desired.

Assume now that $n = 2k+1$ for some natural number k. Recall that, by 10.2.6, $T_{w_0} = \sum_{u \in W} \varepsilon(w_0 u) p_{1,w_0 u} \mathbf{C}_u$. Therefore

$$\begin{aligned} T_{w_0}^{-n}\mathbf{C}_w &= \sum_{u \in W} \varepsilon(w_0 u) p_{1,w_0 u} T_{w_0}^{-n-1} \mathbf{C}_u \mathbf{C}_w \\ &= \sum_{\substack{u,x \in W \\ x \leqslant_L w \text{ and } x \leqslant_R u}} \varepsilon(w_0 u) p_{1,w_0 u} h_{u,w,x} T_{w_0}^{-n-1} \mathbf{C}_x. \end{aligned}$$

This implies that

$$\tau(T_{w_0}^{-n}\mathbf{C}_w) = \sum_{\substack{u,x\in W \\ x\leqslant_L w \text{ and } x\leqslant_R u}} \varepsilon(w_0u)p_{1,w_0u}h_{u,w,x}\tau(T_{w_0}^{-n-1}\mathbf{C}_x),$$

so $$\deg\tau(T_{w_0}^{-n}\mathbf{C}_w) \leqslant \max_{\substack{u,x\in W \\ x\leqslant_L w \text{ and } x\leqslant_R u}} \deg(p_{1,w_0u}h_{u,w,x}\tau(T_{w_0}^{-n-1}\mathbf{C}_x)).$$

Let u and x be two elements of W such that $x \leqslant_L w$ and $x \leqslant_R u$. Since $n+1$ is even and by the previous discussion, we have

$$\deg\tau(T_{w_0}^{-n-1}\mathbf{C}_x) \leqslant -\mathbf{a}(x) - (n+1)\boldsymbol{c}(x).$$

By (P1) and (P4), $\deg p_{1,w_0u} \leqslant -\mathbf{a}(w_0u) \leqslant -\mathbf{a}(w_0x)$. Moreover, $\deg h_{u,w,x} \leqslant \mathbf{a}(x)$. Consequently,

$$\begin{aligned}\deg(p_{1,w_0u}h_{u,w,x}\tau(T_{w_0}^{-n-1}\mathbf{C}_x)) &\leqslant -\mathbf{a}(x) - n\boldsymbol{c}(x)\\ &\leqslant -\mathbf{a}(w) - n\boldsymbol{c}(w).\end{aligned}$$

Moreover, equality holds if and only if $w_0u \in \mathscr{D}$, $\deg h_{u,w,x} = \mathbf{a}(x) = \mathbf{a}(w)$ and $x \in \mathscr{D}$.

We first deduce that

$$\deg\tau(T_{w_0}^{-n}\mathbf{C}_w) \leqslant -\mathbf{a}(w) - n\boldsymbol{c}(w),$$

which is the first assertion of the corollary.

Assume now that $\deg\tau(T_{w_0}^{-n}\mathbf{C}_w) = -\mathbf{a}(w) - n\boldsymbol{c}(w)$. Then there exists u and x in W such that $x \leqslant_L w$, $x \leqslant_R u$, $w_0u \in \mathscr{D}$, $\deg h_{u,w,x} = \mathbf{a}(x) = \mathbf{a}(w)$ and $x \in \mathscr{D}$. Since $\deg h_{u,w,x} = \mathbf{a}(x)$ and $x \in \mathscr{D}$, we deduce from (P2) that $w = u^{-1}$, which shows that $w_0w^{-1} \in \mathscr{D}$.

Conversely, assume that w_0w^{-1} belongs to $\mathscr{D}$. To show that $\deg\tau(T_{w_0}^{-n}\mathbf{C}_w) = -\mathbf{a}(w) - n\boldsymbol{c}(w)$, it is sufficient to show that there is a unique pair (u,x) of elements of W such that $x \leqslant_L w$, $x \leqslant_R u$, $w_0u \in \mathscr{D}$, $\deg h_{u,w,x} = \mathbf{a}(x) = \mathbf{a}(w)$ and $x \in \mathscr{D}$. The existence follows from (P13) (take $u = w^{-1}$ and x to be the unique element of $\mathscr{D}$ belonging to the left cell containing w). Let us now show unicity. Let (u,x) be such a pair. Since $\deg h_{u,w,x} = \mathbf{a}(x)$ and $x \in \mathscr{D}$, we deduce from (P2) that $u = w^{-1}$. Moreover, since $\mathbf{a}(x) = \mathbf{a}(w)$ and $x \leqslant_L w$, we have $x \sim_L w$ by (P9). But, by (P13), x is the unique element of $\mathscr{D}$ belonging to the left cell containing w. ■

If $\delta : W \to W$ is any map, we denote by δ^{op} the map $\delta^{\mathrm{op}} : W \to W$, $w \mapsto \delta(w^{-1})^{-1}$. We are now ready to prove the main result of this chapter.

Mathas–Lusztig's Theorem 17.1.5 ($\mathscr{P}$). *There exists a (unique) sign map η : $W \to \boldsymbol{\mu}_2$, $w \mapsto \eta_w$ and two (unique) involutions ρ and λ of the set W such that, for any two-sided cell Γ and any $w \in \Gamma$, we have*

$$v^{-c(w)}T_{w_0}\mathbf{c}_w = \eta_w \mathbf{c}_{\rho(w)} \quad \textit{and} \quad v^{-c(w)}\mathbf{c}_w T_{w_0} = \eta_w \mathbf{c}_{\lambda(w)}$$

in $\mathcal{H}^{LR}[\Gamma]$. *Note that* $\rho = \lambda^{\mathrm{op}} = \lambda \circ \omega_0 = \omega_0 \circ \lambda$ *and that*

$$\rho(w) \sim_L w \quad \textit{and} \quad \lambda(w) \sim_R w.$$

Moreover,

$$\eta_w = \eta_{w^{-1}} = \eta_{\rho(w)} = \eta_{\lambda(w)} = \eta_{\omega_0(w)}.$$

Proof. Let Γ be a two-sided cell. It follows from Proposition 17.1.2(c) that, if we set $\nu_{x,y} = v^{-c(\Gamma)}\lambda_{x,y}^{(1)}$, then $\nu_{x,y} \in \mathbb{Z}$ for all x, $y \in \Gamma$. As in Section 16.2, we denote by

$$\theta_\Gamma : \mathcal{H}^{\leqslant_{LR}\Gamma} \longrightarrow A$$

the unique A-linear map such that

$$\theta_\Gamma(\mathbf{C}_x) = \delta_{x \in \mathscr{D} \cap \Gamma}$$

for all $x \leqslant_{LR} \Gamma$.

Let $w \in \Gamma$ and let $m = v^{-\mathbf{a}(\Gamma)}\theta_\Gamma(\mathbf{C}_{w^{-1}}v^{-2c(\Gamma)}T_{w_0}^2\mathbf{C}_w)$. By Proposition 17.1.2(c),

$$v^{-2c(\Gamma)}T_{w_0}^2\mathbf{c}_w = \mathbf{c}_w,$$

so it follows from Lemma 16.2.4 that

$$m \equiv n_{\partial_w^L} \mod A_{<0}.$$

On the other hand,

$$v^{-c(\Gamma)}T_{w_0}\mathbf{c}_w = \sum_{x \sim_L w} \nu_{x,w}\mathbf{c}_x.$$

Applying the involution *, we get

$$v^{-c(\Gamma)}\mathbf{c}_{w^{-1}}T_{w_0} = \sum_{x \sim_L w} \nu_{x,w}\mathbf{c}_{x^{-1}}.$$

Then, by Lemma 16.2.4, we get

$$m \equiv n_{\partial_w^L} \sum_{x \sim_L w} \nu_{x,w}^2 \mod A_{<0}.$$

Therefore

$$\sum_{x \sim_L w} \nu_{x,w}^2 = 1.$$

So there exists a unique x such that $x \sim_L w$ and $\nu_{x,w} \neq 0$. It is denoted by $\rho(w)$. We set $\eta_w = \nu_{\rho(w),w}$: then $\eta_w = \pm 1$. This proves the first equality. The fact that ρ is an involution follows by applying twice the multiplication by $v^{-c(\Gamma)}T_{w_0}$ and using Corollary 17.1.3 (and this implies that $\eta_w = \eta_{\rho(w)}$). The fact that $\rho(w) \sim_L w$ follows from the fact that, by construction, $\rho(w) \sim_{LR} w$ and $\rho(w) \leqslant_L w$ and from Lemma 16.1.1.

The second equality follows by applying the automorphism ω_0 (see Corollary 10.1.2), and it also follows that $\rho = \lambda \circ \omega_0 = \omega_0 \circ \lambda$. The fact that $\rho = \lambda^{\mathrm{op}}$ follows by applying the involution *. These arguments also give the last equalities involving the sign function η. ■

We keep the notation λ, ρ and η introduced in Theorem 17.1.5.

17.2. Further Properties of the Pairs (λ, η) and (ρ, η)

The first property is a compatibility of the map λ and ρ with left and right multiplication by w_0.

Proposition 17.2.1 ($\mathscr{P}$). *Let $w \in W$. Then*

$$\lambda(w_0 w) = w_0 \lambda(w), \quad \lambda(w w_0) = \lambda(w) w_0,$$

$$\rho(w_0 w) = w_0 \rho(w), \quad \rho(w w_0) = \rho(w) w_0$$

and

$$\eta_{w w_0} = \varepsilon(w_0 w \rho(w)) \eta_w.$$

Proof. By Theorem 17.1.5 and Corollary 17.1.3, we have

$$v^{-\mathbf{c}(w)} T_{w_0}^{-1} \mathbf{C}_w \equiv \eta_w \mathbf{C}_{\rho(w)} \mod \mathscr{H}^{<_{LR} w}.$$

Applying the automorphism †, we get

$$\varepsilon(w_0) v^{-\mathbf{c}(w)} T_{w_0} \mathbf{C}_w^\dagger \equiv \eta_w \mathbf{C}_{\rho(w)}^\dagger \mod (\mathscr{H}^{<_{LR} w})^\dagger.$$

By Theorem 10.2.3 and Corollary 10.3.4(c), this leads to

$$\varepsilon(w_0 w) v^{\mathbf{c}(w)} \mathbf{D}_{w^{-1} w_0} \equiv \eta_w \varepsilon(\rho(w)) T_{w_0}^{-1} \mathbf{D}_{\rho(w)^{-1} w_0} \mod \mathscr{H}^{\vee, >_{LR} w^{-1} w_0}.$$

In other words,

$$\varepsilon(w_0 w \rho(w)) v^{-\mathbf{c}(w^{-1} w_0)} T_{w_0} \mathbf{D}_{w^{-1} w_0} \equiv \eta_w \mathbf{D}_{\rho(w)^{-1} w_0} \mod \mathscr{H}^{\vee, >_{LR} w^{-1} w_0}.$$

But, by duality (see Proposition 1.2.2 and its proof), we have

$$v^{-\mathbf{c}(u)} T_{w_0} \mathbf{D}_u \equiv \eta_u \mathbf{D}_{\lambda(u)} \mod \mathscr{H}^{\vee, >_{LR} u} \tag{17.2.2}$$

for all $u \in W$. Applying this congruence for $u = w^{-1} w_0$, we get

$$\varepsilon(w_0 w \rho(w)) \eta_{w^{-1} w_0} \mathbf{D}_{\lambda(w^{-1} w_0)} \equiv \eta_w \mathbf{D}_{\rho(w)^{-1} w_0} \mod \mathscr{H}^{\vee, >_{LR} w^{-1} w_0}.$$

This proves that $\lambda(w^{-1}w_0) = \rho(w)^{-1}w_0$ and

$$\eta_{w^{-1}w_0} = \eta_{w_0 w} = \eta_{w w_0} = \varepsilon(w_0 w \rho(w))\eta_w$$

for all $w \in W$. Replacing w by w^{-1} and using that $\lambda = \rho^{\mathrm{op}}$, we get $\lambda(w w_0) = \lambda(w) w_0$, as desired. The other formulas follow immediately. ■

We now give a characterization of the maps λ and ρ in terms of Duflo involutions and leading coefficients $\gamma_{x,y,z}$. As a consequence, we obtain a formula for the action of λ and ρ on $\mathscr{D}$.

Proposition 17.2.3 ($\mathscr{P}$). *Let $x, y \in W$ and $d \in \mathscr{D}$. Then:*

(a) *If $\gamma_{x,y,w_0 d} \neq 0$, then $d = \partial^L_{w_0 x^{-1}}$ and $y = \rho(x)^{-1}$.*
(b) *If $d = \partial^L_{w_0 x^{-1}}$, then $\gamma_{x,\rho(x)^{-1},w_0 d} = \pm 1$ and $\eta_x = \varepsilon(d) n_d \gamma_{x,\rho(x)^{-1},w_0 d}$.*

Proof. Assume that $\gamma_{x,y,w_0 d} \neq 0$. Then $w_0 d \sim_L x^{-1}$ by (P8) and so $d \sim_L w_0 x^{-1}$ by Theorem 10.3.2. Hence $d = \partial^L_{w_0 x^{-1}}$. This proves the first statement of (a).

On the other hand, $y^{-1} \sim_L x$ by (P8). Therefore, $v^{-c(x)} \lambda_{y^{-1},x} = \eta_x \delta_{y = \rho(x)^{-1}}$ by Theorem 17.1.5. But recall from the proof of Proposition 17.1.2 that

$$v^{-c(x)} \lambda^{(1)}_{y^{-1},x} = v^{-c(x)} \sum_{u \in W} \varepsilon(w_0 u) p_{1,w_0 u} h_{u,x,y^{-1}} = v^{-c(x)} \sum_{u \in W} \varepsilon(u) p_{1,u} h_{w_0 u,x,y^{-1}}$$

and that

$$v^{-c(x)} p_{1,u} h_{w_0 u,x,y^{-1}} \equiv \begin{cases} n_u \gamma_{w_0 u,x,y} & \text{if } u = d = \partial^L_{w_0 x^{-1}}, \\ 0 & \text{otherwise,} \end{cases}$$

modulo $A_{<0}$. Therefore, by (P7),

$$\eta_x \delta_{y = \rho(x)^{-1}} = \varepsilon(d) n_d \gamma_{x,y,w_0 d}.$$

This completes the proof of (a) and (b). ■

Corollary 17.2.4 ($\mathscr{P}$). *Let $d \in \mathscr{D}$. Then $\rho(d) = w_0 \partial^L_{w_0 d} = \partial^L_{d w_0} w_0$, $\lambda(d) = w_0 \partial^L_{d w_0} = \partial^L_{w_0 d} w_0$ and $\eta_d = \varepsilon(\partial^L_{w_0 d}) n_{\partial^L_{w_0 d}} n_d$.*

Proof. We have $d = d^{-1}$ by (P6) and, by Proposition 17.2.3(b), $\gamma_{d,\rho(d)^{-1},w_0 \partial^L_{w_0 d}} \neq 0$. So it follows from (P7) that $\gamma_{\rho(d)^{-1},w_0 \partial^L_{w_0 d},d} \neq 0$, and so the first equality follows from (P2). The other equalities follow easily. ■

Theorem 17.2.5 ($\mathscr{P}$). *The pair (λ, η) (resp. (ρ, η)) is strongly left (resp. right) cellular.*

Proof. It is sufficient to prove that (λ, η) is strongly left cellular. First, (LC3) holds by Theorem 17.1.5.

Let x and y be two elements of W such that $x \sim_L y$. Let Γ (resp. C) denote the two-sided (resp. left) cell containing x and y. Then there exist $x = x_0$, $x_1, \dots, x_m = y = y_0, y_1, \dots, y_n = x$ in W and elements $h_1, \dots, h_m, h'_1, \dots, h'_n$ of $\mathcal{H}$ such that $\mathbf{C}_{x_i}$ (resp. $\mathbf{C}_{y_j}$) appears with a non-zero coefficient in the expression of $h_i \mathbf{C}_{x_{i-1}}$ (resp. $h'_j \mathbf{C}_{y_{j-1}}$) in the Kazhdan–Lusztig basis for $1 \leqslant i \leqslant m$ (resp. $1 \leqslant j \leqslant n$). Therefore, $y = x_m \leqslant_L \cdots \leqslant_L x_2 \leqslant_L x_1 = x = y'_n \leqslant_L \cdots \leqslant_L y'_2 \leqslant_L y'_1 = y$ and so $x_i, y_j \in C$. Hence, if we write

$$h_i \mathbf{C}_{x_{i-1}} \equiv \sum_{u \in \Gamma} \beta_{i,u} \mathbf{C}_u \mod \mathcal{H}^{<_{LR}\Gamma},$$

then $\beta_{i,x_i} \neq 0$ and

$$v^{-\mathbf{c}(\Gamma)} h_i \mathbf{C}_{x_{i-1}} T_{w_0} \equiv \sum_{u \in \Gamma} v^{-\mathbf{c}(\Gamma)} \beta_{i,u} \mathbf{C}_u T_{w_0} \mod \mathcal{H}^{<_{LR}\Gamma}.$$

Therefore, by Theorem 17.1.5,

$$\eta_{x_{i-1}} h_i \mathbf{C}_{\lambda(x_{i-1})} \equiv \sum_{u \in \Gamma} \eta_u \beta_{i,u} \mathbf{C}_{\lambda(u)} \mod \mathcal{H}^{<_{LR}\Gamma},$$

and so $\lambda(x_i) \leqslant_L \lambda(x_{i-1})$. This shows that $\lambda(y) \leqslant_L \lambda(x)$ and we can prove similarly that $\lambda(x) \leqslant_L \lambda(y)$. Therefore, $\lambda(C)$ is contained in a unique left cell C'. But, similarly, $\lambda(C')$ is contained in a unique left cell, and contains C. So $\lambda(C) = C'$ is a left cell. This shows (LC1).

Finally, the map $(\lambda, \eta)_C : \mathcal{H}^L[C] \to \mathcal{H}^L[\lambda(C)]$, $\mathbf{c}_w \mapsto \eta_w \mathbf{c}_{\lambda(w)}$, is obtained through *right* multiplication by $v^{-\mathbf{c}(\Gamma)} T_{w_0}$. Since this right multiplication commutes with the *left* action of $\mathcal{H}$, this implies (LC2). ■

Remark 17.2.6 ($\mathscr{P}$). Let C be a left cell and assume here that the sign function η is constant on C (with constant value denoted by η_C). Let

$$C^\rho = \{w \in C \mid \rho(w) = w\}$$

be the set of fixed points of ρ in C. We denote by $[C]$ the $\mathbb{Z}W$-module obtained by specialization of $\mathcal{H}^L[C]$ through the map $A \to \mathbb{Z}$, $v^a \mapsto 1$. Then

$$\mathrm{Tr}(w_0, [C]) = \eta_C |C^\rho|$$

(by Theorem 17.1.5). □

17.3. Some Examples

Example 17.3.1 (Extremal cells). Since the two-sided cells of 1 and w_0 are singletons, we get that $\lambda(1) = \rho(1) = 1$ and $\lambda(w_0) = \rho(w_0) = w_0$. Moreover, $\eta_1 = (-1)^{\ell(w_0)}$ and $\eta_{w_0} = 1$ (see Example 6.7.2 and Proposition 10.5.2).

In particular, if $|S| = 1$, then $\lambda = \rho = \mathrm{Id}_W$. □

Example 17.3.2 (***Split case***). We assume in this example, and only in this example, that $\mathscr{A} = \mathbb{Z}$ and $\varphi = \ell$. It follows from the positivity properties 15.2.1 that $n_d = 1$ and $\gamma_{x,y,z} \geqslant 0$ for all $x, y, z \in W$ and $d \in \mathscr{D}$. It then follows from Proposition 17.2.3 that $\eta_x = \varepsilon(\partial^L_{w_0 x^{-1}})$ for all $x \in W$. But, by Proposition 5.6.4(a), $P_{1,d} = v^{\ell(d)} p_{1,d} \in \mathbb{Z}[v^2]$, so $\mathbf{\Delta}(d) = \mathbf{a}(d) \equiv \ell(d) \mod 2$. Therefore,

$$\eta_x = (-1)^{\mathbf{a}(w_0 x^{-1})}$$

by (P4). This proves in particular that the sign map η is constant on two-sided cells. It will be shown in Example 17.3.4 that this last property does not hold in general in the unequal parameter case. □

Example 17.3.3 (***Rank 2***). If $|S| = 2$, then the maps λ and ρ will be computed in Chapter 21 (see Proposition 21.5.1). In this case again, the map η is constant on two-sided cells. □

Example 17.3.4 (***Type B_3***). Assume here that (W, S) is of type B_3, so that the Coxeter graph of W is given by

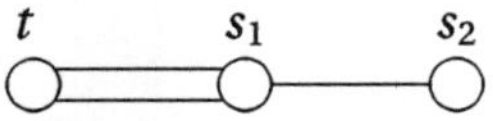

We set $b = \varphi(t)$ and $a = \varphi(s_1) = \varphi(s_2)$. Recall that $a, b > 0$. It follows from computer computations that, if $b \notin \{a, 2a\}$, then the representation afforded by a left cell module is absolutely irreducible (after extension to the fraction field K of A). Since T_{w_0} is central in $\mathscr{H}$, it acts by multiplication by a scalar on all these cell modules, and so $\lambda = \rho = \mathrm{Id}_W$ and the sign function is given by the action of w_0 on the specialization, through $v^a \mapsto 1$, of the corresponding left cell module. In particular, it is constant on two-sided cells.

If $b = a$, then it follows from Example 17.3.2 that the sign function is constant on two-sided cells. However, if $b = 2a$, it may be checked by computer computations that the sign function is not constant on the unique two-sided cell of cardinality 20. □

Notes

Theorem 17.1.5 is due to Mathas in the split case [Mat] and to Lusztig [Lus21] in the general case. A simplification of Lusztig's proof was provided by the author in [Bon5]: it avoids the use of the asymptotic algebra (which is defined in Chapter 19) and the use of (P15), which is sometimes the most difficult conjecture to prove. If W is a Weyl group, Losev [Los] proves that left multiplication by T_{w_0} may be viewed as an equivalence of categories involving the category $\mathscr{O}$.

Chapter 18
Action of the Cactus Group

We first recall the definition of the *cactus group* Cact_W associated with W. We denote by $\mathscr{P}_{\mathrm{ir,sph}}(S)$ the set of spherical subsets I of S such that (W_I, I) is irreducible. If $I \in \mathscr{P}_{\mathrm{sph}}(S)$, we denote by $\boldsymbol{c}_I : W_I \to \mathscr{A}$, $\lambda_I : W_I \to W_I$, $\rho_I : W_I \to W_I$, and $\eta^I : W_I \to \boldsymbol{\mu}_2$ the respective analogues of the maps $\boldsymbol{c}$, λ, ρ and η which were defined in Chapter 17 for the group W (see Theorem 17.1.5), and we set $\omega_I : W_I \to W_I$, $w \mapsto w_I w w_I$. The group Cact_W is the group with the following presentation:

- Generators: $(\sigma_I)_{I \in \mathscr{P}_{\mathrm{ir,sph}}(S)}$;
- Relations: for all $I, J \in \mathscr{P}_{\mathrm{ir,sph}}(S)$, we have:

$$\begin{cases} \text{(C1)} & \sigma_I^2 = 1, \\ \text{(C2)} & [\sigma_I, \sigma_J] = 1 \qquad \text{if } W_{I \cup J} = W_I \times W_J, \\ \text{(C3)} & \sigma_I \sigma_J = \sigma_J \sigma_{\omega_J(I)} \quad \text{if } I \subset J. \end{cases}$$

By construction, the map $\sigma_I \mapsto w_I$ extends to a surjective morphism of groups $\mathrm{Cact}_W \twoheadrightarrow W$ which is not used in this chapter.

Hypothesis. *Recall from the introduction of this Part VI that, in this chapter, we relax the assumption a little: we only assume that* **($\mathscr{P}_{\mathrm{sph}}$)** *holds (and not necessarily* **($\mathscr{P}$)***).*

18.1. Cellularity

The next theorem follows immediately from Theorems 8.4.4, 8.4.8 and 17.2.5:

Theorem 18.1.1 ($\mathscr{P}_{\mathrm{sph}}$). *If $I \in \mathscr{P}_{\mathrm{sph}}(S)$, then the pair (λ_I^L, η_L^I) (resp. (ρ_I^R, η_R^I)) is strongly left (resp. right) cellular.*

C. Bonnafé, *Kazhdan–Lusztig Cells with Unequal Parameters*,
Algebra and Applications 24, https://doi.org/10.1007/978-3-319-70736-5_18

Note that the maps λ_I^L and ρ_I^R depend on the weight function φ, even if this is not clear from the notation. The canonicity of their construction shows that, if $\omega : W \to W$ is an automorphism such that $\omega(S) = S$ and $\varphi \circ \omega = \varphi$, then

$$\omega \circ \lambda_I^L = \lambda_{\omega(I)}^L \circ \omega \quad \text{and} \quad \omega \circ \rho_I^R = \rho_{\omega(I)}^R \circ \omega. \tag{18.1.2}$$

For instance, *if W is finite*, then $\omega_0 : W \to W$ satisfies the above properties and so

$$\omega_0 \circ \lambda_I^L = \lambda_{\omega_0(I)}^L \circ \omega_0 \quad \text{and} \quad \omega_0 \circ \rho_I^R = \rho_{\omega_0(I)}^R \circ \omega_0. \tag{18.1.3}$$

For the next corollary, we recall from the general notation of Chapter 8 that $\omega_{I,L}(w) = \omega_I(\mathrm{pr}_L^I(w))$ and $\omega_{I,R}(w) = \omega_I(\mathrm{pr}_R^I(w))$ (and similarly for $\boldsymbol{c}_{I,L}$ and $\boldsymbol{c}_{I,R}$).

Corollary 18.1.4 ($\mathscr{P}_{\mathrm{sph}}$). *Let $I \in \mathscr{P}_{\mathrm{sph}}(S)$ and let $w \in W$. Then*

$$\eta_{R,w}^I v^{\boldsymbol{c}_{I,R}(w)} T_{w_I} \mathbf{C}_w \equiv \mathbf{C}_{\rho_I^R(w)} \mod \mathscr{H}_I^{<_R \omega_{I,R}(w)} \mathscr{H}$$

and

$$\eta_{L,w}^I v^{\boldsymbol{c}_{I,L}(w)} \mathbf{C}_w T_{w_I} \equiv \mathbf{C}_{\lambda_I^L(w)} \mod \mathscr{H}\mathscr{H}_I^{<_L \omega_{I,L}(w)}.$$

Proof. It is sufficient to prove the second congruence. Let $b \in X_I$ and $y \in W_I$ be such that $w = by$ (so that $y = \mathrm{pr}_L^I(w)$). By 8.1.7,

$$\mathbf{C}_{by} \equiv \sum_{\substack{(a,x) \in X_I \times W_I \\ \text{such that } a \leqslant b \\ \text{and } x \sim_L y}} p_{ax,by}^I T_a \mathbf{C}_x \mod \mathscr{H}\mathscr{H}_I^{<_L y}.$$

If $x \sim_L y$, then $\mathbf{a}_I(x) = \mathbf{a}_I(y)$ and $\mathbf{a}_I(w_I x) = \mathbf{a}_I(w_I y)$ by (P4), so it follows from Theorem 17.1.5 that

$$\eta_y^I v^{\boldsymbol{c}_I(y)} \mathbf{C}_{by} T_{w_I} \equiv \sum_{\substack{(a,x) \in X_I \times W_I \\ \text{such that } a \leqslant b \\ \text{and } x \sim_L y}} \eta_y^I \eta_x^I p_{ax,by}^I T_a \mathbf{C}_{\lambda_I(x)} \mod \mathscr{H}\mathscr{H}_I^{<_L y}.$$

But, by Corollary 8.4.7, $p_{ax,by}^I = \eta_y^I \eta_x^I p_{a\lambda_I(x), b\lambda_I(y)}$, so

$$\eta_{L,w}^I v^{\boldsymbol{c}_{I,L}(w)} \mathbf{C}_w T_{w_I} \equiv \mathbf{C}_{\lambda_I^L(w)} \mod \mathscr{H}\mathscr{H}_I^{<_L y} T_{w_I}.$$

It then remains to notice that $T_{w_I}^{-1} \mathbf{C}_x T_{w_I} = \mathbf{C}_{\omega_I(x)}$ for all $x \in W_I$ (see 10.1.5), so that $\mathscr{H}_I^{<_L y} T_{w_I} = T_{w_I} \mathscr{H}_I^{<_L \omega_I(y)} = \mathscr{H}_I^{<_L \omega_I(y)}$ and the result follows. ■

An important consequence of the previous characterization is the following:

Theorem 18.1.5 ($\mathscr{P}_{\mathrm{sph}}$). *Let $I \in \mathscr{P}_{\mathrm{sph}}(S)$ and let (σ,μ) be a strongly left (resp. right) cellular pair. Then $\sigma \circ \rho_I^R = \rho_I^R \circ \sigma$ (resp. $\sigma \circ \lambda_I^L = \lambda_I^L \circ \sigma$). Moreover, $\eta_{R,\sigma(w)}^I = \mu_w \mu_{\rho_I^R(w)} \eta_{R,w}^I$ (resp. $\eta_{L,\sigma(w)}^I = \mu_w \mu_{\lambda_I^L(w)} \eta_{L,w}^I$) for all $w \in W$.*

Proof. Assume that (σ,μ) is strongly left cellular. Let $w \in W$. By (LC3), we have $\sigma(w) \sim_R w$ and so, by Corollary 8.2.9,

$$(*) \qquad \mathrm{pr}_R^I(\sigma(w)) \sim_R \mathrm{pr}_R^I(w).$$

Now, let us write

$$\eta_{R,w}^I v^{c_{I,R}(w)} T_{w_I} \mathbf{C}_w \equiv \sum_{u \sim_L w} \beta_u \mathbf{C}_u \mod \mathscr{H}^{<_L w},$$

with $\beta_u \in A$. Since (σ,μ) is left cellular, we get

$$\mu_w \eta_{R,w}^I v^{c_{I,R}(w)} T_{w_I} \mathbf{C}_{\sigma(w)} \equiv \sum_{u \sim_L w} \beta_u \mu_u \mathbf{C}_{\sigma(u)} \mod \mathscr{H}^{<_L \sigma(w)}.$$

But, by Corollary 18.1.4, we have

$$\eta_{R,w}^I v^{c_{I,R}(w)} T_{w_I} \mathbf{C}_w \equiv \mathbf{C}_{\rho_I^R(w)} \mod \mathscr{H}_I^{<_R \omega_I(\mathrm{pr}_R^I(w))} \mathscr{H}$$

and $\rho_I^R(w) \sim_L w$ (because $(\rho_I^R, \eta_{I,R})$ is strongly right cellular by Theorem 18.1.1). Therefore, $\beta_{\rho_I^R(w)} = \mu_w \mu_{\rho_I^R(w)}$. Again by Corollary 18.1.4, we get

$$\eta_{R,w}^I v^{c_{I,R}(w)} T_{w_I} \mathbf{C}_{\sigma(w)} \equiv \eta_{R,w}^I \eta_{R,\sigma(w)}^I \mathbf{C}_{\rho_I^R(\sigma(w))} \mod \mathscr{H}_I^{<_R \omega_I(\mathrm{pr}_R^I(w))} \mathscr{H}$$

(by using also $(*)$). Combining these results, we get

$$\mathbf{C}_{\rho_I^R(\sigma(w))} - \eta_{R,w}^I \eta_{R,\sigma(w)}^I \mu_w \mu_{\rho_I^R(w)} \mathbf{C}_{\sigma(\rho_I^R(w))} \in \bigoplus_{z \in \mathscr{E}_1 \cup \mathscr{E}_2 \cup \mathscr{E}_3} A\mathbf{C}_z,$$

where

$$\mathscr{E}_1 = \{\sigma(u) \mid u \sim_L w \text{ and } u \neq \rho_I^R(w)\},$$
$$\mathscr{E}_2 = \{u \in W \mid u <_L \sigma(w)\}$$

and

$$\mathscr{E}_3 = \{u \in W \mid \mathrm{pr}_R^I(u) <_R \omega_I(\mathrm{pr}_R^I(w))\}$$

(we have used the fact that $\mathscr{H}_I^{<_R v} \mathscr{H} = \bigoplus_{\mathrm{pr}_R^I(u) <_R v} A\mathbf{C}_u$ for all $v \in W_I$: see Theorem 8.2.1). So, in order to prove that $\rho_I^R(\sigma(w)) = \sigma(\rho_I^R(w))$ and $\eta_{R,\sigma(w)}^I = \mu_w \mu_{\rho_I^R(w)} \eta_{R,w}^I$, we only need to show that $\sigma(\rho_I^R(w)) \notin \mathscr{E}_1 \cup \mathscr{E}_2 \cup \mathscr{E}_3$.

First, by definition, $\sigma(\rho_I^R(w)) \notin \mathscr{E}_1$. Also, since $(\rho_I^R, \eta_{I,R})$ is strongly right cellular, we get that $\rho_I^R(w) \sim_L w$ by (LC3) and so $\sigma(\rho_I^R(w)) \sim_L \sigma(w)$ because (σ, μ) is left cellular (see (LC1)). So $\sigma(\rho_I^R(w)) \notin \mathscr{E}_2$. Finally, note that $\sigma(\rho_I^R(w)) \sim_R \rho_I^R(w)$ because (σ, μ) is strongly left cellular (see (LC3)). So $\mathrm{pr}_R^I(\sigma(\rho_I^R(w)) \sim_R \mathrm{pr}_R^I(\rho_I^R(w))$ by Corollary 8.2.9. Since $\mathrm{pr}_R^I(\rho_I^R(w)) = \rho_I(\mathrm{pr}_R^I(w)) \sim_{LR} \mathrm{pr}_R^I(w) \sim_{LR} \omega_I(\mathrm{pr}_R^I(w))$ by Lemma 10.3.1, we get $\sigma(\rho_I^R(w)) \notin \mathscr{E}_3$ by (P4), (P8) and (P9). ■

18.2. Action

The key result is the following:

Theorem 18.2.1 ($\mathscr{P}_{\mathrm{sph}}$). *Let $I, J \in \mathscr{P}_{\mathrm{ir,sph}}(S)$. Then:*

(a) $[\lambda_I^L, \rho_J^R] = \mathrm{Id}_W$.
(b) $(\lambda_I^L)^2 = (\rho_I^R)^2 = \mathrm{Id}_W$.
(c) *If $W_{I\cup J} = W_I \times W_J$, then $[\lambda_I^L, \lambda_J^L] = [\rho_I^R, \rho_J^R] = \mathrm{Id}_W$.*
(d) *If $I \subset J$, then $\lambda_I^L \lambda_J^L = \lambda_J^L \lambda_{\omega_J(I)}^L$ and $\rho_I^R \rho_J^R = \rho_J^R \rho_{\omega_J(I)}^R$.*

Proof. (a) follows from Theorem 18.1.5, while (b) is obvious.

(c) Assume that $W_{I\cup J} = W_I \times W_J$. We only need to prove that $[\lambda_I^L, \lambda_J^L] = \mathrm{Id}_W$, the proof of the other equality being similar. Let $w \in W$ and write $w = xw'$, with $x \in X_{I\cup J}$ and $w' \in W_{I\cup J}$. Since $W_{I\cup J} = W_I \times W_J$, there exist $w_1 \in W_I$ and $w_2 \in W_J$ such that $w' = w_1 w_2 = w_2 w_1$. Note also that $xw_1 \in X_J$, $x\lambda_I(w_1) \in X_J$, $xw_2 \in X_I$ and $x\lambda_J(w_2) \in X_I$. Therefore,

$$\lambda_I^L(\lambda_J^L(w)) = \lambda_I^L(xw_1\lambda_J(w_2)) = \lambda_I^L(x\lambda_J(w_2)w_1) = x\lambda_J(w_2)\lambda_I(w_1)$$

and, similarly,

$$\lambda_J^L(\lambda_I^L(w)) = x\lambda_I(w_1)\lambda_J(w_2).$$

So $[\lambda_I^L, \lambda_J^L] = \mathrm{Id}_W$, as desired.

(d) Assume here that $I \subset J$. It is easily checked that we may assume that W is finite and $J = S$. Let $w \in W$. Then

$$\begin{aligned} \lambda_S^L(\lambda_I^L(w)) &= \rho_S^L(\omega_0(\lambda_I^L(w))) && \text{by Theorem 17.1.5.} \\ &= \rho_S \lambda_{\omega_0(I)}^L(\omega_0(w)) && \text{by 18.1.3,} \\ &= \lambda_{\omega_0(I)}^L(\rho_S(\omega_0(w)) && \text{by (a),} \\ &= \lambda_{\omega_0(I)}^L(\lambda_S(w)) && \text{by Theorem 17.1.5.} \end{aligned}$$

This proves the first equality and the second follows from a similar argument. ■

Recall that $\mathfrak{S}_W$ denotes the symmetric group on the set W. The statements (b), (c) and (d) of the previous Theorem 18.2.1 show that there exists a unique morphism of groups

$$\begin{array}{rcl} \mathrm{Cact}_W & \longrightarrow & \mathfrak{S}_W \\ \sigma & \longmapsto & \sigma_\varphi^L \end{array}$$

such that

$$\sigma_{I,\varphi}^L = \lambda_I^L$$

for all $I \in \mathscr{P}_{\mathrm{ir,sph}}(S)$. Note that here we have emphasized the fact that the map depends on φ. The same statements also show that there exists a unique morphism of groups

$$\begin{array}{rcl} \mathrm{Cact}_W & \longrightarrow & \mathfrak{S}_W \\ \sigma & \longmapsto & \sigma_\varphi^R \end{array}$$

such that

$$\sigma_{I,\varphi}^R = \rho_I^R$$

for all $I \in \mathscr{P}_{\mathrm{ir,sph}}(S)$. Moreover, Theorem 18.2.1(a) shows that both actions commute or, in other words, that the map

$$\begin{array}{rcl} \mathrm{Cact}_W \times \mathrm{Cact}_W & \longrightarrow & \mathfrak{S}_W \\ (\sigma_1, \sigma_2) & \longmapsto & \sigma_{1,\varphi}^L \sigma_{2,\varphi}^R \end{array} \tag{18.2.2}$$

is a morphism of groups. Let us summarize the properties of this morphism which have been proved in this chapter:

Theorem 18.2.3 ($\mathscr{P}_{\mathrm{sph}}$). *Let $\sigma \in \mathrm{Cact}_W$. Then there exist two sign maps $\eta_L^{\sigma,\varphi} : W \to \boldsymbol{\mu}_2$ and $\eta_R^{\sigma,\varphi} : W \to \boldsymbol{\mu}_2$ such that the pairs $(\sigma_\varphi^L, \eta_L^{\sigma,\varphi})$ and $(\sigma_\varphi^R, \eta_R^{\sigma,\varphi})$ are strongly left cellular and strongly right cellular respectively.*

Moreover, if $\sigma' \in \mathrm{Cact}_W$, then $[\sigma_\varphi^L, \sigma'^R_\varphi] = \mathrm{Id}_W$.

Note that we do not claim that the sign maps in the above theorem are unique. They are obtained by decomposing σ as a product of the generators and then composing the cellular pairs according to this decomposition: the resulting sign map might depend on the chosen decomposition. It must be added that the maps σ_φ^L and σ_φ^R depend heavily on φ (see, for instance, the case where $|S| = 2$ in Chapter 21).

Remark 18.2.4. It might happen that Lusztig's Conjectures for the triple (W_I, I, φ_I) hold for *some* $I \in \mathscr{P}_{\mathrm{sph}}(S)$, but not for *all* $I \in \mathscr{P}_{\mathrm{sph}}(S)$. The reader may check that the proofs of Theorems 18.1.1 and 18.1.5 and Corollary 18.1.4 remain valid if Lusztig's Conjectures hold only for the triple (W_I, I, φ_I) considered in the statement. Similarly, Theorem 18.2.1 is valid if Lusztig's Conjectures hold only for the triples (W_I, I, φ_I) and (W_J, J, φ_J) involved in the statement.

This remark can be applied, for instance, to the subsets $I \in \mathscr{P}_{\mathrm{sph}}(S)$ such that $|I| = 2$ (as we will prove in Chapter 21 that Lusztig's Conjectures hold for dihedral groups), or to subsets I such that φ_I is constant (by Theorem 15.2.5). For instance, we always obtain an action of the subgroup of the cactus group generated by the σ_I such that $|I| = 2$. □

Remark 18.2.5. If $|I| = 1$, then it follows from Example 17.3.1 that $\lambda_I = \rho_I = \mathrm{Id}_{W_I}$ and so $\lambda_I^L = \rho_I^L = \mathrm{Id}_W$. This shows that the element $\sigma_I \in \mathrm{Cact}_W$ acts trivially on W, for the action defined in Theorem 18.2.3. □

Notes

The main results of this chapter are due to Losev [Los] if (W, S) is a finite Weyl group and to the author [Bon5] in the general case (following a suggestion of Losev [Los, §5.1]). In his particular case, Losev's method provides much more information, as the action of the cactus group is viewed as the combinatorial shadow of wall-crossing functors on the category $\mathscr{O}$. His motivation for this was trying to understand conjectures of Rouquier and the author on the link between Kazhdan–Lusztig cells and Calogero–Moser cells [BoRo1], [BoRo2], [BoRo3].

Chapter 19
Asymptotic Algebra

19.1. Definition

We denote by $\mathscr{J}$ a free $\mathbb{Z}$-module admitting a $\mathbb{Z}$-basis $(\mathbf{t}_w)_{w\in W}$ indexed by W. If X is a subset of W, we denote by $\mathscr{J}[X]$ the $\mathbb{Z}$-submodule

$$\mathscr{J}[X] = \bigoplus_{w\in X} \mathbb{Z}\mathbf{t}_w.$$

Let

$$\begin{array}{ccc} \mathscr{J}\times\mathscr{J} & \longrightarrow & \mathscr{J} \\ (f,g) & \longmapsto & fg \end{array}$$

be defined as the unique $\mathbb{Z}$-bilinear map such that

$$\mathbf{t}_x\mathbf{t}_y = \sum_{z\in W} \gamma_{x,y,z^{-1}}\mathbf{t}_z.$$

It is called the *multiplication* in $\mathscr{J}$.

Theorem 19.1.1 ($\mathscr{P}$). *The multiplication in $\mathscr{J}$ is associative. If, moreover, $\mathscr{D}$ is finite, then $\sum_{d\in\mathscr{D}} n_d\mathbf{t}_d$ is the unit element.*

Proof. It follows from (P8) that, if Γ and Γ' are two-sided cells of W such that $\Gamma\neq\Gamma'$, then

$$\mathscr{J}[\Gamma]\mathscr{J}[\Gamma'] = 0 \qquad \text{and} \qquad \mathscr{J}[\Gamma]\mathscr{J}[\Gamma]\subset\mathscr{J}[\Gamma].$$

Proving the associativity is hence equivalent to proving that if $w, x, y, z\in\Gamma$, then

$$(*) \qquad \sum_{u\in\Gamma}\gamma_{x,y,u^{-1}}\gamma_{u,z,w^{-1}} = \sum_{u\in\Gamma}\gamma_{x,u,w^{-1}}\gamma_{y,z,u^{-1}}.$$

C. Bonnafé, *Kazhdan–Lusztig Cells with Unequal Parameters*,
Algebra and Applications 24, https://doi.org/10.1007/978-3-319-70736-5_19

Indeed, the left-hand side (resp. right-hand side) is the coefficient of $\mathbf{t}_w$ in $(\mathbf{t}_x\mathbf{t}_y)\mathbf{t}_z$ (resp. $\mathbf{t}_x(\mathbf{t}_y\mathbf{t}_z)$). But the associativity property inside $\mathscr{H}$ implies that, for all $w, x, y, z \in W$,

$$\sum_{u\in W} h_{x,y,u}h_{u,z,w} = \sum_{u\in W} h_{x,u,w}h_{y,z,u}.$$

Now, if $h_{x,y,u}h_{u,z,w} \neq 0$, then $u \leqslant_{LR} x$ and $w \leqslant_{LR} u$, so $u \in \Gamma$. Similarly, if $h_{x,u,w}h_{y,z,u} \neq 0$, then $u \in \Gamma$. Therefore,

$$\sum_{u\in \Gamma} h_{x,y,u}h_{u,z,w} = \sum_{u\in \Gamma} h_{x,u,w}h_{y,z,u}.$$

This implies (∗), by looking at the coefficient of $v^{2\mathbf{a}(\Gamma)}$ (see (P4)).

Assume, moreover, that $\mathscr{D}$ is finite and let $e = \sum_{d\in\mathscr{D}} n_d\mathbf{t}_d$. If $x \in W$, then

$$e\mathbf{t}_x = \sum_{\substack{y\in W\\ d\in\mathscr{D}}} n_d\gamma_{d,x,y^{-1}}\mathbf{t}_y.$$

So, by (P7), this gives

$$e\mathbf{t}_x = \sum_{\substack{y\in W\\ d\in\mathscr{D}}} n_d\gamma_{x,y^{-1},d}\mathbf{t}_y$$

and so $e\mathbf{t}_x = \mathbf{t}_x$ by (P2) and (P3). Similarly, one gets $\mathbf{t}_x e = \mathbf{t}_x$. ■

Definition 19.1.2. *The $\mathbb{Z}$-module $\mathscr{J}$ endowed with the above multiplication is called the* ***asymptotic algebra*** *associated with (W, S, φ).*

Note that, contrarily to our usual conventions in this book, $\mathscr{J}$ is not necessarily unital. It is unital if $\mathscr{D}$ is finite (by (P13), this is equivalent to saying that W has only finitely many left cells): note that this is the case, for instance, if W is finite or is an affine Weyl group (see [Lus20, Theorem 18.2]). In this case, we denote by $1_{\mathscr{J}}$ the unit element of $\mathscr{J}$. However, there exist Coxeter groups admitting infinitely many left cells (see Corollary 24.5.2).

Corollary 19.1.3 ($\mathscr{P}$). *We have:*

(a) *If Γ is a two-sided cell, then $\mathscr{J}[\Gamma]$ is also a (not necessarily unital) algebra. If $\mathscr{D}\cap\Gamma$ is finite, then it is unital, with unit given by $e_\Gamma = \sum_{d\in\mathscr{D}\cap\Gamma} n_d\mathbf{t}_d$: in this case, e_Γ is a central idempotent of $\mathscr{J}$ and*

$$\mathscr{J}[\Gamma] = \mathscr{J}e_\Gamma = e_\Gamma\mathscr{J} = e_\Gamma\mathscr{J}e_\Gamma.$$

(b) *If C is a left cell and if d denotes the unique element of $\mathscr{D} \cap C$, then $n_d \mathbf{t}_d$ is an idempotent in $\mathscr{J}$ and*

$$\mathscr{J}[C] = \mathscr{J}\mathbf{t}_d \qquad \text{and} \qquad \mathscr{J}[C^{-1}] = \mathbf{t}_d \mathscr{J}.$$

If C' is another left cell and if d' denotes the unique element of $\mathscr{D} \cap C'$, then

$$\mathscr{J}[C' \cap C^{-1}] = t_d \mathscr{J} t_{d'}.$$

In particular, $\mathscr{J}[C \cap C^{-1}] = \mathbf{t}_d \mathscr{J} \mathbf{t}_d$ is a unital algebra, with unit $n_d \mathbf{t}_d$.

It follows from 14.2.5 that the $\mathbb{Z}$-linear map $^* : \mathscr{J} \to \mathscr{J}$ defined by

$$\mathbf{t}_w^* = \mathbf{t}_{w^{-1}}$$

for all $w \in W$ is an involutive anti-automorphism of the (non-unital) algebra $\mathscr{J}$. If C (resp. Γ) is a left (resp. two-sided) cell, then the involution * stabilizes $\mathscr{J}[C \cap C^{-1}]$ (resp. $\mathscr{J}[\Gamma]$, by (P14)).

19.2. A Morphism from $\mathscr{H}$ to $A \otimes_{\mathbb{Z}} \mathscr{J}$

Hypothesis and notation. *Recall from the general hypotheses of Part VI that we assume that Lusztig's Conjectures* (Pk) *hold for $1 \leqslant k \leqslant 14$. In this section, and only in this section, we assume moreover that Lusztig's Conjecture* (P15) *holds and that $\mathscr{D}$ is* **finite**, *that is, we assume that there are finitely many left cells.*

If B is a commutative A-algebra, recall from the general notation that $B\mathscr{H} = B \otimes_A \mathscr{H}$ and $B\mathscr{J} = B \otimes_{\mathbb{Z}} \mathscr{J}$.

We define two A-linear maps $\phi, \phi^{\mathrm{op}} : \mathscr{H} \longrightarrow A\mathscr{J}$ by the following formulas:

$$\phi(\mathbf{C}_x) = \sum_{\substack{d \in \mathscr{D} \\ y \in W \\ \text{such that } \mathbf{a}(d) = \mathbf{a}(y)}} n_d h_{x,d,y} \mathbf{t}_y$$

and

$$\phi^{\mathrm{op}}(\mathbf{C}_x) = \sum_{\substack{d \in \mathscr{D} \\ y \in W \\ \text{such that } \mathbf{a}(d) = \mathbf{a}(y)}} n_d h_{d,x,y} \mathbf{t}_y.$$

Note that the definitions of ϕ and ϕ^{op} make sense since $\mathscr{D}$ is finite and since, for any $x, x' \in W$, the set of $y \in W$ such that $h_{x,x',y} \neq 0$ is finite.

Let us give another construction of these morphisms. We endow $A\mathscr{J}$ with the structure of a left $\mathscr{H}$-module by identifying the A-module $A\mathscr{J}$ with $\bigoplus_{\Gamma\in\mathrm{Cell}_{LR}(W)}\mathscr{H}^{LR}[\Gamma]$ in the natural way (by sending, for $w\in\Gamma\in\mathrm{Cell}_{LR}(W)$, $\mathbf{c}_w\in\mathscr{H}^{LR}[\Gamma]$ to $\mathbf{t}_w$). If $h\in\mathscr{H}$ and $j\in A\mathscr{J}$, we denote by $h\bullet j\in A\mathscr{J}$ the image of j under this left action of h. Taking into account (P11), we have

(19.2.1) $$\mathbf{C}_x\bullet\mathbf{t}_y=\sum_{\substack{z\in W\\ \text{such that } \mathbf{a}(z)=\mathbf{a}(y)}} h_{x,y,z}\,\mathbf{t}_z.$$

Also, $A\mathscr{J}$ is endowed with the structure of a right $A\mathscr{J}$-module just by right multiplication. Applying $\mathrm{Id}_A\otimes_{\mathbb{Z}}\mathrm{coef}_{\mathbf{a}(y)}: A\otimes_{\mathbb{Z}}A\to A\otimes_{\mathbb{Z}}\mathbb{Z}=A$ to the identity (P15), we see that the left $\mathscr{H}$-module structure and the right $A\mathscr{J}$ structure on $A\mathscr{J}$ commute, so that

(19.2.2) *$A\mathscr{J}$ becomes an $(\mathscr{H},A\mathscr{J})$-bimodule.*

Similarly, we can endow $A\mathscr{J}$ with the structure of a right $\mathscr{H}$-module (still denoted by $\bullet$) and a left $A\mathscr{J}$-module which commute, and

(19.2.3) *$A\mathscr{J}$ becomes an $(A\mathscr{J},\mathscr{H})$-bimodule.*

With this notation, it follows from 19.2.1 that

(19.2.4) $$\phi(h)=h\bullet 1_{\mathscr{J}} \qquad\text{and}\qquad \phi^{\mathrm{op}}(h)=1_{\mathscr{J}}\bullet h$$

for all $h\in\mathscr{H}$.

Theorem 19.2.5 ($\mathscr{P}$). *Assume that $\mathscr{D}$ is **finite** and that Lusztig's Conjecture* (P15) *holds for the triple (W,S,φ). Then the A-linear maps ϕ and ϕ^{op} are morphisms of unital A-algebras.*

Proof. Note that $\phi^{\mathrm{op}}(h)=\phi(h^*)^*$, so that it is sufficient to prove that ϕ is a morphism of algebras. But, by 19.2.2 and 19.2.4, we have

$$\phi(h)\phi(h')=(h\bullet 1_{\mathscr{J}})\phi(h')=h\bullet\phi(h')=h\bullet(h'\bullet 1_{\mathscr{J}})=hh'\bullet 1_{\mathscr{J}}=\phi(hh')$$

for all h, $h'\in\mathscr{H}$. The proof of the theorem is now complete (taking into account that the equality $\phi(\mathbf{C}_1)=1_{\mathscr{J}}$ is immediate). ■

If B is any commutative A-algebra, we denote by ϕ_B and ϕ_B^{op} the morphisms of algebras $B\mathscr{H}\to B\mathscr{J}$ obtained respectively from ϕ and ϕ^{op} by extension of scalars.

Proposition 19.2.6 ($\mathscr{P}$). *Assume that $\mathscr{D}$ is* ***finite*** *and that Lusztig's Conjecture* (P15) *holds for the triple (W,S,φ). Let B be a commutative A-algebra. Then:*

(a) $(\operatorname{Ker}\phi_B)^N = (\operatorname{Ker}\phi_B^{\mathrm{op}})^N = 0$, *where N is the cardinality of the image of* $\mathbf{a}$ *(which is finite because $\mathscr{D}$ is finite).*

(b) *If, moreover, $B = R[\mathscr{A}]$ (where R is a commutative ring) with its natural structure of an A-algebra, then ϕ_B and ϕ_B^{op} are injective. In particular, ϕ and ϕ^{op} are injective.*

Proof. (a) By (P4) and (P11), it is sufficient to show that, if Γ is a two-sided cell, then $(\operatorname{Ker}\phi_B) \cdot (B\mathscr{H}^{\leqslant_{LR}\Gamma}) \subset B\mathscr{H}^{<_{LR}\Gamma}$. In other words, this amounts to proving that, if $h \in \operatorname{Ker}(\phi_B)$, then h acts by 0 on $\bigoplus_{\Gamma \in \mathrm{Cell}_{LR}(W)} B\mathscr{H}^{LR}[\Gamma]$. So let $h \in \operatorname{Ker}(\phi_B)$. Then, through the identification $\bigoplus_{\Gamma \in \mathrm{Cell}_{LR}(W)} B\mathscr{H}^{LR}[\Gamma] \simeq B\mathscr{J}$ used for constructing the $\bullet$-action, we only need to show that $h \bullet j = 0$ for all $j \in B\mathscr{J}$. But $h \bullet j = (h \bullet 1_{\mathscr{J}}) j = \phi_B(h) j = 0$, and the result follows.

(b) Assume now that $B = R[\mathscr{A}]$ (where R is a commutative ring) with its natural structure of an A-algebra. Let $h \in B\mathscr{H}$ and assume that $h \neq 0$. Let us write $h = \sum_{x \in X} \beta_x \mathbf{C}_x$, where X is a non-empty finite subset of W such that $\beta_x \in B \setminus \{0\}$ for all $x \in X$. Let $x_0 \in X$ be such that $\deg(\beta_{x_0})$ is maximal in $\mathscr{A}$, let $a = \mathbf{a}(x_0)$ and $a' = \deg(\beta_{x_0}) \in \mathscr{A}$, let $\beta \in B$ be the coefficient of $\mathbf{t}_{x_0}$ in $\phi(h)$ and let $r = \operatorname{coef}_{a+a'}(\beta) \in R$. Then

$$r = \sum_{\substack{x \in X \\ d \in \mathscr{D} \\ \text{such that } \mathbf{a}(d) = \mathbf{a}(x_0)}} n_d \operatorname{coef}_{a+a'}(\beta_x h_{x,d,x_0}).$$

But, by the choice of x_0, we get

$$r = \sum_{\substack{x \in X \\ d \in \mathscr{D} \\ \text{such that } \mathbf{a}(d) = \mathbf{a}(x_0)}} n_d \operatorname{coef}_{a'}(\beta_x) \gamma_{x,d,x_0^{-1}}.$$

By (P7) and (P2), this gives

$$r = \sum_{\substack{d \in \mathscr{D} \\ \text{such that } \mathbf{a}(d) = \mathbf{a}(x_0)}} n_d \operatorname{coef}_{a'}(\beta_{x_0}) \gamma_{d,x_0^{-1},x_0}.$$

It then follows from (P13) and (P5) that

$$r = \operatorname{coef}_{a'}(\beta_{x_0}) \neq 0,$$

so $\phi_B(h) \neq 0$. ■

Corollary 19.2.7 ($\mathscr{P}$). *Assume that W is finite and that Lusztig's Conjecture* (P15) *holds for the triple (W,S,φ). Let K be a field which is a commutative A-algebra such that $K\mathscr{H}$ is semisimple. Then ϕ_K and ϕ_K^{op} are isomorphisms of algebras.*

Proof. Indeed, by Proposition 19.2.6(a), $\mathrm{Ker}(\phi_K)$ is a nilpotent two-sided ideal so ϕ_K is injective because $K\mathscr{H}$ is semisimple. The corollary now follows by comparison of dimensions. ■

Remark 19.2.8 **($\mathscr{P}$).** The above Corollary 19.2.7 applies, for instance, if K is the fraction field of A, or if K is a field of characteristic not dividing the order of W, viewed as an A-algebra through the specialization $v^a \mapsto 1$ (indeed, in this case, $K\mathscr{H}$ is just the group algebra KW, so it is semisimple). □

19.3. Examples

Example 19.3.1 **(*Type A_2*).** As in Examples 3.12.7, 4.4.13, 5.7.9, 6.7.6, and 8.4.2, we assume here that $S=\{s,t\}$ and $m_{st}=3$, so that (W,S) is of type A_2. We set $\varphi(s)=\varphi(t)=a$ and we assume, following Commentary 6.2.12 and as in Examples 5.7.9, 6.7.6 and 8.4.2, that $a>0$. Here is the corresponding Coxeter graph, together with the values of φ:

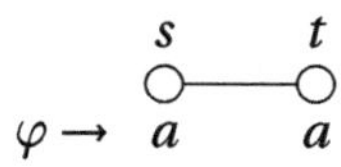

Let $\Gamma=\{s,t,st,ts\}$ be the two-sided cell of cardinality 4. It follows from the computations in Example 5.7.9, that

$$\mathbf{a}(w)=\begin{cases} 0 & \text{if } w=1, \\ a & \text{if } w\in\Gamma, \\ 3a & \text{if } w=w_0. \end{cases}$$

Also, $\mathscr{J}=\mathbb{Z}\mathbf{t}_1\oplus\mathbb{Z}\mathbf{t}_{w_0}\oplus\mathscr{J}[\Gamma]$ (as a direct sum of rings), and $\mathbf{t}_1$ and $\mathbf{t}_{w_0}$ are central idempotents. Finally, it is easily checked that the $\mathbb{Z}$-linear map

$$\begin{array}{rcl} \mathscr{J}[\Gamma] & \longrightarrow & \mathrm{Mat}_2(\mathbb{Z}) \\ k\mathbf{t}_s+l\mathbf{t}_{st}+m\mathbf{t}_{ts}+n\mathbf{t}_t & \longmapsto & \begin{pmatrix} k & l \\ m & n \end{pmatrix} \end{array}$$

is an isomorphism of algebras. □

Example 19.3.2 (Type B_2). As in Examples 3.12.8, 4.4.14, 5.7.10, 6.7.7 and 8.4.3, we assume here that $S=\{s,t\}$ with $m_{st}=4$, so that (W,S) is of type B_2. We set $\varphi(s)=a$ and $\varphi(t)=b$, and assume, following Commentary 6.2.12 and as in Examples 5.7.10, 6.7.7 and 8.4.3, that a, $b>0$. Here is the corresponding Coxeter graph, together with the values of φ:

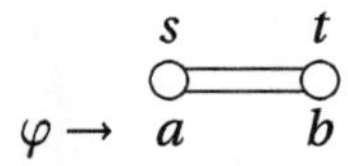

We compute here Lusztig's **a**-function, as well as the asymptotic algebra $\mathscr{J}$. We need to distinguish between the split case and the other cases.

First case: assume that $a=b$. We set $\Gamma=\{s,t,st,ts,sts,tst\}$. It follows from the computations in Example 5.7.10, that

$$\mathbf{a}(w)=\begin{cases}0 & \text{if } w=1,\\ a & \text{if } w\in\Gamma,\\ 4a & \text{if } w=w_0.\end{cases}$$

Moreover, the only interesting algebra $\mathscr{J}[\Gamma']$ is the algebra corresponding to $\Gamma'=\Gamma$. Taking into account the anti-involution * of $\mathscr{J}$ and the fact that exchanging s and t induces an automorphism of $\mathscr{J}$, the multiplication is completely determined by the following rules (here, $w\in\Gamma$):

$$\mathbf{t}_s\mathbf{t}_w=\begin{cases}\mathbf{t}_w & \text{if } \mathscr{L}(w)=\{s\},\\ 0 & \text{otherwise,}\end{cases}\qquad \mathbf{t}_{st}\mathbf{t}_{ts}=\mathbf{t}_s+\mathbf{t}_{sts}\quad\text{and}\quad \mathbf{t}_{sts}^2=\mathbf{t}_s.$$

In particular, if C is the left cell containing s, then

$$\mathscr{J}[C\cap C^{-1}]=\mathbf{t}_s\mathscr{J}\mathbf{t}_s=\mathbb{Z}\mathbf{t}_s\oplus\mathbb{Z}\mathbf{t}_{sts}\simeq\mathbb{Z}\boldsymbol{\mu}_2,$$

where we recall that $\boldsymbol{\mu}_2=\{1,-1\}$ is a cyclic group of order 2.

Second case: assume that $a<b$. Let $\Gamma^<=\{t,st,ts,sts\}$ denote the unique two-sided cell of cardinality 4 (see Example 6.7.7). Lusztig's **a**-function is given by

$$\mathbf{a}(w)=\begin{cases}0 & \text{if } w=1,\\ a & \text{if } w=s,\\ b & \text{if } w\in\Gamma,\\ 2b-a & \text{if } w=tst,\\ 2(a+b) & \text{if } w=w_0.\end{cases}$$

Then $\mathscr{J}$ admits the following decomposition (as a direct sum of algebras)

$$\mathscr{J}=\mathbb{Z}\mathbf{t}_1\oplus\mathbb{Z}\mathbf{t}_s\oplus\mathbb{Z}\mathbf{t}_{tst}\oplus\mathbb{Z}\mathbf{t}_{w_0}\oplus\mathscr{J}[\Gamma^<],$$

and it follows from the computations in Example 5.7.10 that $\mathbf{t}_1$, $\mathbf{t}_s$ and $\mathbf{t}_{w_0}$ are idempotents, that $n_{tst}=-1$, that $\mathbf{t}_{sts}^2=-\mathbf{t}_{sts}$ and that the $\mathbb{Z}$-linear map

$$\begin{aligned} \mathscr{J}[\Gamma^<] &\longrightarrow \mathrm{Mat}_2(\mathbb{Z}) \\ k\mathbf{t}_{sts}+l\mathbf{t}_{st}+m\mathbf{t}_{ts}+n\mathbf{t}_t &\longmapsto \begin{pmatrix} k & l \\ m & n \end{pmatrix} \end{aligned}$$

is an isomorphism of algebras. □

Example 19.3.3 **($\mathscr{P}$).** Assume in this example that W is finite and, moreover, that Lusztig's Conjecture (P15) holds for the triple (W,S,φ). We take the notation of Chapter 17 $(\lambda,\rho,\eta,\ldots)$. Then it follows directly from 19.2.4 and Theorem 17.1.5 that

$$\phi(T_{w_0})=\sum_{d\in\mathscr{D}} n_d\eta_d v^{\mathbf{c}(d)}\mathbf{t}_{\rho(d)}$$

and

$$\phi^{\mathrm{op}}(T_{w_0})=\sum_{d\in\mathscr{D}} n_d\eta_d v^{\mathbf{c}(d)}\mathbf{t}_{\lambda(d)},$$

where $\mathbf{c}(z)=\mathbf{a}(z)-\mathbf{a}(w_0z)$ and η_z have been defined for all $z\in W$ in Section 17.1 and Theorem 17.1.5. Recall that, if $d\in\mathscr{D}$, then

$$\rho(d)=w_0\partial^L_{w_0d} \qquad \text{and} \qquad \eta_d=\varepsilon(\partial^L_{w_0d})n_{\partial^L_{w_0d}}n_d$$

(see Corollary 17.2.4). Similarly, it follows from 19.2.4 and Corollary 17.1.3 that

$$\phi(T_{w_0}^2)=\phi^{\mathrm{op}}(T_{w_0}^2)=\sum_{d\in\mathscr{D}} n_d v^{2\mathbf{c}(d)}\mathbf{t}_d.$$

Now, let

$$\mathfrak{T}=\sum_{d\in\mathscr{D}\cap\Gamma} n_d\eta_d\mathbf{t}_{\rho(d)}\in\mathscr{J}.$$

Then

$$(*) \qquad \mathfrak{T}=\sum_{\Gamma\in\mathrm{Cell}_{LR}(W)} v^{-\mathbf{c}(\Gamma)}\phi(T_{w_0})e_\Gamma,$$

where $e_\Gamma=\sum_{d\in\mathscr{D}\cap\Gamma} n_d\mathbf{t}_d$ is the central idempotent such that $\mathscr{J}[\Gamma]=\mathscr{J}e_\Gamma$. Therefore,

$$(19.3.4) \qquad \mathfrak{T}^2=\sum_{\Gamma\in\mathrm{Cell}_{LR}(W)} v^{2(\mathbf{a}(w_0\Gamma)-\mathbf{a}(\Gamma))}\phi(T_{w_0}^2)e_\Gamma=1_{\mathscr{J}}.$$

In particular, $\mathfrak{T}$ is invertible in $\mathscr{J}$. Note also that

$$(19.3.5) \qquad \mathfrak{T}\mathbf{t}_w=n_w\eta_w\mathbf{t}_{\rho(w)} \qquad \text{and} \qquad \mathbf{t}_w\mathfrak{T}=n_w\eta_w\mathbf{t}_{\lambda(w)}$$

for all $w \in W$. Indeed, if Γ denotes the two-sided cell of W, it follows from $(*)$ that $\mathfrak{T}\mathbf{t}_w = v^{-\mathbf{c}(\Gamma)} T_{w_0} \bullet \mathbf{t}_x$, and the result follows from the Mathas–Lusztig Theorem 17.1.5. In particular, if we denote by $\omega_0^{\mathcal{J}}$ the automorphism of $\mathcal{J}$ induced by the automorphism ω_0 of (W,S) (defined in Chapter 10: $\omega_0(w) = w_0 w w_0$), then

(19.3.6) $$\mathfrak{T} j \mathfrak{T}^{-1} = \omega_0^{\mathcal{J}}(j)$$

for all $j \in \mathcal{J}$, because $\mathfrak{T}^{-1} = \mathfrak{T}$ and $\lambda\rho(w) = w_0 w w_0$. □

Notes

The construction of the asymptotic algebra is due to Lusztig [Lus12, §2] in the split case and again to Lusztig [Lus20] in the general case. See also [Lus13]. Example 19.3.3 is due to Lusztig [Lus21].

Exercises

Exercise 19.1 (Type G_2). As in Exercises 3.1, 4.1, 5.1, 6.1, 8.3, we assume here that $S = \{s,t\}$ with $m_{st} = 6$, so that (W,S) is of type G_2. We set $\varphi(s) = a$ and $\varphi(t) = b$, and assume here that a, $b > 0$. Here is the corresponding Coxeter graph, together with the values of φ:

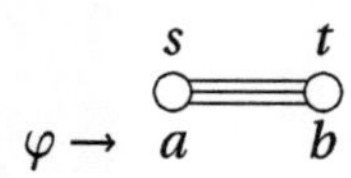

Determine, according to the values of a, b, the structure of the asymptotic algebra $\mathcal{J}$.

Exercise 19.2. Let Γ be a two-sided cell. Let

$$\tau_\Gamma : \mathcal{J}[\Gamma] \longrightarrow \mathbb{Z}$$

be the $\mathbb{Z}$-linear map defined by

$$\tau_\Gamma(\mathbf{t}_w) = \delta_{w \in \mathscr{D}}$$

for all $w \in \Gamma$. Show that

$$\tau_\Gamma(\mathbf{t}_x \mathbf{t}_y) = \delta_{xy=1}$$

for all x, $y \in \Gamma$. In particular, if Γ is finite, prove that $\mathcal{J}[\Gamma]$ is a symmetric algebra.

Prove similar results for the algebra $\mathcal{J}[C \cap C^{-1}]$, where C is a left cell.

Chapter 20
Automorphisms

Hypothesis. *In this chapter, and only in this chapter, we fix a group G acting on the Coxeter system (W,S): this means that G acts by automorphisms of the group W stabilizing S. We also assume that the weight function φ is G-invariant.*

Following Theorem 3.10.6, we denote by $(S/G)_{\mathrm{sph}}$ the set of G-orbits I in S such that W_I is finite, we set

$$S_G = \{w_I \mid I \in (S/G)_{\mathrm{sph}}\}$$

and we recall that (W^G, S_G) is a Coxeter system. Extending φ to a braided map $\varphi : W \to \mathbb{N}$, we denote by φ_G the restriction of φ to S_G. It is a weight function by Corollary 3.10.7.

Let $\mathscr{H}_G$ denote the Hecke algebra $\mathscr{H}(W^G, S_G, \varphi_G)$. If $w \in W^G$, we denote by T_w^G the corresponding standard basis element of $\mathscr{H}_G$. We denote by $(\mathbf{C}_w^G)_{w \in W^G}$ the Kazhdan–Lusztig basis of $\mathscr{H}_G$ and by $\leqslant_?^G$ and $\sim_?^G$ the respective analogues of $\leqslant_?$ and $\sim_?$ for the triple (W^G, S_G, φ_G). The structure constants of $\mathscr{H}_G$ in the Kazhdan–Lusztig basis are denoted by $h_{x,y,z}^G$. Finally, we set $\boldsymbol{\Delta}_G(z) = -\deg(p_{1,z}^G)$, where $\mathbf{C}_y^G = \sum_{x \in W^G} p_{x,y}^G T_x^G$.

We work under the hypothesis that the following variant of Hypothesis **($\mathscr{P}$)** holds (see the introduction of this Part VI).

($\mathscr{P}_G$) For all subgroups H of G, Conjecture 14.2.1 holds for the triple (W^H, S_H, φ_H), so that Lusztig's **a**-function $\mathbf{a}_H : W^H \to \mathscr{A}$, the numbers $\gamma_{x,y,z}^H \in \mathbb{Z}$ (which we define as the analogues of $\mathbf{a}$ and $\gamma_{x,y,z}$ but for W^H) and the set $\mathscr{D}_H = \{z \in W \mid \boldsymbol{\Delta}_H(z) = \mathbf{a}_H(z)\}$ are well-defined. Moreover, Lusztig's Conjecture (Pk) holds for the triple (W^H, S_H, φ_H) and for all $1 \leqslant k \leqslant 14$.

C. Bonnafé, *Kazhdan–Lusztig Cells with Unequal Parameters*,
Algebra and Applications 24, https://doi.org/10.1007/978-3-319-70736-5_20

20.1. Statement

Theorem 20.1.1 ($\mathscr{P}_G$). *Assume that G is **solvable** and let $x, y \in W^G$. Then:*

(a) $\mathbf{a}_G(x) = \mathbf{a}(x)$.
(b) $\mathscr{D}_G = \mathscr{D} \cap W^G$ *(i.e. $\mathscr{D}_G = \mathscr{D}^G$).*
(c) $x \sim_L y$ *(resp. $x \sim_R y$) if and only if $x \sim_L^G y$ (resp. $x \sim_R^G y$).*
(d) *If $x \sim_{LR}^G y$, then $x \sim_{LR} y$.*

Remark 20.1.2. If W is finite, if (W,S) is irreducible and if G is non-trivial, then it follows from the classification that necessarily φ is constant (so we may assume that $\varphi = \ell$) and there exists an element $\sigma \in G$ such that $W^\sigma = W^G$. So, in this case, the above Theorem 20.1.1 is already contained in Proposition 15.3.3. □

Remark 20.1.3. In [Bon3, Theorem A(d)], the author claimed that, if $x, y \in W^G$, then $x \sim_{LR}^G y$ if and only if $x \sim_{LR} y$. However, the proof of the converse of Theorem 20.1.1(d) given in [Bon3] is incorrect, so we still do not know if it holds. □

The proof is given in Section 20.3. Here, we just provide some easy reductions. First, we may assume that G acts faithfully on W. Since $W = \langle S \rangle$, this shows that G acts faithfully on S and, since S is finite, this forces G to be finite. Since G is assumed to be solvable, it admits a normal subgroup H of prime index. It is then enough to prove Theorem 20.1.1 for the quadruple (W,S,φ,H) and for the quadruple $(W^H, S_H, \varphi_H, G/H)$. So, by an easy induction argument, we see that it is enough to prove Theorem 20.1.1 whenever G is a finite p-group, where p is a prime number.

Hypothesis. *We fix a prime number p and, until the end of this chapter, we assume that G is a **finite p-group**.*

20.2. Brauer Quotient

20.2.A. Definition

For all the facts contained in this subsection, the reader may refer to [Thé, §11]: even though this reference deals only with Λ-algebras (where Λ is a commutative complete local noetherian $\mathbb{Z}_p$-algebra) which are Λ-modules of finite type, the proofs can be applied almost word for word to our more general situation.

Let R be a commutative ring and let M be an RG-module. If H is a subgroup of G, we

$$\begin{aligned}\mathrm{Tr}_H^G : M^H &\longrightarrow M^G\\ m &\longmapsto \sum_{\sigma\in[G/H]} \sigma(m).\end{aligned}$$

We also define

$$\mathrm{Tr}(M)=\sum_{H<G}\mathrm{Tr}_H^G(M^H),$$

where $H < G$ means that H is a proper subgroup of G. This is an R-submodule of M^G, containing pM^G. The *Brauer quotient* $\mathrm{Br}_G(M)$ is then defined by

$$\mathrm{Br}_G(M)=M^G/\mathrm{Tr}(M)$$

and we denote by $\mathrm{br}_G : M^G \to \mathrm{Br}_G(M)$ the canonical map (it is called the *Brauer morphism*).

Lemma 20.2.1. *Assume that $pR \neq R$ and that M admits an R-basis $\mathscr{B}$ which is permuted by the action of G. Then $\mathrm{Br}_G(M)$ is a free R/pR-module with basis $(\mathrm{br}_G(b))_{b\in\mathscr{B}^G}$.*

If M is an R-algebra and if G acts on M by algebra automorphisms, then $\mathrm{Tr}(M)$ is a two-sided ideal of M^G and so $\mathrm{Br}_G(M)$ is an R-algebra. Of course, br_G is a morphism of algebras in this case. We recall the following result:

Lemma 20.2.2. *Assume that $pR\neq R$, that M is an R-algebra, that G acts on M by algebra automorphisms, that M admits an R-basis $\mathscr{B}$ which is permuted by G and let us write $ab=\sum_{c\in\mathscr{B}}\lambda_{a,b,c}c$ for $a, b\in\mathscr{B}$. If $a, b\in\mathscr{B}^G$, then*

$$\mathrm{br}_G(a)\mathrm{br}_G(b)=\sum_{c\in\mathscr{B}^G}\pi(\lambda_{a,b,c})\mathrm{br}_G(c),$$

where $\pi : R\to R/pR$ is the canonical morphism.

20.2.B. Applications to Hecke Algebras

Since G stabilizes S and φ, it acts on $\mathscr{H}$ by A-algebra automorphisms which permute the standard basis $(T_w)_{w\in W}$. Let $\mathbb{F}_p$ denote the finite field with p elements. Since $A/pA=\mathbb{F}_p[\mathscr{A}]$, it follows from Lemma 20.2.1 that:

Corollary 20.2.3. $(\mathrm{br}_G(T_w))_{w\in W^G}$ *is an $\mathbb{F}_p[\mathscr{A}]$-basis of the $\mathbb{F}_p[\mathscr{A}]$-algebra* $\mathrm{Br}_G(\mathscr{H})$.

Now, let

$$\mathrm{can}_G : \mathscr{H}_G \longrightarrow \mathrm{Br}_G(\mathscr{H})$$

be the unique A-linear map such that

$$\mathrm{can}_G(T_w^G) = \mathrm{br}_G(T_w)$$

for all $w \in W^G$. The main result of this subsection is the following:

Proposition 20.2.4. *The map* $\mathrm{can}_G : \mathscr{H}_G \longrightarrow \mathrm{Br}_G(\mathscr{H})$ *is a surjective morphism of A-algebras whose kernel is* $p\mathscr{H}_G$.

Proof. It follows from Corollary 20.2.3 that the map can_G is surjective and that $\mathrm{Ker}(\mathrm{can}_G) = p\mathscr{H}_G$. It remains to show that can_G is a morphism of algebras. First, note that if $x, y \in W^G$ satisfy $\ell_G(xy) = \ell_G(x) + \ell_G(y)$, then $\ell(xy) = \ell(x) + \ell(y)$ (by Theorem 3.10.6) and so

$$\begin{aligned}\mathrm{can}_G(T_x^G T_y^G) &= \mathrm{can}_G(T_{xy}^G) = \mathrm{br}_G(T_{xy})\\ &= \mathrm{br}_G(T_x T_y) = \mathrm{br}_G(T_x)\mathrm{br}_G(T_y) = \mathrm{can}_G(T_x^G)\mathrm{can}_G(T_y^G).\end{aligned}$$

So it remains to show that if $I \in (S/G)_{\mathrm{sph}}$, then

$$\mathrm{br}_G((T_{w_I} - v^{\varphi(w_I)})(T_{w_I} + v^{-\varphi(w_I)})) = 0. \tag{?}$$

Since w_I is the longest element of W_I, we have, by 10.5.1,

$$\mathbf{C}_{w_I} = \sum_{w \in W_I} v^{\varphi(w) - \varphi(w_I)} T_w$$

and, by Proposition 10.5.2(c),

$$(T_{w_I} - v^{\varphi(w_I)})\mathbf{C}_{w_I} = 0.$$

But $(W_I)^G = \{1, w_I\}$. Since $\varphi(w) = \varphi(\sigma(w))$ for all $w \in W_I$ and all $\sigma \in G$, we have

$$\mathbf{C}_{w_I} \equiv T_{w_I} + v^{-\varphi(w_I)} \mod \mathrm{Tr}(\mathscr{H}).$$

Indeed, if $w \in W_I \setminus W_I^G$, if $G \cdot w$ denotes its G-orbit and H denotes its stabilizer in G, then $H < G$ and

$$\sum_{w' \in G \cdot w} v^{\varphi(w') - \varphi(w_I)} T_{w'} = \mathrm{Tr}_H^G(v^{\varphi(w) - \varphi(w_I)} T_w).$$

This completes the proof of (?). ■

Corollary 20.2.5. $\mathbb{F}_p \otimes_{\mathbb{Z}} \mathscr{H}_G \simeq \mathrm{Br}_G(\mathscr{H})$.

Corollary 20.2.6. *If $h \in \mathscr{H}_G$ and $h' \in \mathscr{H}$ are such that* $\mathrm{can}_G(h) = \mathrm{br}_G(h')$, *then* $\tau_G(h) \equiv \tau(h') \mod pA$.

Proposition 20.2.7. *If $w \in W^G$, then* $\mathrm{can}_G(\mathbf{C}_w^G) = \mathrm{br}_G(\mathbf{C}_w)$.

Proof. Let $C = \mathrm{can}_G(\mathbf{C}_w^G) - \mathrm{br}_G(\mathbf{C}_w)$. Then

$$\overline{C} = C$$

where $^{-}: \mathrm{Br}_G(\mathscr{H}) \to \mathrm{Br}_G(\mathscr{H})$ is defined by $\overline{\mathrm{br}_G(h)} = \mathrm{br}_G(\overline{h})$ for all $h \in \mathscr{H}^G$. Moreover, $C \in \mathrm{Br}_G(\mathscr{H}_{<0})$ so $C = 0$ by Lusztig's Lemma. ∎

Corollary 20.2.8. *If $x, y, z \in W^G$, then $h_{x,y,z} \equiv h_{x,y,z}^G \mod pA$ and $\tau(\mathbf{C}_z) \equiv \tau_G(\mathbf{C}_z^G) \mod pA$.*

Proof. This follows immediately from Proposition 20.2.7, from Lemma 20.2.2 and from Corollary 20.2.6. ∎

20.3. Proof of Theorem 20.1.1

Theorem 20.1.1 follows immediately from the next theorem by the easy induction argument explained in Section 20.1.

Theorem 20.3.1 ($\mathscr{P}_G$). *Assume that G is a finite p-group and let x and y be two elements of W^G. Then:*

(a) $\mathbf{a}_G(x) = \mathbf{a}(x)$.
(b) $\mathscr{D}_G = \mathscr{D} \cap W^G$ $(= \mathscr{D}^G)$.
(c) *$x \sim_L^G y$ (resp. $x \sim_R y$) if and only if $x \sim_L y$ (resp. $x \sim_R y$).*
(d) *If $x \sim_{LR}^G y$, then $x \sim_{LR} y$.*

Proof. (a) By Corollary 20.2.8, we have, for all $x, y, z \in W^G$:

(1) If $\gamma_{x,y,z^{-1}} \not\equiv 0 \mod p$, then $\mathbf{a}(z) \leqslant \mathbf{a}_G(z)$.
(2) If $\gamma_{x,y,z^{-1}}^G \not\equiv 0 \mod p$, then $\mathbf{a}_G(z) \leqslant \mathbf{a}(z)$.

Now let $z \in W^G$. By (P3), there exists a unique $d \in \mathscr{D}$ such that $\gamma_{z^{-1},z,d} \neq 0$. From the uniqueness, we get that $d \in \mathscr{D}^G \subseteq W^G$. By (P5) and (P7), we get that $\gamma_{z,d,z^{-1}} = \pm 1$. So $\mathbf{a}(z) \leqslant \mathbf{a}_G(z)$ by (1).

The same argument shows that there exists a $d \in \mathscr{D}_G$ such that $\gamma_{z,d,z^{-1}}^G = \pm 1$, so (2) can be applied to get that $\mathbf{a}_G(z) \leqslant \mathbf{a}(z)$. The proof of (a) is complete.

Before going further, let us state the following consequence of (a):

Corollary 20.3.2 ($\mathscr{P}_G$). *If $x, y, z \in W^G$, then $\gamma_{x,y,z} \equiv \gamma_{x,y,z}^G \mod p$.*

Proof. This follows from Theorem 20.3.1(a) and Corollary 20.2.8. ∎

(b) Let $d \in \mathscr{D}^G$. By (P5), we have $n_d = \pm 1$. Moreover, by Corollary 20.2.8, we have

$$\tau(\mathbf{C}_d) \equiv \tau_G(\mathbf{C}_d^G) \mod pA.$$

This shows that the coefficient of $v^{-\mathbf{\Delta}(d)}$ in $\tau_G(\mathbf{C}_d^G)$ is non-zero. So $\mathbf{\Delta}_G(d) \leqslant \mathbf{\Delta}(d)$. But, by (P1),

$$\mathbf{a}_G(d) \leqslant \mathbf{\Delta}_G(d) \leqslant \mathbf{\Delta}(d) = \mathbf{a}(d).$$

So $\mathbf{a}_G(d) = \mathbf{\Delta}_G(d) = \mathbf{\Delta}(d) = \mathbf{a}(d)$ by (a). In particular, $d \in \mathscr{D}_G$.

The same argument shows that, if $d \in \mathscr{D}_G$, then $\mathbf{\Delta}(d) \leqslant \mathbf{\Delta}_G(d)$ and again we get similarly that $d \in \mathscr{D}$. The proof of (b) is complete.

(c) Let d (resp. e) be the unique element of $\mathscr{D}$ such that $\gamma_{x^{-1},x,d} = \pm 1$ (resp. $\gamma_{y^{-1},y,e} = \pm 1$). By uniqueness and (b), we have $d, e \in \mathscr{D}^G = \mathscr{D}_G$. By Corollary 20.3.2, we also get $\gamma^G_{x^{-1},x,d} \neq 0$ and $\gamma^G_{y^{-1},y,e} \neq 0$. Therefore, by (P8), we have

$$x \sim_L d, \quad x \sim_L^G d, \quad y \sim_L e \quad \text{and} \quad y \sim_L^G e.$$

But, by (P13), we have $x \sim_L y$ (resp. $x \sim_L^G y$) if and only if $d = e$. This proves (c).

(d) Since $\sim_{LR}^G$ is the equivalence relation generated by $\sim_L^G$ and $\sim_R^G$ (see Lemma 16.1.1(b)), the statement (d) follows from (c). ■

20.4. Asymptotic Algebra

Let $\mathscr{J}_G$ denote the asymptotic algebra of $\mathscr{H}_G$, whose standard basis is denoted by $(\mathbf{t}_w^G)_{w \in W}$. Then it follows immediately from Corollary 20.3.2 and from Lemma 20.2.2 that:

Theorem 20.4.1 ($\mathscr{P}_G$). *Assume that G is a finite p-group. Then we have an isomorphism of algebras*

$$\mathbb{F}_p \otimes_{\mathbb{Z}} \mathscr{J}_G \simeq \mathrm{Br}_G(\mathscr{J})$$

which sends $1_{\mathbb{F}_p} \otimes_{\mathbb{Z}} \mathbf{t}_w^G$ *to* $\mathrm{br}_G(\mathbf{t}_w)$.

Remark 20.4.2 **($\mathscr{P}_G$).** If Γ is a two-sided cell of W meeting W^G and if C is a left cell of W meeting W^G, then Theorem 20.4.1 can be easily extended as follows: the $\mathbb{F}_p$-linear map which send $1_{\mathbb{F}_p} \otimes_{\mathbb{Z}} \mathbf{t}_w^G$ to $\mathrm{br}_G(\mathbf{t}_w)$ induces isomorphisms of algebras

$$\mathbb{F}_p \otimes_{\mathbb{Z}} \mathscr{J}_G[\Gamma^G] \xrightarrow{\sim} \mathrm{Br}_G(\mathscr{J}[\Gamma])$$

and
$$\mathbb{F}_p \otimes_{\mathbb{Z}} \mathscr{J}_G[C^G \cap (C^G)^{-1}] \simeq \mathrm{Br}_G(\mathscr{J}[C \cap C^{-1}]).$$

They glue together to induce the isomorphism of algebras $\mathbb{F}_p \otimes_{\mathbb{Z}} \mathscr{J}_G \xrightarrow{\sim} \mathrm{Br}_G(\mathscr{J})$ described in Theorem 20.4.1. □

Part VII
Examples

In this part we treat several examples, where the methods developed in the previous can be successful. Chapter 21 deals with finite dihedral groups: we give the description of cells and prove Lusztig's Conjectures (and all the conjectures stated in this book), and we also describe the cellular maps which are used for in bigger groups. In Chapter 22, we describe the partition into cells of the symmetric group. Affine Weyl groups are particularly interesting because of their links with Lie theory: we have decided to include the description of the partition of an affine Weyl group of type $\tilde{A}_2$, with a proof. Finally, Chapter 24 is entirely devoted to the particular case of free Coxeter groups (due to Shi and Yang): partition into cells is obtained, as well as a proof of Lusztig's Conjectures. The particular case where $|S| = 2$ is the infinite dihedral group: we describe the asymptotic algebra in terms of the representation rings of $\mathbf{SL}_2(\mathbb{C})$ or $\mathbf{PGL}_2(\mathbb{C})$. Finally, Chapter 25 contains pictures describing the partition into cells of finite, affine and some hyperbolic groups of rank 3 (and a few bigger ones): it does not contain any proofs.

Hypothesis. *Recall that S is finite and* $\varphi > 0$.

Chapter 21
Finite Dihedral Groups

Hypothesis and notation. *In this chapter, and only in this chapter, we assume that* $S = \{s, t\}$ *with* $s \neq t$*, and we set* $m = m_{st}$*. We assume that* $3 \leqslant m < \infty$*. If* m *is odd (resp. even), we write* $m = 2l + 1$ *(resp.* $m = 2l + 2$*) where* $l \geqslant 1$*. We denote by* $\omega : W \to W$ *the unique automorphism of* W *such that* $\omega(s) = t$ *and* $\omega(t) = s$*: it is involutive. If* r *is a natural number, we set*

$$s_r = \mathrm{Prod}(s,t;r)^{-1} \quad \textit{and} \quad t_r = \mathrm{Prod}(t,s;r)^{-1}.$$

We fix a primitive m*-th root of unity* ζ *in* $\mathbb{C}^\times$*. Finally, we set*

$$a = \varphi(s) \quad \textit{and} \quad b = \varphi(t),$$

and we assume that $a, b > 0$.

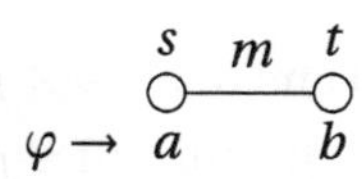

The aim of this chapter is to describe the Kazhdan–Lusztig basis, the different partitions into cells, as well as the structure of the cell modules for all choices of parameters. The first important consequence is the construction of left, right and two-sided cellular maps (which were used in Chapters 9 and 12). The second important consequence is the proof of Lusztig's Conjectures (Pk) for $1 \leqslant k \leqslant 15$. We also compute the maps $\lambda, \rho : W \to W$ and $\eta : W \to \boldsymbol{\mu}_2$ induced by the multiplication by T_{w_0} (see Theorem 17.1.5): this leads to a new cellular map whenever $a \neq b$.

C. Bonnafé, *Kazhdan–Lusztig Cells with Unequal Parameters*,
Algebra and Applications 24, https://doi.org/10.1007/978-3-319-70736-5_21

21.1. The Structure of W

21.1.A. About Elements

Let us recall some elementary facts. First,

$$|W| = 2m. \tag{21.1.1}$$

Note also that

$$\begin{cases} s_1 = s, \\ t_1 = t, \\ t_r s = s_{r+1}, \\ s_r t = t_{r+1} \end{cases} \tag{21.1.2}$$

if $r \geqslant 0$ and that

$$\ell(s_r) = \ell(t_r) = r \tag{21.1.3}$$

if, moreover, $r \leqslant m$. Also

$$s_m = t_m = w_0. \tag{21.1.4}$$

The list of elements of W is given by

$$W = \{1, s_1, s_2, \dots, s_{m-1}, t_1, t_2, \dots, t_{m-1}, w_0\}. \tag{21.1.5}$$

The Bruhat order is determined as follows: if $0 \leqslant i, j \leqslant m-1$, then

$$\begin{cases} s_i \leqslant s_j \text{ if and only if } i \leqslant j, \\ t_i \leqslant t_j \text{ if and only if } i \leqslant j, \\ s_i \leqslant t_j \text{ if and only if } i \leqslant j-1, \\ t_i \leqslant s_j \text{ if and only if } i \leqslant j-1. \end{cases} \tag{21.1.6}$$

The inverses of elements of W are given by

$$s_r^{-1} = \begin{cases} s_r & \text{if } r \text{ is odd,} \\ t_r & \text{if } r \text{ is even,} \end{cases} \quad \text{and} \quad t_r^{-1} = \begin{cases} t_r & \text{if } r \text{ is odd,} \\ s_r & \text{if } r \text{ is even.} \end{cases} \tag{21.1.7}$$

The right descent sets are given by

(21.1.8)
$$\begin{cases} \mathcal{R}(1)=\varnothing, \\ \mathcal{R}(s_r)=\{s\} & \text{if } 1 \leqslant r \leqslant m-1, \\ \mathcal{R}(t_r)=\{t\} & \text{if } 1 \leqslant r \leqslant m-1, \\ \mathcal{R}(w_0)=\{s,t\}, \end{cases}$$

while the left descent sets are given (for $1 \leqslant r \leqslant m-1$) by

(21.1.9)
$$\begin{cases} \mathcal{L}(s_r)=\begin{cases} \{s\} & \text{if } r \text{ is odd,} \\ \{t\} & \text{if } r \text{ is even,} \end{cases} \\ \mathcal{L}(t_r)=\begin{cases} \{t\} & \text{if } r \text{ is odd,} \\ \{s\} & \text{if } r \text{ is even,} \end{cases} \end{cases}$$

and

(21.1.10)
$$\mathcal{L}(w_0)=\{s,t\}.$$

21.1.B. About Characters

We denote by $\mathbf{1}_W$ the trivial character of W. Let $W^+=\langle st\rangle$. Then W^+ is cyclic of order m,

(21.1.11)
$$W^+=\mathrm{Ker}(\varepsilon) \quad \text{and} \quad W=W^+ \rtimes \langle s\rangle = W^+ \rtimes \langle t\rangle.$$

If $i \in \mathbb{Z}$, we set

$$\begin{aligned} \xi_i:\ W^+ &\longrightarrow \mathbb{C}^\times \\ (st)^j &\longmapsto \zeta^{ij}. \end{aligned}$$

Then ξ_i is an irreducible (linear) character of W^+, $\xi_{i+m}=\xi_i$ and

(21.1.12)
$$\mathrm{Irr}(W^+)=\{\xi_0,\xi_1,\dots,\xi_{m-1}\}.$$

We then set

$$\chi_i=\mathrm{Ind}_{W^+}^W \xi_i.$$

Note that

(21.1.13)
$$\chi_i=\chi_{i+m}=\chi_{-i}=\chi_{m-i}$$

for all $i \in \mathbb{Z}$. The following lemma is easy and well-known:

Lemma 21.1.14. *The irreducible characters of W are given as follows:*

(a) *If m is odd (recall that $m=2l+1$), then:*

(a$_1$) *If $1 \leqslant i \leqslant l$, then χ_i is irreducible.*

(a$_2$) $\chi_0 = \mathbf{1}_W + \varepsilon$.
(a$_3$) $\mathrm{Irr}(W) = \{1, \varepsilon, \chi_1, \chi_2, \ldots, \chi_l\}$.

(b) *If m is even (recall that $m = 2l+2$), then:*

(b$_1$) *If $1 \leqslant i \leqslant l$, then χ_i is irreducible.*
(b$_2$) $\chi_0 = \mathbf{1}_W + \varepsilon$ *and* $\chi_{l+1} = \varepsilon_s + \varepsilon_t$.
(b$_3$) $\mathrm{Irr}(W) = \{1, \varepsilon, \varepsilon_s, \varepsilon_t, \chi_1, \chi_2, \ldots, \chi_l\}$.

In this lemma, if m is even, then ε_s and ε_t denote the linear characters of W defined in Exercise 3.5. They are determined by the following conditions:

$$\begin{cases} \varepsilon_s(s) = -1, \\ \varepsilon_s(t) = 1, \end{cases} \qquad \text{and} \qquad \begin{cases} \varepsilon_t(s) = 1, \\ \varepsilon_t(t) = -1. \end{cases}$$

21.2. The Kazhdan–Lusztig Basis

If $y \in W$, we set

$$\mathbf{C}_y^= = \sum_{x \leqslant y} v^{\varphi(x)-\varphi(y)} T_x.$$

Recall that, since $a, b > 0$, then

(21.2.1) $$\mathbf{C}_s^= = T_s + v^{-a} = \mathbf{C}_s \qquad \text{and} \qquad \mathbf{C}_t^= = T_t + v^{-b} = \mathbf{C}_t.$$

Lemma 21.2.2. *Let $r \in \mathbb{N}$. Then*

$$\mathbf{C}_{t_r}^= \mathbf{C}_s = \begin{cases} \mathbf{C}_{s_{r+1}}^= & \text{if } r \in \{0,1\}, \\ \mathbf{C}_{s_{r+1}}^= + v^{a-b} \mathbf{C}_{s_{r-1}}^= & \text{if } 2 \leqslant r \leqslant m-1, \end{cases}$$

and

$$\mathbf{C}_{s_r}^= \mathbf{C}_t = \begin{cases} \mathbf{C}_{t_{r+1}}^= & \text{if } r \in \{0,1\}, \\ \mathbf{C}_{t_{r+1}}^= + v^{b-a} \mathbf{C}_{t_{r-1}}^= & \text{if } 2 \leqslant r \leqslant m-1. \end{cases}$$

Proof. We only need to prove the first equality (the other follows by applying the automorphism ω) and we may assume that $r \geqslant 2$ (since the case where $r \in \{0,1\}$ is straightforward). First, note that

$$\mathbf{C}_{s_{r+1}}^= = \sum_{i=0}^{r} v^{\varphi(t_i)-\varphi(t_r)} T_{t_i} \mathbf{C}_s,$$

and
$$\mathbf{C}^{=}_{s_{r-1}} = \sum_{i=0}^{r-2} v^{\varphi(t_i)-\varphi(t_{r-2})} T_{t_i}\mathbf{C}_s.$$

Since $\varphi(t_r) = a+b+\varphi(t_{r-2})$, we get

$$\mathbf{C}^{=}_{s_{r+1}} + v^{a-b}\mathbf{C}^{=}_{s_{r-1}} = (T_{t_r} + v^{\varphi(t_{r-1})-\varphi(t_r)}T_{t_{r-1}})\mathbf{C}_s + \sum_{i=0}^{r-2} v^{\varphi(t_i)-\varphi(t_r)}(1+v^{2a})T_{t_i}\mathbf{C}_s.$$

But $\mathbf{C}_s^2 = (v^a+v^{-a})\mathbf{C}_s$, so

$$\mathbf{C}^{=}_{s_{r+1}} + v^{a-b}\mathbf{C}^{=}_{s_{r-1}} = \Big(T_{t_r} + v^{\varphi(t_{r-1})-\varphi(t_r)}T_{t_{r-1}} + \sum_{i=0}^{r-2} v^{a+\varphi(t_i)-\varphi(t_r)}T_{t_i}\mathbf{C}_s\Big)\mathbf{C}_s.$$

Now, $T_{t_i}\mathbf{C}_s = T_{s_{i+1}} + v^{-a}T_{t_i}$ and $a+\varphi(t_i) = \varphi(s_{i+1})$, so

$$\begin{aligned} T_{t_r} &+ v^{\varphi(t_{r-1})-\varphi(t_r)}T_{t_{r-1}} + \sum_{i=0}^{r-2} v^{a+\varphi(t_i)-\varphi(t_r)}T_{t_i}\mathbf{C}_s \\ &= \sum_{i=0}^{r} v^{\varphi(t_i)-\varphi(t_r)}T_{t_i} + \sum_{i=0}^{r-2} v^{\varphi(s_{i+1})-\varphi(t_r)}T_{s_{i+1}} = \mathbf{C}^{=}_{t_r}, \end{aligned}$$

and the result follows. ■

If $m = 2l+2$ is even and $w \in W$, let us define $\mathbf{C}^{<}_w$ and $\mathbf{C}^{>}_w$ by the following formulas:

$$\begin{cases} \mathbf{C}^{<}_w = \mathbf{C}^{>}_w = \mathbf{C}^{=}_w & \text{if } \ell(w) \text{ is even,} \\ \mathbf{C}^{<}_w = \mathbf{C}^{>}_w = \mathbf{C}^{=}_w & \text{if } w \in \{s,t\}, \\ \mathbf{C}^{<}_{s_{2r+1}} = \mathbf{C}^{=}_{s_{2r+1}} + v^{a-b}\mathbf{C}^{=}_{s_{2r-1}} & \text{if } 1 \leqslant r \leqslant l, \\ \mathbf{C}^{<}_{t_{2r+1}} = \sum_{i=0}^{r} (-1)^i v^{i(a-b)}\mathbf{C}^{=}_{t_{2(r-i)+1}} & \text{if } 1 \leqslant r \leqslant l, \\ \mathbf{C}^{>}_{s_{2r+1}} = \sum_{i=0}^{r} (-1)^i v^{i(b-a)}\mathbf{C}^{=}_{s_{2(r-i)+1}} & \text{if } 1 \leqslant r \leqslant l, \\ \mathbf{C}^{>}_{t_{2r+1}} = \mathbf{C}^{=}_{s_{2r+1}} + v^{b-a}\mathbf{C}^{=}_{s_{2r-1}} & \text{if } 1 \leqslant r \leqslant l. \end{cases}$$

Lemma 21.2.3. *Assume that m is even and that $0 \leqslant r \leqslant m-1$. Then:*

(a) $\mathbf{C}^{<}_{t_r}\mathbf{C}_s = \mathbf{C}^{<}_{s_{r+1}}$.

(b) $$\mathbf{C}^{<}_{s_r}\mathbf{C}_t = \begin{cases} \mathbf{C}^{<}_{t_{r+1}} & \textit{if } r \in \{0,1\}, \\ \mathbf{C}^{<}_{t_{r+1}} + (v^{b-a}+v^{a-b})\mathbf{C}^{<}_{t_{r-1}} & \textit{if } r \in \{2,3\}, \\ \mathbf{C}^{<}_{t_{r+1}} + (v^{b-a}+v^{a-b})\mathbf{C}^{<}_{t_{r-1}} + \mathbf{C}^{<}_{t_{r-3}} & \textit{if } 4 \leqslant r \leqslant m-1. \end{cases}$$

(a′) $\mathbf{C}_{s_r}^{>}\mathbf{C}_t = \mathbf{C}_{t_{r+1}}^{>}$.

(b′) $\mathbf{C}_{t_r}^{<}\mathbf{C}_s = \begin{cases} \mathbf{C}_{s_{r+1}}^{<} & \text{if } r \in \{0,1\}, \\ \mathbf{C}_{s_{r+1}}^{<} + (v^{b-a}+v^{a-b})\mathbf{C}_{s_{r-1}}^{<} & \text{if } r \in \{2,3\}, \\ \mathbf{C}_{s_{r+1}}^{<} + (v^{b-a}+v^{a-b})\mathbf{C}_{s_{r-1}}^{<} + \mathbf{C}_{s_{r-3}}^{<} & \text{if } 4 \leqslant r \leqslant m-1. \end{cases}$

Proof. We prove only (a) and (b), as (a′) and (b′) can be deduced from (a) and (b) by applying the automorphism ω. The cases where $r \in \{0,1,2,3\}$ are straightforward. So we assume that $4 \leqslant r \leqslant m-1$.

(a) Let us first assume that r is odd. Write $r = 2r'+1$. Then

$$\mathbf{C}_{t_r}^{<} = \sum_{i=0}^{r'} (-1)^i v^{i(a-b)} \mathbf{C}_{t_{2(r'-i)+1}}^{=}.$$

So it follows from Lemma 21.2.2 that

$$\begin{aligned} \mathbf{C}_{t_r}^{<}\mathbf{C}_s &= \sum_{i=0}^{r'-1} (-1)^i v^{i(a-b)} (\mathbf{C}_{s_{2(r'-i)+2}}^{=} + v^{a-b}\mathbf{C}_{s_{2(r'-i)}}^{=}) + (-1)^{r'} v^{r'(a-b)} \mathbf{C}_{s_2} \\ &= \sum_{i=0}^{r'} (-1)^i v^{i(a-b)} \mathbf{C}_{s_{2(r'-i)+2}}^{=} + \sum_{i=0}^{r'-1} (-1)^i v^{(i+1)(a-b)} \mathbf{C}_{s_{2(r'-i)}}^{=} \\ &= \sum_{i=0}^{r'} (-1)^i v^{i(a-b)} \mathbf{C}_{s_{2(r'-i)+2}}^{=} + \sum_{i=1}^{r'} (-1)^{i-1} v^{i(a-b)} \mathbf{C}_{s_{2(r'-i)+2}}^{=} \\ &= \mathbf{C}_{s_{2r'+2}}^{=} \\ &= \mathbf{C}_{s_{r+1}}^{<}, \end{aligned}$$

as expected.

Assume now that r is even. Then $\mathbf{C}_{t_r}^{<} = \mathbf{C}_{t_r}^{=}$ and so, by Lemma 21.2.2, we get

$$\mathbf{C}_{t_r}^{<}\mathbf{C}_s = \mathbf{C}_{s_{r+1}}^{=} + v^{a-b}\mathbf{C}_{s_{r-1}}^{=} = \mathbf{C}_{s_{r+1}}^{<}$$

(the last equality follows from the fact that $r+1$ is odd). The proof of (a) is complete.

(b) Let us first assume that r is odd. Then $\mathbf{C}_{s_r}^{<} = \mathbf{C}_{s_r}^{=} + v^{a-b}\mathbf{C}_{s_{r-2}}^{=}$. So it follows from Lemma 21.2.2 that

$$\mathbf{C}_{s_r}^{<}\mathbf{C}_t = \mathbf{C}_{t_{r+1}}^{=} + v^{b-a}\mathbf{C}_{t_{r-1}}^{=} + v^{a-b}(\mathbf{C}_{t_{r-1}}^{=} + v^{b-a}\mathbf{C}_{t_{r-3}}^{=}).$$

Since $r+1$, $r-1$ and $r-3$ are even, we get from the definition that

$$\mathbf{C}_{s_r}^{<}\mathbf{C}_t = \mathbf{C}_{t_{r+1}}^{<} + (v^{b-a}+v^{a-b})\mathbf{C}_{t_{r-1}}^{<} + \mathbf{C}_{t_{r-3}}^{<},$$

as desired.

Assume now that r is even. Write $r = 2r'$. Then $\mathbf{C}^{<}_{s_r} = \mathbf{C}^{=}_{s_r}$, so it follows from Lemma 21.2.2 that

$$(*) \qquad \mathbf{C}^{<}_{s_r}\mathbf{C}_t = \mathbf{C}^{=}_{t_{r+1}} + v^{a-b}\mathbf{C}^{=}_{t_{r-1}}.$$

But,

$$\begin{aligned}\mathbf{C}^{<}_{t_{r+1}} + v^{a-b}\mathbf{C}^{<}_{t_{r-1}} &= \sum_{i=0}^{r'} (-1)^i v^{i(a-b)}\mathbf{C}^{=}_{2(r'-i)+1} + \sum_{i=1}^{r'}(-1)^{i-1} v^{i(a-b)}\mathbf{C}^{=}_{2(r'-i)+1} \\ &= \mathbf{C}_{t_{r+1}}.\end{aligned}$$

So it follows from $(*)$ that

$$\mathbf{C}^{<}_{s_r}\mathbf{C}_t = \mathbf{C}^{<}_{t_{r+1}} + v^{a-b}\mathbf{C}^{<}_{t_{r-1}} + v^{b-a}(\mathbf{C}^{<}_{t_{r-1}} + v^{a-b}\mathbf{C}^{<}_{t_{r-3}}),$$

as expected. ■

The description of the Kazhdan–Lusztig basis follows from the above computations.

Theorem 21.2.4. *Let $w \in W$. Then*

$$\mathbf{C}_w = \begin{cases} \mathbf{C}^{<}_w & \textit{if } a < b, \\ \mathbf{C}^{=}_w & \textit{if } a = b, \\ \mathbf{C}^{>}_w & \textit{if } a > b. \end{cases}$$

Proof. Note that

$$(\equiv) \qquad \mathbf{C}^{=}_w \equiv T_w \mod \mathcal{H}_{<0}$$

(because $a, b > 0$). Also Lemma 21.2.3 together with an easy induction argument on the length of w implies that

$$(-) \qquad \overline{\mathbf{C}^{<}_w} = \mathbf{C}^{<}_w \quad \text{and} \quad \overline{\mathbf{C}^{>}_w} = \mathbf{C}^{>}_w.$$

Assume first that $a < b$. By $(\equiv)$, it follows from the definition that $\mathbf{C}^{<}_w \equiv T_w \mod \mathcal{H}_{<0}$. Now, using $(-)$, we get $\mathbf{C}^{<}_w = \mathbf{C}_w$, as desired. The case where $a > b$ follows by applying the automorphism ω.

Assume now that $a = b$. Then Lemma 21.2.2 together with an easy induction argument implies that

$$\overline{\mathbf{C}^{=}_w} = \mathbf{C}^{=}_w.$$

So $\mathbf{C}^{=}_w = \mathbf{C}_w$ by $(\equiv)$. ■

21.3. Cells, Cell Modules

We follow the notation of Chapter 12. Since $|S|=2$, we have

$$\Gamma_s^{\text{uni}}=\{w\in W \mid \mathscr{R}(w)=\{s\}\}, \qquad \Gamma_t^{\text{uni}}=\{w\in W \mid \mathscr{R}(w)=\{t\}\}$$

and
$$\Gamma^{\text{uni}}=\Gamma_s^{\text{uni}}\,\dot\cup\,\Gamma_t^{\text{uni}}.$$

Note that
$$\Gamma_s^{\text{uni}}=\{s_r \mid 1\leqslant r\leqslant m-1\}=\{s_1,s_2,\dots,s_{m-1}\}$$

and
$$\Gamma_t^{\text{uni}}=\{t_r \mid 1\leqslant r\leqslant m-1\}=\{t_1,t_2,\dots,t_{m-1}\}$$

Now, we define three maps $\sigma_<$, $\sigma_>$ and $\sigma_=$ as follows:

$$\sigma_<(w)=\begin{cases} w & \text{if } w\in\{1,s,w_0s,w_0\},\\ ws & \text{if } w\notin\{1,s,w_0s,w_0\},\end{cases}$$

$$\sigma_>(w)=\begin{cases} w & \text{if } w\in\{1,t,w_0t,w_0\},\\ wt & \text{if } w\notin\{1,t,w_0t,w_0\},\end{cases}$$

and
$$\sigma_=(w)=\begin{cases} w & \text{if } w\in\{1,w_0\},\\ \omega(w)w_0 & \text{if } w\notin\{1,w_0\}.\end{cases}$$

These are involutions of W.

We finally introduce a last notation. If C and C' are two left cells, then we write $C \overset{L}{\dashrightarrow} C'$ if the following two conditions are satisfied:

- $C' <_L C$;
- If C'' is a left cell such that $C' \leqslant_L C'' \leqslant_L C$, then $C''\in\{C,C'\}$.

We similarly define the relations $\overset{R}{\dashrightarrow}$ and $\overset{LR}{\dashrightarrow}$. With this notation, we can state the main result of this section:

Theorem 21.3.1. *Recall that $|S|=2$, W is finite and $a,b>0$. Then:*

(a) *If $a<b$, then m is even and:*

(a$_1$) *The left cells of W and the elementary relations $\overset{L}{\dashrightarrow}$ are given by the following diagram:*

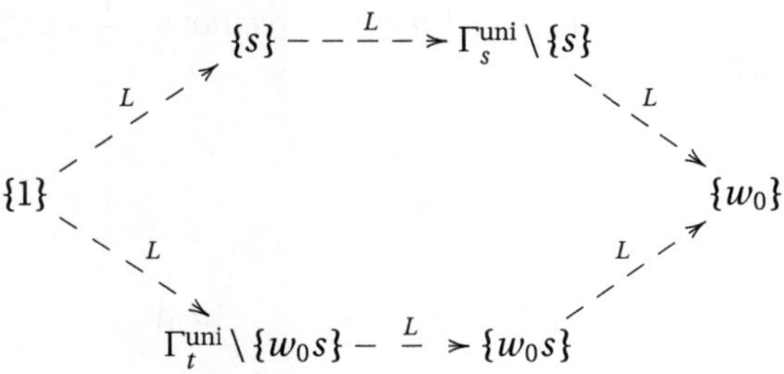

Recall that $s = s_1$ *and* $w_0 s = t_{m-1}$.

(a$_2$) *The two-sided cells of* W *and the elementary relations* $\overset{LR}{\dashrightarrow}$ *are given by the following diagram:*

$$\{1\} \overset{LR}{\dashrightarrow} \{s\} \overset{LR}{\dashrightarrow} \Gamma^{\mathrm{uni}} \setminus \{s, w_0 s\} \overset{LR}{\dashrightarrow} \{w_0 s\} \overset{LR}{\dashrightarrow} \{w_0\}.$$

(a$_3$) *The map* $\sigma_< : W \to W$ *is left cellular. It stabilizes all the left cells of cardinality* 1 *and exchanges the two other left cells (which have cardinality* $m-1$*). It is not two-sided cellular.*

(a′) *If* $a > b$, *then* m *is even and:*

(a$'_1$) *The left cells of* W *and the elementary relations* $\overset{L}{\dashrightarrow}$ *are given by the following diagram:*

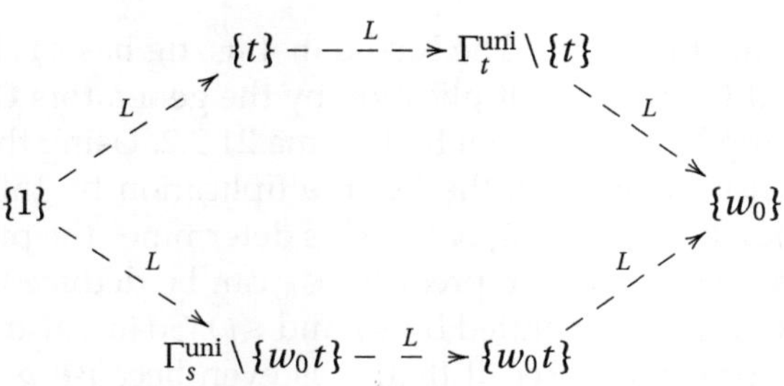

Recall that $t = t_1$ *and* $w_0 t = s_{m-1}$.

(a$'_2$) *The two-sided cells of* W *and the elementary relations* $\overset{LR}{\dashrightarrow}$ *are given by the following diagram:*

$$\{1\} \overset{LR}{\dashrightarrow} \{t\} \overset{LR}{\dashrightarrow} \Gamma^{\mathrm{uni}} \setminus \{t, w_0 t\} \overset{LR}{\dashrightarrow} \{w_0 t\} \overset{LR}{\dashrightarrow} \{w_0\}.$$

(a$'_3$) *The map* $\sigma_> : W \to W$ *is left cellular. It stabilizes all the left cells of cardinality* 1 *and exchanges the two other left cells (which have cardinality* $m-1$*). It is not two-sided cellular.*

(b) *If* $a = b$, *then:*

(b$_1$) *The left cells of W and the elementary relations* $\overset{L}{\dashrightarrow}$ *are given by the following diagram:*

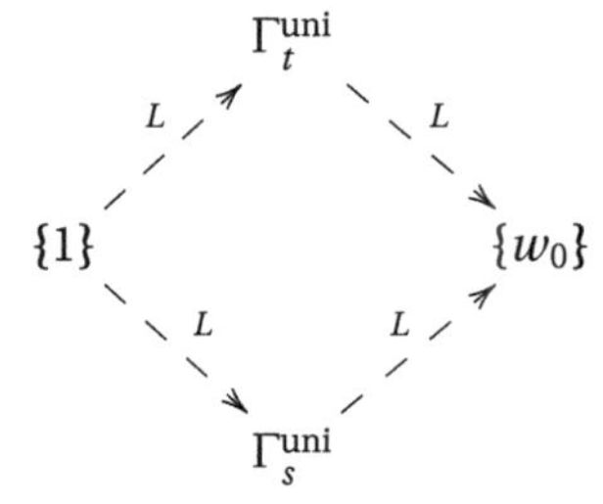

(b$_2$) *The two-sided cells of W and the elementary relations* $\overset{LR}{\dashrightarrow}$ *are given by the following diagram:*

$$\{1\} \overset{LR}{\dashrightarrow} \Gamma^{\mathrm{uni}} \overset{LR}{\dashrightarrow} \{w_0\}.$$

(b$_3$) *The map* $\sigma_= : W \longrightarrow W$ *is left cellular and*

$$\sigma_=(\Gamma_s^{\mathrm{uni}}) = \begin{cases} \Gamma_t^{\mathrm{uni}} & \textit{if } m \textit{ is odd,} \\ \Gamma_s^{\mathrm{uni}} & \textit{if } m \textit{ is even,} \end{cases} \quad \textit{and} \quad \sigma_=(\Gamma_t^{\mathrm{uni}}) = \begin{cases} \Gamma_s^{\mathrm{uni}} & \textit{if } m \textit{ is odd,} \\ \Gamma_t^{\mathrm{uni}} & \textit{if } m \textit{ is even.} \end{cases}$$

If, moreover, m is even, then $\sigma_=$ *is two-sided cellular.*

Proof. (a) Assume that $a < b$. The Kazhdan–Lusztig basis is known by Theorem 21.2.4 and the right multiplication by the generators $\mathbf{C}_s$ and $\mathbf{C}_t$ in the Kazhdan–Lusztig basis is known by Lemma 21.2.2. Using the antiautomorphism $*$, this also determines the left multiplication by the generators $\mathbf{C}_s$ and $\mathbf{C}_t$ in the Kazhdan–Lusztig basis. This determines the preorder $\leqslant_L$ (and so (a$_1$) follows easily), and the preorder $\leqslant_R$ can be deduced, as well as the preorder $\leqslant_{LR}$ (which is generated by $\leqslant_L$ and $\leqslant_R$). So (a$_2$) also follows easily.

Let us now prove (a$_3$). Recall that m is even because $a \neq b$. Therefore, $\Gamma_s^{\mathrm{uni}} \setminus \{s\} = \{s_2, s_3, \dots, s_{m-1}\}$ and $\Gamma_t^{\mathrm{uni}} \setminus \{w_0 s\} = \{t_1, t_2, \dots, t_{m-2}\}$ and right multiplication by s induces a bijection between these two left cells by 21.1.2. This shows that $\sigma_<$ satisfies Property (LC1) of Definition 8.4.1. To show Property (LC2), it is now sufficient to notice (thanks to Lemma 21.2.3(a)) that the map $\mathscr{H}^L[\Gamma_t^{\mathrm{uni}} \setminus \{w_0 s\}] \to \mathscr{H}^L[\Gamma_s^{\mathrm{uni}} \setminus \{s\}]$, $\mathbf{c}_w \mapsto \mathbf{c}_{\sigma_=(w)}$, is just given by the right multiplication by $\mathbf{C}_s$ inside $\mathscr{H}^{LR}[\Gamma^{\mathrm{uni}} \setminus \{s, w_0 s\}]$, so it is a morphism of left $\mathscr{H}$-modules. This completes the proof of (a).

Statement (a′) follows from (a) by applying the automorphism ω.

(b) Assume now that $a = b$. Again, (b$_1$) and (b$_2$) follow from the knowledge of the Kazhdan–Lusztig basis by Theorem 21.2.4 and the knowledge of the right (and so, the left) multiplication by $\mathbf{C}_s$ and $\mathbf{C}_t$ in the Kazhdan–Lusztig basis by Lemma 21.2.2.

Let us now prove (b$_3$). Two cases may occur:

• *Assume first that m is odd.* Then $\sigma_=$ exchanges the two left cells Γ_s^{uni} and Γ_t^{uni} and is the involution given by the following diagram:

$$\begin{array}{ccccccc}
\Gamma_s^{\mathrm{uni}} & & s_1 & s_2 & \cdots & s_{m-2} & s_{m-1} \\
\sigma_= \updownarrow & & \updownarrow & \updownarrow & & \updownarrow & \updownarrow \\
\Gamma_t^{\mathrm{uni}} & & t_{m-1} & t_{m-2} & \cdots & t_2 & t_1
\end{array}$$

So it is clear that $\sigma_=$ satisfies (LC1). To show Property (LC2), it is now sufficient to notice that the matrices of the left multiplication by $\mathbf{C}_s$ and $\mathbf{C}_t$, expressed in the A-basis $(\mathbf{c}_{s_1}, \mathbf{c}_{s_2}, \dots, \mathbf{c}_{s_{m-1}})$ of $\mathscr{H}^L[\Gamma_s^{\mathrm{uni}}]$, are equal to the corresponding matrices which give the action of $\mathbf{C}_s$ and $\mathbf{C}_t$, expressed in the A-basis $(\mathbf{c}_{t_1}, \mathbf{c}_{t_2}, \dots, \mathbf{c}_{t_{m-1}})$ of $\mathscr{H}^L[\Gamma_t^{\mathrm{uni}}]$. These matrices can be computed thanks to Lemma 21.2.2.

• *Assume now that m is even.* Then $\sigma_=$ stabilizes all the left cells and is the involution given by the following diagrams:

$$\begin{array}{ccccccc}
\Gamma_s^{\mathrm{uni}} & & s_1 & s_2 & \cdots & s_{m-2} & s_{m-1} \\
\sigma_= \updownarrow & & \updownarrow & \updownarrow & & \updownarrow & \updownarrow \\
\Gamma_s^{\mathrm{uni}} & & s_{m-1} & s_{m-2} & \cdots & s_2 & s_1
\end{array}$$

$$\begin{array}{ccccccc}
\Gamma_t^{\mathrm{uni}} & & t_1 & t_2 & \cdots & t_{m-2} & t_{m-1} \\
\sigma_= \updownarrow & & \updownarrow & \updownarrow & & \updownarrow & \updownarrow \\
\Gamma_t^{\mathrm{uni}} & & t_{m-1} & t_{m-2} & \cdots & t_2 & t_1
\end{array}$$

So it is clear that $\sigma_=$ satisfies (LC1). To show Property (LC2), it is now sufficient to notice that the multiplication by $\mathbf{C}_s$ and $\mathbf{C}_t$, expressed in the A-basis $(\mathbf{c}_{s_1}, \mathbf{c}_{s_2}, \dots, \mathbf{c}_{s_{m-1}})$ of $\mathscr{H}^L[\Gamma_s^{\mathrm{uni}}]$ (resp. $\mathscr{H}^L[\Gamma_t^{\mathrm{uni}}]$), commutes with the permutation matrix given by $\sigma_=$: this follows again from Lemma 21.2.2. ■

Remark 21.3.2. We provide only a few details for the proof of the fact that $\sigma_=$ is left cellular (in the split case) because another proof will be given later: indeed, $\sigma_=$ coincides with the map λ induced by right multiplication by T_{w_0} (see Proposition 21.5.1) and the sign map $\eta : W \to \mu_2$ given by Theorem 17.1.5 is constant on two-sided cells (see Example 17.3.2). □

Corollary 21.3.3. *If $|S| = 2$, then Conjectures 6.5.2 and 6.5.3 hold. The semicontinuity Conjecture 7.2.1 also holds.*

Proof. The first two conjectures are straightforward using Theorem 21.3.1. For the semicontinuity Conjecture 7.2.1, it follows easily from Theorem 21.3.1 for non-zero parameters, and the general case is left to the reader (see Exercise 21.1). ■

If C is a left cell, we denote by $[C]$ the character of the $\mathbb{C}W$-module $\mathbb{C}\otimes_A \mathscr{H}^L[C]$, where the morphism $A \to \mathbb{C}$ is given by $v^a \mapsto 1$ for all $a \in \mathscr{A}$. The character $[C]$ is called the *cellular character* associated with C.

Corollary 21.3.4. *The cellular characters are given as follows:*

(a) *If $a < b$, then $m = 2l+2$ is even and*

$$\begin{cases} [1] = \varepsilon, \\ [s] = \varepsilon_t, \\ [w_0 s] = \varepsilon_s, \\ [w_0] = \mathbf{1}_W, \\ [\Gamma_s^{\mathrm{uni}} \setminus \{s\}] = [\Gamma_t^{\mathrm{uni}} \setminus \{w_0 s\}] = \chi_1 + \chi_2 + \cdots + \chi_l. \end{cases}$$

(a′) *If $a > b$, then $m = 2l+2$ is even and*

$$\begin{cases} [1] = \varepsilon, \\ [t] = \varepsilon_s, \\ [w_0 t] = \varepsilon_t, \\ [w_0] = \mathbf{1}_W, \\ [\Gamma_s^{\mathrm{uni}} \setminus \{w_0 t\}] = [\Gamma_t^{\mathrm{uni}} \setminus \{t\}] = \chi_1 + \chi_2 + \cdots + \chi_l. \end{cases}$$

(b) *If $a = b$ and $m = 2l+1$ is odd, then*

$$\begin{cases} [1] = \varepsilon, \\ [w_0] = \mathbf{1}_W, \\ [\Gamma_s^{\mathrm{uni}}] = [\Gamma_t^{\mathrm{uni}}] = \chi_1 + \chi_2 + \cdots + \chi_l. \end{cases}$$

(c) *If $a = b$ and $m = 2l+2$ is even, then*

$$\begin{cases} [1] = \varepsilon, \\ [w_0] = \mathbf{1}_W, \\ [\Gamma_s^{\mathrm{uni}}] = \varepsilon_t + \chi_1 + \chi_2 + \cdots + \chi_l, \\ [\Gamma_t^{\mathrm{uni}}] = \varepsilon_s + \chi_1 + \chi_2 + \cdots + \chi_l. \end{cases}$$

Proof. The semisimplicity of the algebra $\mathbb{C}W$ implies that

$$(*) \qquad \sum_{C \in \mathrm{Cell}_L(W)} [C] = \begin{cases} \mathbf{1}_W + \varepsilon + 2(\chi_1 + \chi_2 + \cdots \chi_l) & \text{if } m \text{ is odd,} \\ \mathbf{1}_W + \varepsilon + \varepsilon_s + \varepsilon_t + 2(\chi_1 + \chi_2 + \cdots \chi_l) & \text{if } m \text{ is even.} \end{cases}$$

(a) Assume that $a < b$. The result for cells of cardinality 1 is straightforward. Then it follows from $(*)$ that

$$[\Gamma_s^{\mathrm{uni}} \setminus \{s\}] + [\Gamma_t^{\mathrm{uni}} \setminus \{w_0 s\}] = 2(\chi_1 + \chi_2 + \cdots \chi_l),$$

and it follows from the fact that $\sigma_<$ is left cellular that

$$[\Gamma_s^{\mathrm{uni}} \setminus \{s\}] = [\Gamma_t^{\mathrm{uni}} \setminus \{w_0 s\}].$$

This implies (a), while (a′) follows by applying the automorphism ω.

(b) Assume that $a = b$ and m is odd. The result for cells of cardinality 1 is straightforward. Then it follows from $(*)$ that

$$[\Gamma_s^{\mathrm{uni}}] + [\Gamma_t^{\mathrm{uni}}] = 2(\chi_1 + \chi_2 + \cdots \chi_l),$$

and it follows from the fact that $\sigma_=$ is left cellular that

$$[\Gamma_s^{\mathrm{uni}}] = [\Gamma_t^{\mathrm{uni}}].$$

This implies (b).

(c) Assume that $a = b$ and $m = 2l + 2$ is even. The result for cells of cardinality 1 is straightforward. Then it follows from $(*)$ that

$$[\Gamma_s^{\mathrm{uni}}] + [\Gamma_t^{\mathrm{uni}}] = \varepsilon_s + \varepsilon_t + 2(\chi_1 + \chi_2 + \cdots \chi_l).$$

But $\Gamma_s^{\mathrm{uni}} = w_0 \Gamma_t^{\mathrm{uni}}$, so it follows from Proposition 10.4.1 that

$$[\Gamma_s^{\mathrm{uni}}] = \varepsilon [\Gamma_t^{\mathrm{uni}}].$$

Since $\varepsilon \varepsilon_s = \varepsilon_t$ and $\varepsilon \chi_i = \chi_i$ for $1 \leqslant i \leqslant l$, this implies that

$$[\Gamma_s^{\mathrm{uni}}] = \gamma + \chi_1 + \chi_2 + \cdots \chi_l \quad \text{and} \quad [\Gamma_t^{\mathrm{uni}}] = \varepsilon\gamma + \chi_1 + \chi_2 + \cdots \chi_l,$$

where $\gamma \in \{\varepsilon_s, \varepsilon_t\}$. So it remains to prove that $\gamma = \varepsilon_t$.

From Corollary 5.5.6, we get that $\mathrm{Tr}(T_s, \mathscr{H}^L[\Gamma_s^{\mathrm{uni}}]) = (l+1)v^a - l v^{-a}$ so, after specializing $v^a \mapsto 1$ for all $a \in \mathscr{A}$, this gives

$$\gamma(s) + \sum_{i=1}^{l} \chi_i(s) = 1.$$

But $\chi_i(s) = 0$, so $\gamma(s) = 1$. This proves that $\gamma = \varepsilon_t$. ∎

21.4. Lusztig's a-Function and Lusztig's Conjectures

Proposition 21.4.1. *Lusztig's* **a** *function is given by:*

(a) *If $a<b$, then m is even and*

$$\mathbf{a}(z)=\begin{cases} 0 & \text{if } z=1,\\ a & \text{if } z=s,\\ b & \text{if } z\in\Gamma^{\mathrm{uni}}\setminus\{s,w_0s\},\\ b+\dfrac{m}{2}(b-a) & \text{if } z=w_0s,\\ \dfrac{m}{2}(a+b) & \text{if } z=w_0.\end{cases}$$

(a′) *If $a>b$, then m is even and*

$$\mathbf{a}(z)=\begin{cases} 0 & \text{if } z=1,\\ b & \text{if } z=t,\\ a & \text{if } z\in\Gamma^{\mathrm{uni}}\setminus\{t,w_0t\},\\ a+\dfrac{m}{2}(a-b) & \text{if } z=w_0t,\\ \dfrac{m}{2}(a+b) & \text{if } z=w_0.\end{cases}$$

(b) *If $a=b$, then*

$$\mathbf{a}(z)=\begin{cases} 0 & \text{if } z=1,\\ a=b & \text{if } z\in\Gamma^{\mathrm{uni}},\\ ma & \text{if } z=w_0.\end{cases}$$

(c) *In all cases, if $x,y,z\in W$ are such that* $\deg(h_{x,y,z})=\mathbf{a}(z)$, *then $x\sim_{LR} y\sim_{LR} z$.*

Proof. Using the automorphism ω, we may assume that $a\leqslant b$. Let us first prove (b). So assume that $a=b$. Then it follows from Theorem 15.2.5 that Lusztig's Conjecture (Pk) holds for $1\leqslant k\leqslant 15$. Since $s,t\in\Gamma^{\mathrm{uni}}$, (P4) and (P12) imply that $\mathbf{a}(w)=\mathbf{a}(s)=\mathbf{a}_s(s)=a$ for all $w\in\Gamma^{\mathrm{uni}}$, as desired. Now, $\mathbf{a}(1)=0$ and $\mathbf{a}(w_0)=ma$ by Example 14.2.10. This proves (b). Note also that, in the case $a=b$, then (c) follows from Lusztig's Conjecture (P8).

Assume now that $a<b$. Note that this forces m to be even, $m=2l+2$. Let us prove (a) and (c) in this case. We denote by $\mathbf{a}':W\to\mathscr{A}$ the map defined by the expected formula for **a** given in (a). Note that the statements (a) and (c) are easily checked for $z\in\{1,s\}$ in this case. If $z=w_0$, then (a) and (c) follow from Example 14.2.10. Also, it follows from Lemma 21.2.3 that $\mathbf{C}_s\mathbf{C}_{w_0s}=\mathbf{C}_{w_0}$ and $\mathbf{C}_t\mathbf{C}_{w_0s}=(v^b+v^{-b})\mathbf{C}_{w_0s}$. Therefore, the following equalities in $\mathscr{H}^L[w_0s]$ hold:

(2) $$\begin{cases} T_s \mathbf{c}_{w_0 s} = -v^{-a} \mathbf{c}_{w_0 s}, \\ T_t \mathbf{c}_{w_0 s} \equiv v^{b} \mathbf{c}_{w_0 s}. \end{cases}$$

So $T_w \mathbf{c}_{w_0 s} = (-1)^{\ell_s(w)} v^{\ell_t(w)b - \ell_s(w)a} \mathbf{c}_{w_0 s}$. This proves that $\deg(h_{w,w_0 s,w_0 s}) \leqslant (l+1)b - la$, with equality only if $w = w_0 s$. A similar result holds for $h_{w_0 s,w,w_0 s}$. This proves (a) and (c) for $z = w_0 s$.

Finally, assume that $z \in \Gamma^{\mathrm{uni}} \setminus \{s, w_0 s\}$ and let $x, y \in W$. If $z = t_i$ with $1 \leqslant i \leqslant m-2$, then $h_{z,t,z} = v^b + v^{-b}$, so $\mathbf{a}(z) \geqslant b$. If $z = s_i$ with $2 \leqslant i \leqslant m-1$, then $\mathbf{C}_{s_i} = \mathbf{C}_{t_{i-1}} \mathbf{C}_s$ by Lemma 21.2.3 and so $h_{t_{i-1},ts,s_i} = v^b + v^{-b}$. Again, $\mathbf{a}(z) \geqslant b$. Therefore,

$$\mathbf{a}(z) \geqslant b = \mathbf{a}'(z)$$

for all $z \in \Gamma^{\mathrm{uni}} \setminus \{s, w_0 s\}$.

Now, if x or y belongs to $\{w_0 s, w_0\}$, then it follows from Theorem 21.3.1(a) that $h_{x,y,z} = 0$. If x or y belongs to $\{1, s\}$, then $\deg(h_{x,y,z}) \leqslant a < b = \mathbf{a}'(z)$. So, if $\deg(h_{x,y,z}) = \mathbf{a}(z) \geqslant \mathbf{a}'(z)$, then $x \sim_{LR} y \sim_{LR} z$. This proves (c). We now prove by induction on $(\ell(y), \ell(x))$ (ordered lexicographically) that $\deg(h_{x,y,z}) \leqslant b$. We have just seen that we may assume that $x \sim_{LR} y \sim_{LR} z$. The result is obvious if $\ell(y) \in \{1,2\}$. Two cases remain to be treated:

• If $y = s_i$ with $3 \leqslant i \leqslant m-1$, then $\mathbf{C}_x \mathbf{C}_y = \mathbf{C}_x \mathbf{C}_{t_{i-1}} \mathbf{C}_s$ by Lemma 21.2.3 and $\mathbf{C}_x \mathbf{C}_{t_{i-1}} \equiv \sum_{j=1}^{m-2} \alpha_j \mathbf{C}_{t_j} \mod \mathscr{H}^{\leqslant_{LR} w_0 s}$ because $\mathscr{R}(t_{i-1}) = \{t\}$. But $\deg(\alpha_j) \leqslant b$ by the induction hypothesis, so $\mathbf{C}_x \mathbf{C}_{s_i} \equiv \sum_{j=1}^{m-2} \alpha_j \mathbf{C}_{s_{j+1}} \mod \mathscr{H}^{\leqslant_{LR} w_0 s}$ again by Lemma 21.2.3, and the result follows.

• If $y = t_i$ with $3 \leqslant i \leqslant m-2$, then $\mathbf{C}_{t_i} = \mathbf{C}_{t_{i-2}} h - \delta_{i>4} \mathbf{C}_{t_{i-4}}$, where $h = \mathbf{C}_{st} - v^{b-a} - v^{a-b}$ (see Lemma 21.2.3). By the induction hypothesis, $\mathbf{C}_x \mathbf{C}_{t_{i-2}} \equiv \sum_{j=1}^{m-2} \alpha_j \mathbf{C}_{t_j} \mod \mathscr{H}^{\leqslant_{LR} w_0 s}$ with $\deg(\alpha_j) \leqslant b$, and so

$$\mathbf{C}_x \mathbf{C}_{t_i} \equiv \sum_{j=1}^{m-2} \alpha_j \mathbf{C}_{t_j} h - \delta_{i>4} \mathbf{C}_x \mathbf{C}_{t_{i-4}} \mod \mathscr{H}^{\leqslant_{LR} w_0 s}.$$

So it remains to show that $\mathbf{C}_{t_j} h \in \bigoplus_{k=1}^m \mathbb{Z} \mathbf{C}_{t_k}$ for $1 \leqslant j \leqslant m-2$. But this follows from Lemma 21.2.3. ■

Corollary 21.4.2. *If $z \in W$, then $\mathbf{a}(z) \leqslant \boldsymbol{\Delta}(z)$ and*

$$\mathscr{D} = \begin{cases} \{1, s, t, sts, w_0 s, w_0\} & \text{if } a < b, \\ \{1, s, t, tst, w_0 t, w_0\} & \text{if } a > b, \\ \{1, s, t, w_0\} & \text{if } a = b. \end{cases}$$

Proof. Assume first that $a < b$ and recall that this forces m to be even, $m = 2l+2$. Recall that $w_0 \in \mathscr{D}$ by Example 14.2.10. Also, an easy computation shows that 1, s, t, $sts \in \mathscr{D}$. By definition,

$$p_{1,w_0 s} = p_{1,t_{2l+1}} = \sum_{i=0}^{l} (-1)^i v^{i(a-b)} v^{-(l-i)a-(l-i+1)b}$$
$$= \sum_{i=0}^{l} (-1)^i v^{(2i-l)a-(l+1)b}.$$

So $\deg(p_{1,w_0 s}) = l\,a-(l+1)b = -\mathbf{a}(w_0 s)$, so $w_0 s \in \mathscr{D}$. This proves that

$$\{1, s, t, sts, w_0 s, w_0\} \subset \mathscr{D}.$$

Now, let $z \in \Gamma^{\mathrm{uni}} \setminus \{s, w_0 s\}$. It remains to show that $\mathbf{a}(z) \leqslant \boldsymbol{\Delta}(z)$, with equality if and only if $z \in \{t, sts\}$. If $tz < z$, then $p_{1,z} = v^{-b} p_{t,z}$ by Proposition 5.6.1. So $\boldsymbol{\Delta}(z) \geqslant b = \mathbf{a}(z)$, with equality if and only if $z = t$. Similarly, if $zt < z$, then $\boldsymbol{\Delta}(z) \geqslant b = \mathbf{a}(z)$, with equality if and only if $z = t$. Therefore, we may assume that $z = s_{2i+1}$, with $1 \leqslant i \leqslant l$. But, by definition,

$$p_{1,s_{2i+1}} = v^{-(i+1)a-ib} + v^{a-b} v^{-ia-(i-1)b} = v^{-(i+1)a-ib}(1+v^{2a}).$$

So $\boldsymbol{\Delta}(s_{2i+1}) = b+(i-1)a+(i-1)b \geqslant b$, with equality if and only if $i=1$.

Assume now that $a = b$. By Theorem 21.2.4, we get that $\boldsymbol{\Delta}(z) = \varphi(z)$. So the result follows from the computation of the $\mathbf{a}$-function given by Proposition 21.4.1. ■

Theorem 21.4.3. *If* $|S| = 2$, *then Lusztig's Conjecture* (Pk) *holds for* $1 \leqslant k \leqslant 15$.

Proof. The case where $a = b$ is the split case, so it is a particular case of Theorem 15.2.5. The case where $a > b$ follows from the case $a < b$ by using the automorphism ω. So we may, and we will, assume that $a < b$.

(P1) has been proved in Corollary 21.4.2.

(P4), (P9), (P10) and (P11) follow from Theorem 21.3.1 and Proposition 21.4.1.

(P6) follows from the description of $\mathscr{D}$ given in Corollary 21.4.2.

(P14) follows from the description of two-sided cells given in Theorem 21.3.1.

We now prove four preliminary facts. First, since (P1) holds, it follows from Lemma 14.3.2(a) that

$$(\clubsuit) \qquad \sum_{d \in \mathscr{D}} \gamma_{x,y,d}\, n_d = \delta_{xy=1}$$

for all $x, y \in W$. Also,

$$(\diamondsuit) \qquad \text{if } \gamma_{x,y,z} \neq 0, \text{ then } z^{-1} \sim_R x \text{ and } z^{-1} \sim_L y.$$

Indeed, if $\gamma_{x,y,z} \neq 0$, then $h_{x,y,z^{-1}} \neq 0$ and $\deg(h_{x,y,z^{-1}}) = \mathbf{a}(z^{-1})$. So $x \sim_{LR} y \sim_{LR} z^{-1}$ by Proposition 21.4.1(c). Moreover, $z^{-1} \leqslant_R x$ and $z^{-1} \leqslant_L y$. So (♢) follows from the description of left and two-sided cells given in Theorem 21.3.1.

The descriptions of $\mathscr{D}$ and of the left cells show that

(♡) *if C is a left cell, then* $|C \cap \mathscr{D}| = 1$.

Together with (♢), this implies that:

(♠) if $x, y \in W$, then there is at most one d $\in \mathscr{D}$ such that $\gamma_{x,y,d} \neq 0$.

(P2), (P3) and (P5): Let x, $y \in W$. Let d denote the unique element of $\mathscr{D}$ such that $d = d^{-1} \sim_L y$ (by (♡)). It follows from (♣) and (♠) that

$$\gamma_{x,y,d}\, n_d = \delta_{xy=1}.$$

This proves (P2), (P3) and (P5).

(P7) follows from (P2), (P3), (P4), (P5) and Proposition 14.3.3(d).

(P8) follows from (P7) and Proposition 14.3.3(e).

(P12) follows from (P3), (P4) and (P8) for W and W_I (for any $I \subsetneq S$, then $|I| = 0$ or 1 and things become obvious) and Proposition 14.3.3(i).

(P13) The first statement is (♡) and the second one follows from (P3), (♢) and (♠).

(P15) By Lemma 14.3.5, it is sufficient to prove that, if s', $t' \in S$ and $y \in W$ are such that $s'y > y$ and $yt' > y$, then

$$(*) \qquad \mathbf{C}_{s'}(\hat{\mathbf{c}}_y \mathbf{C}_{t'}) = (\mathbf{C}_{s'} \hat{\mathbf{c}}_y)\mathbf{C}_{t'}.$$

But if $s' = s$, then the structure constants $\lambda_{y',y}^{s'}$ belong to $\mathbb{Z}$, and so $(*)$ follows from the fact that $\mathbf{C}_{s'}(\mathbf{c}_y \mathbf{C}_{t'}) = (\mathbf{C}_{s'} \mathbf{c}_y)\mathbf{C}_{t'}$ (as in the proof of (P15) in the split case, see the proof of Theorem 15.2.5). Similarly, if $t' = s$, then $(*)$ holds. In other words, we may assume that $s' = t' = t$, which forces $y = s_{2r+1}$, with $0 \leqslant r \leqslant m-1$. But this then follows from Lemma 21.2.3(b). ■

21.5. Action of T_{w_0}

Since Lusztig's Conjectures (Pk) hold for $1 \leqslant k \leqslant 15$, it follows from the results of Chapter 17 that there exist involutions λ and ρ of W and a sign map $\eta : W \to \mu_2$ such that

$$T_{w_0}\mathbf{C}_w \equiv \eta_w \mathbf{C}_{\rho(w)} \mod \mathscr{H}^{<_{LR} w}$$

and
$$\mathbf{C}_w T_{w_0} \equiv \eta_w \mathbf{C}_{\lambda(w)} \mod \mathscr{H}^{<_{LR} w}$$

for all $w \in W$ (see Theorem 17.1.5). The explicit description of λ, ρ and η is given by the following proposition:

Proposition 21.5.1. *We have:*

(a) *If $a < b$, then $m = 2l+2$ is even and*

$$\begin{cases} \lambda(w) = \rho(w) = w & \text{if } w \in \{1, s, w_0 s, w_0\}, \\ \lambda(s_{2i}) = \rho(s_{2i}) = s_{m-2i} & \text{if } 1 \leqslant i \leqslant l, \\ \lambda(s_{2i+1}) = \rho(s_{2i+1}) = s_{m+1-2i} & \text{if } 1 \leqslant i \leqslant l, \\ \lambda(t_{2i}) = \rho(t_{2i}) = t_{m-2i} & \text{if } 1 \leqslant i \leqslant l, \\ \lambda(t_{2i-1}) = \rho(t_{2i-1}) = t_{m-1-2i} & \text{if } 1 \leqslant i \leqslant l. \end{cases}$$

Moreover,

$$\eta_w = \begin{cases} 1 & \text{if } w \in \{1, w_0\}, \\ (-1)^{l+1} & \text{if } w \in \{s, w_0 s\}, \\ -1 & \text{if } w \notin \{1, s, w_0 s, w_0\}. \end{cases}$$

(a′) *If $a > b$, then $m = 2l+2$ is even and*

$$\begin{cases} \lambda(w) = \rho(w) = w & \text{if } w \in \{1, t, w_0 t, w_0\}, \\ \lambda(s_{2i}) = \rho(s_{2i}) = s_{m-2i} & \text{if } 1 \leqslant i \leqslant l, \\ \lambda(s_{2i-1}) = \rho(s_{2i-1}) = s_{m-1-2i} & \text{if } 1 \leqslant i \leqslant l, \\ \lambda(t_{2i}) = \rho(t_{2i}) = t_{m-2i} & \text{if } 1 \leqslant i \leqslant l, \\ \lambda(t_{2i+1}) = \rho(t_{2i+1}) = t_{m+1-2i} & \text{if } 1 \leqslant i \leqslant l. \end{cases}$$

Moreover,

$$\eta_w = \begin{cases} 1 & \text{if } w \in \{1, w_0\}, \\ (-1)^{l+1} & \text{if } w \in \{t, w_0 t\}, \\ -1 & \text{if } w \notin \{1, t, w_0 t, w_0\}. \end{cases}$$

(b) *If $a = b$, then $\lambda = \sigma_{=}, \rho = \sigma_{\_}$ and*

$$\eta_w = \begin{cases} (-1)^m & \text{if } w = 1, \\ 1 & \text{if } w = w_0, \\ -1 & \text{if } w \notin \{1, w_0\}. \end{cases}$$

Proof. In all cases, the result is easy if w belongs to a cell of cardinality 1. So we may assume that w belongs to the unique two-sided cell of cardinality >1.

(a) Assume first that $a<b$. Then $m=2l+2$ is even and so w_0 is central in W. In particular, $\lambda=\rho$. We now prove by induction on $i\in\{1,2,\dots,l\}$ that

$$(\mathscr{E}_i)\qquad\qquad \lambda(t_{2i-1})=\rho(t_{2i-1})=t_{m-1-2i}\quad\text{and}\quad \eta_{t_{2i-1}}=-1.$$

Since $t\in\mathscr{D}$, it follows from Corollary 21.4.2 that

$$\lambda(t)=w_0\partial^L_{w_0t}=w_0s_3=t_{m-3}\quad\text{and}\quad \eta_t=-1.$$

This proves $(\mathscr{E}_1)$ and also $(\mathscr{E}_l)$ because λ is an involution (and this proves the result if $l\in\{1,2\}$).

Now assume that $l\geqslant 3$ (and so $m\geqslant 8$). Let $h=\mathbf{C}_{ts}-v^{a-b}-v^{b-a}$. Then $\mathbf{C}_{t_3}=h\mathbf{C}_t$ and, since (λ,η) is left cellular, $\eta_{t_3}\mathbf{C}_{\lambda(t_3)}\equiv -h\mathbf{C}_{t_{m-3}} \mod \mathscr{H}^{\leqslant_{LR} t_{m-1}}$. But $h\mathbf{C}_{t_{m-3}}=\mathbf{C}_{t_{m-1}}+\mathbf{C}_{t_{m-5}}$ by Lemma 21.2.3(b), so $\eta_{t_3}\mathbf{C}_{\lambda(t_3)}\equiv-\mathbf{C}_{t_{m-5}} \mod \mathscr{H}^{\leqslant_{LR} t_{m-1}}$. Hence $\lambda(t_3)=t_{m-5}$ and $\eta_{t_3}=-1$, as expected. This proves $(\mathscr{E}_2)$ and $(\mathscr{E}_{l-1})$ (and this proves the result if $l\in\{3,4\}$).

So we can assume that $l\geqslant 5$ and $3\leqslant i\leqslant l-2$. Then $\mathbf{C}_{t_{2i-1}}=h\mathbf{C}_{t_{2i-3}}-\mathbf{C}_{t_{2i-5}}$ by Lemma 21.2.3(b). By the induction hypothesis and the left cellularity of (λ,η), we get

$$\eta_{t_{2i-1}}\mathbf{C}_{\lambda(t_{2i-1})}\equiv -h\mathbf{C}_{t_{m+1-2i}}-\mathbf{C}_{t_{m+3-2i}} \mod \mathscr{H}^{\leqslant_{LR} t_{m-1}}.$$

But $h\mathbf{C}_{t_{m+1-2i}}=\mathbf{C}_{t_{m+3-2i}}-\mathbf{C}_{t_{m-1-2i}}$, and so $\lambda(t_{2i-1})=t_{m-1-2i}$ and $\eta_{t_{2i-1}}=-1$.

Now, it follows from Lemma 21.2.3(a) that $\mathbf{C}_{t_{2i}}=\mathbf{C}_s\mathbf{C}_{t_{2i-1}}$, so the left cellularity of (λ,η) forces again $\lambda(t_{2i})=t_{m-2i}$ and $\eta_{t_{2i}}=-1$.

Finally, the map $\sigma_<$ is left cellular, so it commutes with $\rho=\lambda$ by Theorem 18.1.5. So, for all $w\in\Gamma^{\mathrm{uni}}\setminus\{s,w_0s\}$, we have $\lambda(ws)=\lambda(w)s$ and $\eta_{ws}=\eta_w$. This completes the proof of (a).

(a′) follows from (a) by applying the automorphism ω.

(b) As for (a), we get that $\lambda(s)=w_0\partial^L_{w_0s}=w_0t=\sigma_=(s)$ and similarly $\lambda(t)=\sigma_=(t)$. We then follow the same induction argument as in (a), taking into account the left cellularity of (λ,η) and the fact that $\mathbf{C}_s\mathbf{C}_{t_i^{-1}}=\mathbf{C}_{s_{i+1}^{-1}}+\mathbf{C}_{s_{i-1}^{-1}}$ and $\mathbf{C}_t\mathbf{C}_{s_i^{-1}}=\mathbf{C}_{t_{i+1}^{-1}}+\mathbf{C}_{t_{i-1}^{-1}}$. ■

Corollary 21.5.2. *If $|S|=2$, then the sign map $\eta: W\to\mu_2$ is constant on two-sided cells.*

Remark 21.5.3. Note that, to compute the action of T_{w_0} in the dihedral group case, we have used that Lusztig's Conjectures hold (by Theorem 21.4.3), so that one could obtain directly the image of the Duflo involutions. A direct proof would require much more computation. □

Exercises

Exercise 21.1. Assume here that $a = 0$ and $b > 0$ (in particular, m is even). Using the results of this chapter, compute the partition into cells and the structure of the cell modules.

Complete the proof of Corollary 21.3.3.

Exercise 21.2.* Assume here that $\mathscr{A} = \mathbb{Z}$ and $a = b = 1$ (split case). Prove directly, without using the results of Elias–Williamson [ElWi] which were recalled in Chapter 15, that $p_{x,y} \in \mathbb{N}[v^{-1}]$ and $h_{x,y,z} \in \mathbb{N}[v, v^{-1}]$ for all x, y and $z \in W$.

Chapter 22
The Symmetric Group

> **Hypothesis and notation.** *In this chapter, and only in this chapter, we fix a natural number $n \geqslant 1$ and we assume that $S = \{s_1, s_2, \ldots, s_{n-1}\}$ and that (W,S) is of type A_{n-1}, with Coxeter graph given by*
>
> 
>
>
> *As all the elements of S are conjugate in W, we may, and we will, assume that we are in the split case (i.e. $\varphi(s) = 1$ for all $s \in S$).*
>
> *We identify W with the symmetric group $\mathfrak{S}_n$ on the set $\{1, 2, \ldots, n\}$, by viewing s_i as the transposition $(i, i+1)$.*

We will describe the partition of $W = \mathfrak{S}_n$ into left, right and two-sided cells in terms of the Robinson–Schensted–Knuth correspondence. This will be obtained as an application of the methods developed in Chapters 8 and 9 (parabolic induction, cellular maps, coplactic/Vogan classes). It turns out that, in this case, coplactic classes coincide with Vogan left classes, so that they both give the partition into left cells.

22.1. Some Combinatorics

The reader may refer, for instance, to [Ful] or [Sag] for more details on results stated in this section. We will only provide a proof for Lemma 22.1.9, which will be used as a preliminary result for the main theorem of this chapter.

C. Bonnafé, *Kazhdan–Lusztig Cells with Unequal Parameters*,
Algebra and Applications 24, https://doi.org/10.1007/978-3-319-70736-5_22

22.1.A. Tableaux

We denote by **Part** the set of partitions, namely the set of sequences $\lambda = (\lambda_i)_{i \geqslant 1}$ where $\lambda_i \geqslant \lambda_{i+1}$ for all i and $\lambda_i = 0$ for some i. For simplification, we will sometimes write $\lambda = (\lambda_1, \lambda_2, \dots, \lambda_r)$ whenever $\lambda_{r+1} = 0$: if, moreover, $\lambda_r \neq 0$, then r will be called the *length* of the partition λ. We set $|\lambda| = \sum_{i \geqslant 1} \lambda_i$: the number $|\lambda|$ is called the *size* of λ. We denote by **Part**(n) the set of *partitions of* n, namely the set of partitions of size n. If $\lambda \in$ **Part**, we denote by $Y(\lambda)$ its *Young diagram*, namely

$$Y(\lambda) = \{(i,j) \in \mathbb{N}^* \times \mathbb{N}^* \mid 1 \leqslant i \leqslant r \text{ and } 1 \leqslant j \leqslant \lambda_i\}.$$

We denote by λ^* the dual partition of λ, that is, the partition such that

$$Y(\lambda^*) = \{(j,i) \mid (i,j) \in Y(\lambda)\}.$$

A *tableau of shape* λ is a map $P : Y(\lambda) \to \mathbb{Z}$. The tableau P is called *semistandard* if the following conditions are satisfied:

- if $i' \geqslant i$ and $j' \geqslant j$, then $P(i',j') \geqslant P(i,j)$;
- if $i' > i$ and $j' \geqslant j$, then $P(i',j') > P(i,j)$.

It is called standard if it is semistandard and if, moreover,

- the image of P is exactly $\{1, 2, \dots, |\lambda|\}$.

A *tableau* is a tableau P of shape λ for some partition λ: then λ will be denoted by $\mathrm{sh}(P)$ (the *shape* of P) and $|\lambda|$ is called the *size* of P and will be denoted by $|P|$. We denote by $\mathbf{ST}_2(n)$ the set of ordered pairs of standard tableaux of the same shape and of size n.

22.1.B. The Robinson–Schensted–Knuth Correspondence

If P is a semistandard tableau and $i \in \mathbb{Z}$, we define a new semistandard tableau $P \leftarrow i$ of size $|P|+1$ as follows (*row-insertion algorithm,* by induction on $|P|$):

- If i is larger than or equal to all the entries in the first row of P, then $P \leftarrow i$ is just obtained by adding a box with entry i at the end of the first row.
- Otherwise, let P' be the tableau obtained from P by removing the first row and let i' be the left-most entry of the first row of P such that $i' > i$.

Then this entry i' is replaced by i (which gives a new row R), and we obtain $P \leftarrow i$ by adding as a first row R to the semistandard tableau $P' \leftarrow i'$.

If $w \in W = \mathfrak{S}_n$, we set

$$\mathbf{P}(w) = (\dots((\varnothing \leftarrow w(1)) \leftarrow w(2)) \leftarrow \cdots) \leftarrow w(n).$$

Then $\mathbf{P}(w)$ is a standard tableau of size n. We set $\mathrm{sh}(w) = \mathrm{sh}(\mathbf{P}(w))$. It turns out that

$$\mathrm{sh}(w) = \mathrm{sh}(w^{-1}). \tag{22.1.1}$$

So, if we set

$$\mathbf{Q}(w) = \mathbf{P}(w^{-1}),$$

we can define a map

$$\begin{array}{rccc} \mathrm{RSK}\colon & \mathfrak{S}_n & \longrightarrow & \mathbf{ST}_2(n) \\ & w & \longmapsto & (\mathbf{P}(w), \mathbf{Q}(w)). \end{array}$$

Then (see, for instance, [Ful, §4.1]):

Theorem 22.1.2 (Robinson–Schensted–Knuth correspondence). *The map*

$$\mathrm{RSK}\colon \mathfrak{S}_n \longrightarrow \mathbf{ST}_2(n)$$

is bijective.

Corollary 22.1.3. *Let $w \in \mathfrak{S}_n$. Then $w^2 = 1$ if and only if $\mathbf{P}(w) = \mathbf{Q}(w)$.*

22.1.C. Descent Set

If P is a standard tableau of size n and if $k \in \{1, 2, \dots, n\}$, we denote by $(\mathbf{i}_P(k), \mathbf{j}_P(k))$ the element $P^{-1}(k) \in Y(\lambda)$. We define the *descent set* $\mathscr{D}(P)$ of P by

$$\mathscr{D}(P) = \{s_k \mid \mathbf{i}_P(k) < \mathbf{i}_P(k+1)\}.$$

Then (see, for instance, [Sag, Chapter 5, Exercise 22])

$$\mathscr{L}(w) = \mathscr{D}(\mathbf{P}(w)) \qquad \text{and} \qquad \mathscr{R}(w) = \mathscr{D}(\mathbf{Q}(w)) \tag{22.1.4}$$

for all $w \in W = \mathfrak{S}_n$.

22.1.D. Dominance Order

Let λ and μ be two partitions of n. We will write $\lambda \trianglelefteq \mu$ if

$$\sum_{i=1}^{k} \lambda_i \leqslant \sum_{i=1}^{k} \mu_i$$

for all $k \geqslant 1$. This defines a partial order on $\mathbf{Part}(n)$, called the *dominance order*. Recall that

$$\lambda \trianglelefteq \mu \Longleftrightarrow \mu^* \trianglelefteq \lambda^*. \tag{22.1.5}$$

We set

$$I_\lambda = S \setminus \{s_k \mid \exists\, i,\ k = \lambda_1 + \lambda_2 + \cdots + \lambda_i\}$$

and we denote by $\mathfrak{S}_\lambda$ the *Young subgroup* of $\mathfrak{S}_n$ defined as the standard parabolic subgroup W_{I_λ}. It is naturally isomorphic to $\mathfrak{S}_{\lambda_1} \times \mathfrak{S}_{\lambda_2} \times \cdots \times \mathfrak{S}_{\lambda_r}$. For simplification, we set $X_\lambda = X_{I_\lambda}$ and $w_\lambda = w_{I_\lambda}$ (the longest element of $\mathfrak{S}_\lambda$). We also set

$$P_\lambda = \mathbf{P}(w_{\lambda^*}) = \mathbf{Q}(w_{\lambda^*}).$$

Then it is clear and well-known that

$$\mathrm{sh}(w_{\lambda^*}) = \mathrm{sh}(P_\lambda) = \lambda \tag{22.1.6}$$

and that

$$P_\lambda(i,j) = \lambda_1^* + \cdots + \lambda_{j-1}^* + i \tag{22.1.7}$$

for all $(i,j) \in Y(\lambda)$.

Example 22.1.8. If $\lambda = (4,4,1) \in \mathbf{Part}(9)$, then $\lambda^* = (3,2,2,2)$, $w_{\lambda^*} = (1,3)(4,5)(6,7)(8,9)$ and

$$P_\lambda = \begin{array}{|c|c|c|c|}\hline 1 & 4 & 6 & 8 \\ \hline 2 & 5 & 7 & 9 \\ \hline 3 \\ \cline{1-1}\end{array}.$$

Note also that $I_{\lambda^*} = \{s_1, s_2, s_4, s_6, s_8\}$. □

Lemma 22.1.9. *Let* $\lambda \in \mathbf{Part}(n)$ *and let* $w \in X_{\lambda^*} w_{\lambda^*}$. *Then:*

(a) $\mathrm{sh}(w) \trianglelefteq \lambda$.
(b) *If* $\mathrm{sh}(w) = \lambda$, *then* $\mathbf{Q}(w) = P_\lambda$.

Proof. We will prove both (a) and (b) by induction on n. So we will need to emphasize the role of n by denoting by $\mathbf{P}_n(w)$ and $\mathbf{Q}_n(w)$ the tableaux $\mathbf{P}(w)$ and $\mathbf{Q}(w)$, respectively. Note first that (a) and (b) are obvious for $n \in \{1,2\}$. So we assume that $n \geqslant 3$ and that (a) and (b) hold when n is replaced by $n-1$. We set $S' = \{s_1, s_2, \ldots, s_{n-2}\}$, $X' = X_{S'}$ and $W' = W_{S'} = \mathfrak{S}_{n-1,1}$: we will identify W' with $\mathfrak{S}_{n-1}$. If $1 \leqslant k \leqslant n$, we denote by c_k the cycle $(k, k+1, \ldots, n)$, so that

$$X' = \{c_1, c_2, \ldots, c_n\}.$$

There exists a unique pair $(a, w') \in X' \times W'$ such that $w = aw'$. Then $a = c_k$ where $k = w(n)$ and

(22.1.10) $\quad \mathbf{Q}_{n-1}(w')$ *is obtained from* $\mathbf{Q}_n(w)$ *by removing the box containing* n.

We set $\mu = \mathrm{sh}(w) \in \mathbf{Part}(n)$ and $\mu' = \mathrm{sh}(w') \in \mathbf{Part}(n-1)$.

The length of λ^* is equal to λ_1 and we denote by λ' the partition of $n-1$ such that

$$\lambda'^* = (\lambda_1^*, \lambda_2^*, \ldots, \lambda_{\lambda_1-1}^*, \lambda_{\lambda_1}^* - 1).$$

In other words, if we denote by d the largest natural number such that $\lambda_d = \lambda_1$, then

$$\lambda' = (\lambda_1, \lambda_2, \ldots, \lambda_{d-1}, \lambda_d - 1, \lambda_{d+1}, \lambda_{d+2}, \ldots).$$

Again, there exists a unique pair $(b, u) \in X' \times W'$ such that $w_{\lambda^*} = bu$. It turns out that $b = c_{\lambda_1^* + \cdots + \lambda_{\lambda_1 - 1}^* + 1}$ and $u = w_{\lambda'^*}$. By hypothesis, we have $\mathscr{R}(w_{\lambda^*}) \subset \mathscr{R}(w)$ and so $\mathscr{R}(w_{\lambda'^*}) \subset \mathscr{R}(w')$, which means exactly that $w' \in (X_{\lambda'^*} \cap W')w_{\lambda'^*}$. So, by the induction hypothesis, we have:

(a′) $\mathrm{sh}(w') \trianglelefteq \lambda'$;

(b′) If $\mathrm{sh}(w') = \lambda'$, then $\mathbf{Q}_{n-1}(w') = P_{\lambda'}$.

Let μ' denote the shape of w'. Then there exists a unique $k_0 \geqslant 1$ such that $\mu_i = \mu'_i$ for $i \neq k_0$ and $\mu_{k_0} = \mu'_{k_0} + 1$. Let s denote the length of μ. We have

$$\mu'_1 + \mu'_2 + \cdots + \mu'_k \leqslant \lambda'_1 + \lambda'_2 + \cdots + \lambda'_k$$

for all $k \geqslant 1$ and

$$\mu'_1 + \mu'_2 + \cdots + \mu'_s = n - 1 = \lambda'_1 + \lambda'_2 + \cdots + \lambda'_s.$$

Now, if $\mathrm{sh}(w) \ntrianglelefteq \lambda$, this means that there exists a k such that

$$\mu_1 + \mu_2 + \cdots + \mu_k > \lambda_1 + \lambda_2 + \cdots + \lambda_k.$$

This forces $k < d$ and

$$\mu_1 + \mu_2 + \cdots + \mu_{k-1} = \lambda_1 + \lambda_2 + \cdots + \lambda_{k-1}$$

and

$$\mu_1 + \mu_2 + \cdots + \mu_k = \lambda_1 + \lambda_2 + \cdots + \lambda_k + 1.$$

So $\mu_1 \geqslant \mu_k = \lambda_k + 1 = \lambda_1 + 1$. But $\mu'_1 \leqslant \lambda'_1$. This shows that $k = k_0 = 1$ and $d > 1$. But, if $d > 1$, then $s_{n-1} \in \mathscr{R}(w_{\lambda^*})$. On the other hand, if $k_0 = 1$, this means that n occurs in the first row of $\mathbf{Q}_n(w)$, so $s_{n-1} \notin \mathscr{D}(\mathbf{Q}_n(w)) = \mathscr{R}(w)$, which is impossible.

So we have proved that $\mathrm{sh}(w) \trianglelefteq \lambda$. Now assume, moreover, that $\mu = \mathrm{sh}(w) = \lambda$. Since $\mu' \trianglelefteq \lambda'$ and $Y(\mu')$ (resp. $Y(\lambda')$) is obtained from $Y(\lambda)$ by removing one inner corner (i_μ, j_μ) (resp. (i_λ, j_λ)), this shows that $i_\mu \geqslant i_\lambda$. But, by construction, i_λ is maximal amongst the elements i such that there exists a j such that $(i,j) \in Y(\lambda)$. So $i_\mu = i_\lambda$, and so $(i_\mu, j_\mu) = (i_\lambda, j_\lambda)$. In particular, $\mu' = \mathrm{sh}(w') = \lambda'$ so $\mathbf{Q}_{n-1}(w') = \mathbf{P}_{\lambda'}$ by the induction hypothesis, and so $\mathbf{Q}_n(w) = \mathbf{P}_\lambda$ by 22.1.10. ■

22.2. Kazhdan–Lusztig Cells

22.2.A. *Knuth Relations*

Let $\mathscr{K}(n)$ denote the subgroup of Cact_W generated by the σ_I, where $I \in \mathscr{P}_{\mathrm{ir,sph}}(S)$ has cardinality 2 (in other words, $I = \{s_i, s_{i+1}\}$ for some $1 \leqslant i \leqslant n-2$: we set $\sigma_i = \sigma_{s_i, s_{i+1}}$). Any element $\sigma \in \mathscr{K}(n)$ induces two permutations σ^L and σ^R of the set W (see 18.2.2) and we recall the following facts (see Theorem 18.2.3: note that we do not need sign maps here because we are in the split case and the sign is constant on two-sided cells by Example 17.3.2):

Lemma 22.2.1. *Let σ and τ be two elements of $\mathscr{K}(n)$. Then:*

(a) *σ^L is a strongly left cellular map, while τ^R is a strongly right cellular map.*
(b) $[\sigma^L, \tau^R] = 1$.
(c) *$\sigma^R(w) = \sigma^L(w^{-1})^{-1}$ for all $w \in W$.*

We denote by $\mathscr{K}^L(n)$ (resp. $\mathscr{K}^R(n)$) the image of $\mathscr{K}(n)$ in $\mathfrak{S}_W$ under the map $\sigma \mapsto \sigma^L$ (resp. $\sigma \mapsto \sigma^R$). It follows from Lemma 22.2.1(b) that

(22.2.2) $$[\mathscr{K}^L(n), \mathscr{K}^R(n)] = 1.$$

It is readily seen that the Knuth relations can be defined in these terms: $x \smile^{\varphi}_{L} y$ if and only if there exists an $i \in \{1, 2, \dots, n-2\}$ such that $y = \sigma_i^R(x)$. In particular,

(22.2.3) $x \smile^{\varphi}_{L} y$ if and only if x and y are in the same $\mathscr{K}^R(n)$-orbit.

The following result is due to Knuth (see, for instance, [Sag, Theorem 3.4.3]):

Theorem 22.2.4. *Let x and y be two elements of $W = \mathfrak{S}_n$. Then:*

(a) *The elements x and y are in the same $\mathscr{K}^R(n)$-orbit if and only if $\mathbf{Q}(x) = \mathbf{Q}(y)$.*

(b) *The elements x and y are in the same $\mathscr{K}^L(n)$-orbit if and only if* $\mathbf{P}(x)=\mathbf{P}(y)$.
(c) *The elements x and y are in the same $(\mathscr{K}^L(n)\times\mathscr{K}^R(n))$-orbit if and only if* $\mathrm{sh}(x)=\mathrm{sh}(y)$.

Corollary 22.2.5. *Let x and y be two elements of $W=\mathfrak{S}_n$, let $\sigma\in\mathscr{K}^L(n)$ and let $\tau\in\mathscr{K}^R(n)$. Then:*

(a) $\mathbf{Q}(x)=\mathbf{Q}(y)$ *if and only if* $\mathbf{Q}(\sigma(x))=\mathbf{Q}(\sigma(y))$.
(b) $\mathbf{P}(x)=\mathbf{P}(y)$ *if and only if* $\mathbf{P}(\tau(x))=\mathbf{P}(\tau(y))$.

Proof. This follows from Theorem 22.2.4 and from the fact that σ (resp. τ) permutes the $\mathscr{K}^R(n)$-orbits (resp. $\mathscr{K}^L(n)$-orbits) by 22.2.2. ■

22.2.B. Vogan Classes

It is readily seen from the definitions that $\mathscr{K}^L(n)$ is the group $\mathscr{V}_{\nabla_2}$ defined in Section 9.3 and so, if $x, y\in W=\mathfrak{S}_n$, then

(22.2.6) $\quad \tau_x^{(\nabla_2,\mathscr{R})}=\tau_y^{(\nabla_2,\mathscr{R})}$ *if and only if* $\mathscr{R}(\sigma(x))=\mathscr{R}(\sigma(y))$ *for all* $\sigma\in\mathscr{K}^L(n)$

(see Definition 9.3.1 and Example 9.3.6).

22.2.C. Cells

The following theorem describes the partition of $W=\mathfrak{S}_n$ into left, right and two-sided cells.

Theorem 22.2.7. *Let x and y be two elements of $\mathfrak{S}_n$. Then:*

(a) $x\sim_L y$ *if and only if* $\mathbf{Q}(x)=\mathbf{Q}(y)$.
(b) $x\sim_R y$ *if and only if* $\mathbf{P}(x)=\mathbf{P}(y)$.
(c) $x\sim_{LR} y$ *if and only if* $\mathrm{sh}(x)=\mathrm{sh}(y)$.

Proof. It is clear that (a) is equivalent to (b) and we will first prove that (c) follows from (a). Indeed, by Lemma 16.1.1, the relation $\sim_{LR}$ is generated by $\sim_L$ and $\sim_R$. So, if we assume (a), it is clear that if $x\sim_{LR} y$, then $\mathrm{sh}(x)=\mathrm{sh}(y)$ by 22.1.1. Conversely, if $\mathrm{sh}(x)=\mathrm{sh}(y)$, write $(P,Q)=\mathrm{RSK}(x)$ and $(P',Q')=\mathrm{RSK}(y)$. Then $(P,Q')\in\mathbf{ST}_2(n)$ so that we can define $z=\mathrm{RSK}^{-1}(P,Q')$. If we assume (a), then $x\sim_R z\sim_L y$, so $x\sim_{LR} y$.

Let us now prove (a). By Propositions 9.2.2 and 9.3.3 and by Theorem 22.2.4, it is sufficient to prove that the following are equivalent:

(1) $\mathbf{Q}(x)=\mathbf{Q}(y)$;
(2) x and y have the same Vogan $\tau^{(\nabla_2,\mathscr{R})}$-invariant (i.e. $\mathscr{R}(\sigma(x))=\mathscr{R}(\sigma(y))$ for all $\sigma\in\mathscr{K}^L(n)$).

So, let us first assume that $\mathbf{Q}(x)=\mathbf{Q}(y)$ and let $\sigma\in\mathscr{K}^L(n)$. It follows from Corollary 22.2.5 that $\mathbf{Q}(\sigma(x))=\mathbf{Q}(\sigma(y))$, and so $\mathscr{R}(\sigma(x))=\mathscr{R}(\sigma(y))$ by 22.1.4. This proves that (1) implies (2).

Conversely, assume that $\mathscr{R}(\sigma(x))=\mathscr{R}(\sigma(y))$ for all $\sigma\in\mathscr{K}^L(n)$. Let $\lambda=\mathrm{sh}(x)$ and $\mu=\mathrm{sh}(y)$. By Theorem 22.2.4(c), there exists $(\sigma,\tau)\in\mathscr{K}^L(n)\times\mathscr{K}^R(n)$ such that $w_{\lambda^*}=\sigma\tau(x)=\tau\sigma(x)$. Note also that $\mathrm{sh}(\sigma\tau(y))=\mathrm{sh}(y)=\mu$. Now, $\mathbf{Q}(w_{\lambda^*})=\mathbf{Q}(\sigma(x))$ and $\mathbf{Q}(\sigma\tau(y))=\mathbf{Q}(\sigma(y))$ by Theorem 22.2.4(a). So $\mathscr{R}(\sigma(x))=\mathscr{R}(w_{\lambda^*})$ by 22.1.4 and so $\mathscr{R}(\sigma(y))=\mathscr{R}(w_{\lambda^*})$ by hypothesis. Therefore, $\sigma(y)\in X_{\lambda^*}w_{\lambda^*}$. It then follows from Lemma 22.1.9(b) that $\mu=\mathrm{sh}(\sigma(y))\trianglelefteq\lambda$. By symmetry, we can prove that $\lambda\trianglelefteq\mu$, which proves that $\lambda=\mu$.

Therefore, $\sigma(y)\in X_{\lambda^*}w_{\lambda^*}$ and $\mathrm{sh}(\sigma(y))=\lambda$, so Lemma 22.1.9(c) implies that $\mathbf{Q}(\sigma(y))=\mathbf{Q}(w_{\lambda^*})=\mathbf{Q}(\sigma(x))$. So $\mathbf{Q}(x)=\mathbf{Q}(y)$ by Corollary 22.2.5(a). ■

22.2.D. Complements

First, note that the left cells are left-connected (because they are $\mathscr{K}^R(n)$-orbits by Theorem 22.2.7(a)), and so Conjecture (L5) holds in type A_{n-1}. It also follows from Theorem 22.2.7(a) that left cells in $W=\mathfrak{S}_n$ are parametrized by standard tableaux of size n: if P is a standard tableau, let $C(P)$ denote the associated left cell, that is

$$C(P)=\{w\in\mathfrak{S}_n \mid \mathbf{Q}(w)=P\}.$$

Then it is a consequence of the Robinson–Schensted correspondence that:

Proposition 22.2.8. *Let P be a standard tableau. Then the left cell $C(P)$ contains a unique involution, namely* $\mathrm{RSK}^{-1}(P,P)$.

22.2.E. Example: Type A_3

The case where $n=2$ (type A_1) is trivial while the case where $n=3$ (type A_2) is easy (see Example 6.7.6). We provide in Figure 22.a the description of the left and two-sided cells in the case where $n=4$ (type A_3), following the convention of Section 6.6.

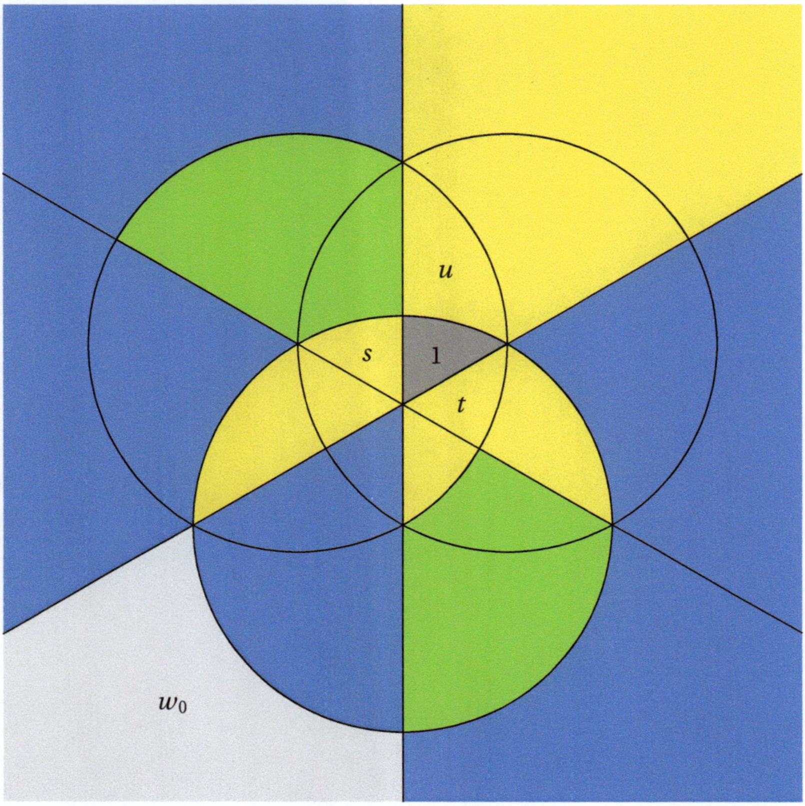

Fig. 22.a. Cells in type A_3

Notes

The partition of the symmetric group into cells was obtained by Kazhdan and Lusztig in their original paper [KaLu1]. We mainly followed the treatment made by Ariki [Ari].

Exercises

Exercise 22.1* (Hohlweg [Hoh]). Let λ be a partition of n. We denote by $\mathscr{P}_\lambda(S)$ the set of subsets I of S such that I is conjugate to I_λ in W. Show that the elements of minimal length in the two-sided cell $\mathrm{sh}^{-1}(\lambda)$ are the elements w_I, for $I \in \mathscr{P}_{\lambda^*}(S)$.

Exercise 22.2* (Schützenberger [Sch]). Show that all the involutions in a two-sided cell of $W = \mathfrak{S}_n$ are conjugate in W (compare with Conjecture 6.5.3).

Chapter 23
Affine Weyl Groups of Type $\tilde{A}_2$

Hypothesis and notation. *We assume in this chapter, and only in this chapter, that (W,S) is a tame Coxeter group of type $\tilde{A}_2$. We write $S=\{s,t,u\}$, so that the Coxeter graph of (W,S) is*

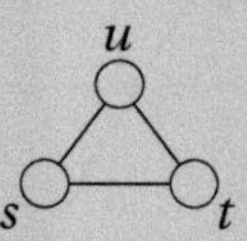

As all the elements of S are conjugate in W, we may, and we will, restrict to the split case (i.e. we assume throughout this chapter that $\mathscr{A}=\mathbb{Z}$ and $\varphi=\ell$).

We aim to describe the partition of W into left (and so, right) and two-sided cells. However, we will not give a self-contained proof, as we will rely on the following result of Lusztig [Lus12, Theorem 7.2] (see Remark 14.1.6), which holds because (W,S) is tame:

$$\forall\, x,y,z \in W,\ \deg(t_{x,y,z}) \leqslant \max_{I \in \mathscr{P}_{\mathrm{sph}}(S)} \varphi(w_I) = 3.$$

By Lemma 14.1.7, this implies that

$$\mathbf{a}(z) \leqslant 3$$

for all $z \in W$ (Figure 23.a).

The result will be displayed according to the conventions of Section 6.6: it is given by Figure 23.b. To have a better idea about which elements correspond to alcoves, we recall from Figure 23.d that the first elements of (W,S) are given by the following picture:

C. Bonnafé, *Kazhdan–Lusztig Cells with Unequal Parameters*,
Algebra and Applications 24, https://doi.org/10.1007/978-3-319-70736-5_23

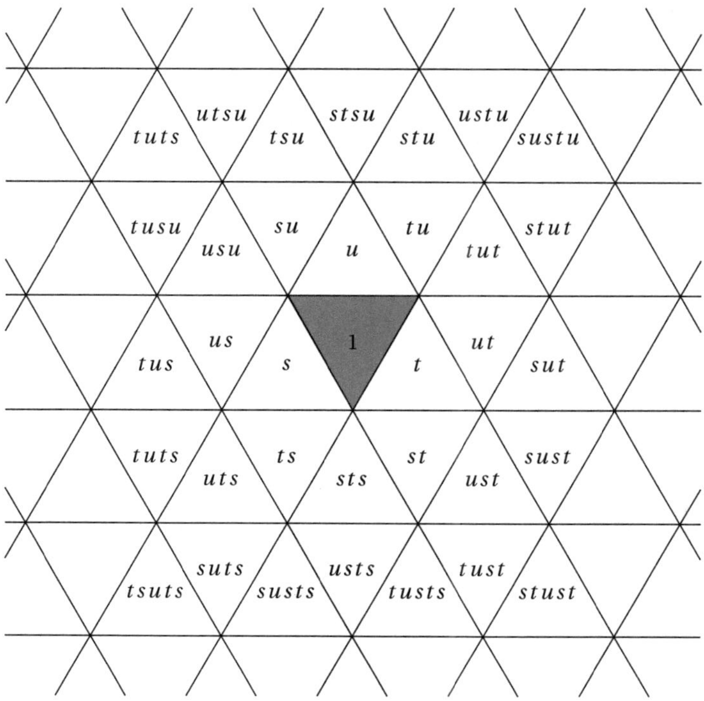

Fig. 23.a. Small length elements in type $\tilde{A}_2$

23.1. The Partition into Cells

Theorem 23.1.1. *The partition of* (W,S) *into left and two-sided cells is given by Figure 23.b.*

The rest of this section is devoted to the proof of Theorem 23.1.1.

23.1.A. Submaximal Cells

As we are in the split case, we will use the notation and results of Chapter 12. Recall that Γ^{uni} denotes the set of $W\setminus\{1\}$ which have a unique reduced expression and, if $s_0 \in S$, we set

$$\Gamma_{s_0}^{\mathrm{uni}} = \{w \in \Gamma^{\mathrm{uni}} \mid \mathscr{R}(w) = \{s_0\}\}.$$

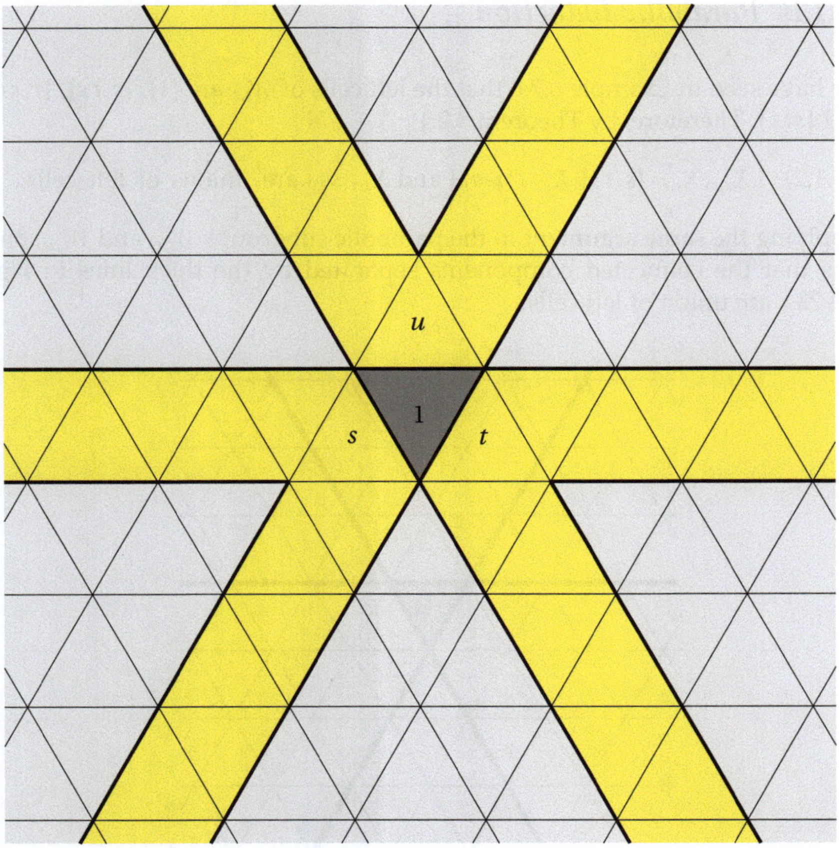

Fig. 23.b. Cells in type $\tilde{A}_2$

It follows from Theorem 12.2.3 that Γ^{uni} is a two-sided cell and that Γ^{uni}_s, $\Gamma^{\mathrm{uni}}_\tau$ and Γ^{uni}_u are the left cells of W contained in Γ^{uni}.

Now, the set Γ^{uni} is in bijection with the set $\mathscr{U}nbr(W,S)$ of unbraided paths in the Coxeter graph $\mathscr{C}ox(W,S)$ (see 12.1.10 and 12.1.11). As $m_{s_0 s_1} = 3$ for all $s_0 \neq s_1 \in S$, these unbraided paths are easily described (just turn around the diagram, in one way or the other), and it follows that the set Γ^{uni} is the set of yellow elements in Figure 23.b, and that its left-connected components are the left cells (see also Corollary 12.2.4). So we have proved that:

Fact 1. *The set of yellow elements form a two-sided cell (the unique submaximal one) and its left-connected components are left cells (the submaximal ones).*

23.1.B. *Parabolic Induction*

We have seen in Example 6.7.6 that the left cells of $W_{s,t}$ are $\{1\}$, $\{s,ts\}$, $\{t,st\}$ and $\{sts\}$. Therefore, by Theorem 8.2.1,

(23.1.2) $X_{s,t}, X_{s,t}\cdot\{s,ts\}, X_{s,t}\cdot\{t,st\}$ and $X_{s,t}\ sts$ are unions of left cells.

Applying the same argument to the parabolic subgroups $W_{s,u}$ and $W_{t,u}$, one gets that the connected components separated by the thick lines in Figure 23.c are union of left cells.

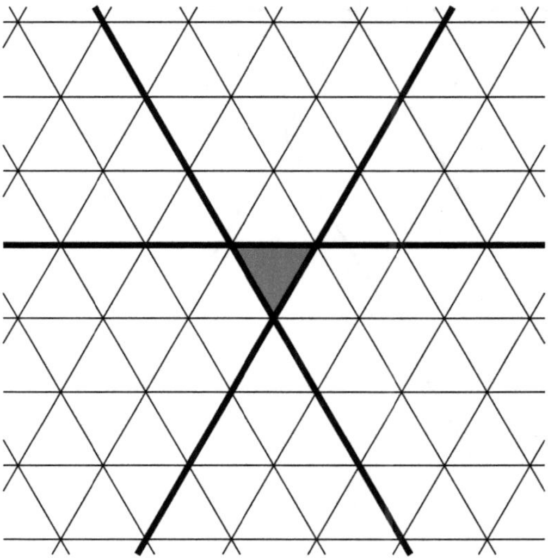

Fig. 23.c. Parabolic induction in type $\tilde{A}_2$

23.1.C. *Cellular Maps (∗-Operation)*

Recall from Example 8.4.2 that the permutation $\sigma_{s,u} : W_{s,u} \to W_{s,u}$ equal to $(s,su)(u,us)$ (written as a product of transpositions) is a left cellular map. We denote by $\bigstar_{s,u}$ its left extension (it is an example of the ∗-operation defined by Kazhdan–Lusztig [KaLu1]).

Lemma 23.1.3. *If* $w \in X_{s,t}sts$*, then* $\bigstar_{s,u} = wu$*.*

Proof. As it is readily seen from the pictures, $X_{s,t}sts$ (which is the bottom blue triangle in Figure 23.c) is contained in $X_{s,u}s$, so the result follows from the definition. ■

Lemma 23.1.3 shows that $X_{s,t}stsu$ is a union of left cells. But, on the picture, $X_{s,t}stsu$ is obtained as the symmetric image of $X_{s,t}sts$ through the affine line defining the reflection u (the thick horizontal line of Figure 23.c). It is described in Figure 23.d (in this figure, the bottom blue cone is $X_{s,t}sts$ and the top blue cone is its image through $\bigstar_{s,u}$).

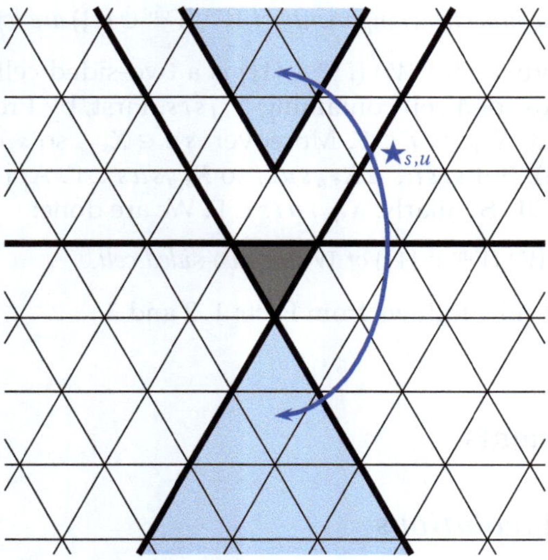

Fig. 23.d. Action of $\bigstar_{s,u}$ on $X_{s,t}sts$

Applying the symmetry between s, t and u, one gets:

Corollary 23.1.4. *The partition described by Figure 23.b is coarser than the partition into left cells.*

To prove the statement about left cells in Theorem 23.1.1, it remains to show that the connected components of the same colour in Figure 23.b are contained in a single left cell.

23.1.D. Lowest Two-sided Cell

Lemma 23.1.5. *The subset $X_{s,t}sts$ is a left cell.*

Proof. Let $x \in X_{s,t}$. Since $\ell(xsts) = \ell(x) + \ell(sts)$, we get that $xsts \leqslant_L sts$. So it follows from (P4), which holds because we are in the split case, that

$\mathbf{a}(xsts) \geqslant \mathbf{a}(sts)$. By (P12), $\mathbf{a}(sts)$ can be computed in $W_{s,t}$, which is of type A_2, and one gets that $\mathbf{a}(sts) = 3$. So $\mathbf{a}(xsts) \geqslant 3$, but it follows from the inequality stated in the introduction of this chapter that $\mathbf{a}(xsts) \leqslant 3$. So $\mathbf{a}(xsts) = 3 = \mathbf{a}(sts)$, and it follows from (P9) that $xsts \sim_L sts$. ■

Corollary 23.1.6. *$X_{s,t}stsu$ is a left cell.*

Proof. This follows from Lemma 23.1.5 and Lemma 23.1.3. ■

Applying again the symmetry in s, t and u, one gets:

Fact 2. *The six left-connected components of $W \setminus (\Gamma^{\text{uni}} \cup \{1\})$ are left cells.*

It remains to prove that $W \setminus (\Gamma^{\text{uni}} \cup \{1\})$ is a two-sided cell. For this, let Γ be the unique two-sided cell containing $X_{s,t}sts$. First, by Proposition 16.3.1 and Lemma 23.1.3, $X_{s,t}stsu \subset \Gamma$. Moreover, $su \in X_{s,t}$, so $susts \in \Gamma$ and so $stsus \in \Gamma$ by (P14). But $stsus \in X_{s,u}sus$, so $X_{s,u}sus \subset \Gamma$. Applying $\bigstar_{s,t}$, one gets that $X_{s,u}sus \subset \Gamma$. Similarly, $X_{s,u}tuts \subset \Gamma$. We are done:

Fact 3. *The subset $W \setminus (\Gamma^{\text{uni}} \cup \{1\})$ of W is a two-sided cell.*

Theorem 23.1.1 now follows from Facts 1, 2 and 3.

23.2. Complements

23.2.A. Duflo Involutions

It follows from the results of the previous sections that:

$$\mathbf{a}(z) = \begin{cases} 0 & \text{if } z = 1, \\ 1 & \text{if } z \in \Gamma^{\text{uni}}, \\ 3 & \text{if } z \in \Gamma. \end{cases} \tag{23.2.1}$$

Using the fact that there is a unique element of $\mathscr{D}$ in each left cell by (P13), one gets easily that

$$\mathscr{D} = \{1, s, t, u, sts, tut, sus, ustsu, stuts, tsust\}. \tag{23.2.2}$$

23.2.B. Other Conjectures

It follows easily from the description of cells in W that Conjecture (Lk) holds in (W, S) for $1 \leqslant k \leqslant 8$. Also:

Proposition 23.2.3. *Conjecture 6.5.3 holds in type $\tilde{A}_2$.*

Proof. It follows from Theorem 3.9.1 that $\{1, s\}$ is a set of representatives of conjugacy classes of involutions in W, so the result follows. ■

Notes

The partition into cells of affine Weyl groups of type $\tilde{A}_2$, $\tilde{B}_2$ and $\tilde{G}_2$ was obtained by Lusztig [Lus12] in the split case.

Chapter 24
Free Coxeter Groups

Hypothesis and notation. *We assume in this chapter, and only in this chapter, that* (W,S) *is a* ***free Coxeter system****. That is, we assume that* $m_{st}=\infty$ *for all* $s,\ t\in S$ *such that* $s\neq t$*. We set* $a_0=0$ *and we write*

$$\varphi(S)=\{a_1,a_2,\ldots,a_l\},\ \textit{with}\ l\geqslant 1\ \textit{and}\ 0=a_0<a_1<a_2<\cdots<a_l$$

and we define

$$S_k=\varphi^{-1}(a_k)\quad\textit{and}\quad I_k=\{s\in S\mid \varphi(s)\leqslant a_k\}$$

for $0\leqslant k\leqslant l$*. In particular,* $S_0=I_0=\varnothing$*. Also* I_{-1} *denotes a formal symbol such that* $W_{I_{-1}}=\varnothing$*. Note that* $I_k=S_0\,\dot\cup\,S_1\,\dot\cup\,\cdots\,\dot\cup\,S_k$ *if* $k\geqslant 0$*.*

Remark.— A free Coxeter group is often called a *universal Coxeter group*, as in [ShYa1] (from which this chapter is largely inspired). □

We describe the partition of W into cells and prove Lusztig's Conjectures (P1), (P2),..., (P15) in this case, whatever the weight function φ is. Recall that we assume throughout this part that S is finite but the reader can check that all the results (except the finiteness result of Corollary 24.5.2) remain valid if only $\varphi(S)$ is assumed to be finite.

C. Bonnafé, *Kazhdan–Lusztig Cells with Unequal Parameters*,
Algebra and Applications 24, https://doi.org/10.1007/978-3-319-70736-5_24

24.1. Group-Theoretic Properties

24.1.A. Reduced Expressions

First, note that every element of W has a unique reduced expression by Lemma 12.1.6. In other words, $\Gamma^{\text{uni}} = W \setminus \{1\}$. Therefore, $\mathscr{L}(x)$ and $\mathscr{R}(x)$ are singletons for all $x \in \Gamma^{\text{uni}}$ (as in Chapter 12, we write $\mathscr{L}(x) = s$ or $\mathscr{R}(x) = s$ if $\mathscr{L}(x) = \{s\}$ or $\mathscr{R}(x) = \{s\}$, respectively). It also follows from Lemma 12.1.6 that

$$\text{(24.1.1)} \qquad \textit{If } \mathscr{R}(x) \neq \mathscr{L}(y), \textit{ then } \ell(xy) = \ell(x) + \ell(y).$$

24.1.B. Involutions

Recall from Proposition 12.1.13 that all the non-trivial involutions in W are reflections. More precisely:

Proposition 24.1.2. *The set $S \cup \{1\}$ is a set of representatives of conjugacy classes of involutions of W. If $s \in S$, then*

$$C_W(s) = \langle s \rangle.$$

Proof. It follows from Lemma 3.2.7 that if s, $t \in S$ are conjugate in W, then $s = t$. This proves the first assertion.

Now, let $s \in S$ and $w \in W$ be such that $ws = sw$. We wish to prove that $w \in \langle s \rangle$. By replacing w by sw, ws or sws if necessary, we may assume that $w \in X_{\{s\},\{s\}}$. If $w \neq 1$, let $(s_1, s_2, \dots, s_r)$ denote its reduced expression (with $r \geqslant 1$). Then $(s, s_1, s_2, \dots, s_r)$ and $(s_1, s_2, \dots, s_r, s)$ are reduced expressions of $sw = ws > w$, so they are equal. This shows that $s = s_1$ and so $sw < w$, which contradicts the fact that $w \in X_{\{s\},\{s\}}$. So $w = 1$, as desired. ■

Proposition 24.1.2 implies that Deodhar's Lemma simplifies drastically in a free Coxeter group:

Deodhar's Lemma 24.1.3. *Let I be a subset of S, let $s \in S$ and let $x \in X_I$ be such that $sx \notin X_I$. Then $s \in I$ and $x = 1$.*

Proof. By Deodhar's Lemma 3.6.12, there exists a $t \in I$ such that $sx = xt$. So $s = xtx^{-1}$ and hence $s = t$ and $x \in \langle s \rangle$ by Proposition 24.1.2. So $s \in I$ and $x = 1$ (because $x \in X_I$). ■

Corollary 24.1.4. *Let I and J be two subsets of S and let $x \in X_{IJ} \setminus \{1\}$. Then*

$$I \cap {}^xJ = \varnothing \quad \textit{and} \quad W_I \cap {}^xW_J = \{1\}.$$

Proof. Recall that the second equality follows from the first one by Corollary 3.6.9. So let I and J be two subsets of S and let $x \in X_{IJ}$ be such that $I \cap {}^xJ \neq \varnothing$. Let $s \in I \cap {}^xJ$. If we write $t = x^{-1}sx \in J$, then $sx = xt \notin X_J$ and $x \in X_J$. So, by Deodhar's Lemma 24.1.3, we get that $x = 1$. ■

24.1.C. Some Partitions of W

If $0 \leqslant k \leqslant l$, we set

$$\Gamma_k = W_{I_k} \setminus W_{I_{k-1}}.$$

So, by the convention explained in the introduction of this chapter, we have $\Gamma_0 = \{1\}$. Since $\varnothing = W_{I_{-1}} \subset \{1\} = W_{I_0} \subset W_{I_1} \subset \cdots \subset W_{I_l} = W$, it is clear that

(24.1.5) $$W = \dot{\bigcup_{0 \leqslant k \leqslant l}} \Gamma_k.$$

If $w \in W$, we denote by $\mathbf{k}(w)$ the unique element of $\{0, 1, \ldots, l\}$ such that $w \in \Gamma_{\mathbf{k}(w)}$. For instance, $\mathbf{k}(1) = 0$ and, if $s \in S$, then $s \in S_{\mathbf{k}(s)}$.

We denote by $\mathscr{C}ell'(W,S)$ the set of pairs (s, w) such that $s \in S$ and $w \in W_{I_{\mathbf{k}(s)-1}}$. If $(s, w) \in \mathscr{C}ell'(W,S)$, then

(24.1.6) $$sw > w \qquad \text{and} \qquad \mathscr{L}(sw) = \{s\}.$$

We set $\mathscr{C}ell(W,S) = \mathscr{C}ell'(W,S) \cup \{(1,1)\}$. If $(s, w) \in \mathscr{C}ell(W,S)$, we set

$$C(s,w) = X_s^{I_{\mathbf{k}(s)}} \cdot sw \quad \subset W_{I_{\mathbf{k}(s)}}$$

where, by convention, we set $X_1^{I_0} = \{1\}$. Recall that the notation $X_I^J = X^J \cap W_I$ (for $I \subset J \subset S$) has been introduced in Definition 3.6.6. If $(s, w) \in \mathscr{C}ell(W,S)$ and $x \in X_s$, then

(24.1.7) $$\ell(xsw) = \ell(x) + \ell(sw)$$

by 24.1.1 (here, again by convention, we set $X_1 = W$). If $1 \leqslant k \leqslant l$, we denote by $\mathscr{C}ell_k(W,S)$ the set of pairs $(s, w) \in \mathscr{C}ell(W,S)$ such that $\mathbf{k}(s) = k$. For instance, $\mathscr{C}ell_0(W,S) = \{(1,1)\}$. Then

(24.1.8) $$\Gamma_k = \dot{\bigcup_{(s,w) \in \mathscr{C}ell_k(W,S)}} C(s,w).$$

Proof. We may assume that $k \geqslant 1$. Let $(s,w) \in \mathscr{C}ell_k(W,S)$ and let $x \in C(s,w)$. Then it follows from 24.1.1 that

$$w = \mathrm{pr}_L^{I_{k-1}}(x) \qquad \text{and} \qquad s = \mathscr{L}(xw^{-1}). \tag{24.1.9}$$

So, if $C(s,w) = C(s',w')$, then $(s,w) = (s',w')$. Conversely, if $x \in \Gamma_k$, then we set $w = \mathrm{pr}_L^{I_{k-1}}(x)$ and $s = \mathscr{L}(xw^{-1})$. Then, by construction, $(s,w) \in \mathscr{C}ell_k(W,S)$ and $x \in C(s,w)$. ■

24.2. Boundedness

Lemma 24.2.1. *Let x, $y \in W$, let $(s_m, s_{m-1}, \dots, s_2, s_1)$ be the reduced expression of x and let $(t_1, t_2, \dots, t_n)$ be the reduced expression of y. We denote by d the biggest natural number such that $(s_1, s_2, \dots, s_d) = (t_1, t_2, \dots, t_d)$ (if $s_1 \neq t_1$, we set $d = 0$). Then*

$$T_x T_y = T_{xy} + \sum_{i=1}^{d} (v^{\varphi(s_i)} - v^{-\varphi(s_i)}) T_{s_m s_{m-1} \cdots s_{i+1} s_i t_{i+1} \cdots t_{n-1} t_n},$$

and $(s_m, s_{m-1}, \dots, s_{i+1}, s_i, t_{i+1}, \dots, t_{n-1}, t_n)$ is a reduced expression if $1 \leqslant i \leqslant d$.

Proof. We may assume that $m, n \geqslant 1$, for otherwise the result is obvious. We will prove this lemma by induction on $\ell(x) = m \geqslant 1$. If $m = 1$, then the result is just the multiplication rule in $\mathscr{H}$. So assume now that $m \geqslant 2$. Two cases may occur:

- If $s_1 \neq t_1$, then $T_x T_y = T_{xy}$ by 24.1.1 and the result follows.
- If $s_1 = t_1$, then we set $x' = x s_1 < x$ and so

$$T_x T_y = T_{x'} T_{s_1} T_y = T_{x'}(T_{s_1 y} + (v^{\varphi(s_1)} - v^{-\varphi(s_1)}) T_y).$$

But $\mathscr{R}(x') = s_2 = \mathscr{L}(s_1 y) \neq \mathscr{L}(y) = s_1$. So

$$T_{x'} T_y = T_{x'y} = T_{s_m s_{m-1} \cdots s_2 s_1 t_2 \cdots t_{n-1} t_n}$$

by 24.1.1 and so, by the induction hypothesis,

$$T_{x'} T_{s_1 y} = T_{x' s_1 y} + \sum_{i=2}^{d} (v^{\varphi(s_i)} - v^{-\varphi(s_i)}) T_{s_m s_{m-1} \cdots s_{i+1} s_i t_{i+1} \cdots t_{n-1} t_n}.$$

So the result also follows in this case. ■

Corollary 24.2.2. *Let x, y and z be three elements of W. Then*

$$\deg(t_{x,y,z}) \leqslant \min(a_{\mathbf{k}(x)}, a_{\mathbf{k}(y)}, a_{\mathbf{k}(z)}).$$

Corollary 24.2.3. *If $0 \leqslant k \leqslant l$, then $W \setminus W_{I_k}$ is a two-sided ideal of W and any left, right or two-sided cell of W_{I_k} is respectively a left, right or two-sided cell of W.*

Proof. This follows from Corollaries 24.2.2 and 11.2.4. ■

Proposition 24.2.4. *Lusztig's* **a**-*function* $\mathbf{a}: W \to \mathscr{A}$ *is well-defined and*

$$\mathbf{a}(z) = a_{\mathbf{k}(z)},$$

for all $z \in W$. Moreover, if $z \in W$, then

$$\gamma_{x,y,z^{-1}} = \operatorname{coef}_{\mathbf{a}(z)}(t_{x,y,z})$$

for all $x, y \in W_{I_{\mathbf{k}(z)}}$.

Proof. Once the first formula is proved, the second follows from Remark 14.2.7 applied to the group $W_{I_{\mathbf{k}(z)}}$. Let x, y, $z \in W$ and let $k = \mathbf{k}(z)$. If $h_{x,y,z} \neq 0$, then it follows from Corollary 24.2.3 that x, $y \in W_{I_k}$, so that $\mathbf{a}(z)$ (if it exists) can be computed in W_{I_k}. By Lemma 14.1.7 and Corollary 24.2.2, $\deg(h_{x,y,z}) \leqslant a_k$ for all $x, y \in W_{I_k}$.

It remains to prove that there exist $x, y \in W_{I_k}$ such that $\deg(h_{x,y,z}) \geqslant a_k$. Let $(s_1, s_2, \dots, s_n)$ be the reduced expression of z. By the minimality of k, there exists an $i \in \{1,2,\dots,n\}$ such that $s_i \in S_k$. Let $x = s_1 s_2 \cdots s_i$ and $y = s_i s_{i+1} \cdots s_n$. Then, as in the proof of Lemma 14.1.7, we get

$$h_{x,y,z} = \sum_{\substack{x' \leqslant x \\ y' \leqslant y \\ z' \in W}} p_{x',y} p_{y',y} q_{z,z'} t_{x',y',z'}.$$

But, if $x' \leqslant x$, $y' \leqslant y$ and $z' \in W$ are such that $t_{x',y',z'} \neq 0$, then x', $y' \in W_{I_k}$ (because $x, y \in W_{I_k}$) and so $z' \in W_{I_k}$. Consequently,

$$h_{x,y,z} = \sum_{\substack{x' \leqslant x \\ y' \leqslant y \\ z' \in W_{I_k}}} p_{x',y} p_{y',y} q_{z,z'} t_{x',y',z'}$$

and it follows from Corollary 24.2.2 that

$$\operatorname{coef}_{a_k}(h_{x,y,z}) = \operatorname{coef}_{a_k}(t_{x,y,z}).$$

Since $t_{x,y,z} = v^{a_k} - v^{-a_k}$ by Lemma 24.2.1, the result follows. ■

Corollary 24.2.5. *Let x, y and z be three elements of W such that $\gamma_{x,y,z} \neq 0$. Then* $\mathbf{k}(x) = \mathbf{k}(y) = \mathbf{k}(z)$ *and* $\gamma_{x,y,z} = 1$.

Proof. This follows from Proposition 24.2.4 (together with its proof) and Lemma 24.2.1. ■

24.3. About Kazhdan–Lusztig Basis Elements

The computation of the structure constants $t_{x,y,z}$ in Lemma 24.2.1 has the following consequence:

Lemma 24.3.1. *Let x, $y \in W$ and let $s \in S$ be such that $sy > y$ and $\mathbf{k}(s) > \mathbf{k}(x)$. Then*

$$\mathbf{C}_{xsy} = \mathbf{C}_x \mathbf{C}_{sy}.$$

Proof. We have

$$\mathbf{C}_x \mathbf{C}_{sy} = \sum_{\substack{x' \leqslant x \\ y' \leqslant sy}} p_{x',x} p_{y',sy} T_{x'} T_{y'}.$$

Let $x' \leqslant x$ and $y' \leqslant sy$. We first prove the following fact:

$$(*) \qquad \text{If } (x', y') \neq (x, sy), \text{ then } p_{x',x} p_{y',sy} T_{x'} T_{y'} \in \mathcal{H}_{<0}.$$

Two cases may occur:

• If $sy' > y'$, then $p_{y',sy} = v^{-\varphi(s)} p_{sy',sy} \in A_{\leqslant -\varphi(s)}$ (see Proposition 5.6.1(a)) and $T_{x'} T_{y'} \in v^{a_{\mathbf{k}(x)}} \mathcal{H}_{\leqslant 0}$ (see Corollary 24.2.2). So $p_{x',x} p_{y',sy} T_{x'} T_{y'} \in \mathcal{H}_{<0}$ because $\varphi(s) > a_{\mathbf{k}(x)}$ by hypothesis.

• If $sy' < y'$, then $\mathcal{L}(y') = s \neq \mathcal{R}(x') \in I_{\mathbf{k}(x)}$. So $T_{x'} T_{y'} = T_{x'y'}$ by 24.1.1 and so $p_{x',x} p_{y',sy} T_{x'} T_{y'} \in \mathcal{H}_{<0}$ since $p_{x',x} p_{y',sy} \in A_{<0}$. This completes the proof of $(*)$.

Now, let $C = \mathbf{C}_x \mathbf{C}_{sy}$. By $(*)$, we have $C \equiv T_{xsy} \mod \mathcal{H}_{<0}$. Since $\overline{C} = C$, we get that $C = \mathbf{C}_{xsy}$. ■

If $s \in S$ and x, $y \in W$, we denote by $\alpha_{s,y}$ the unique element of A such that

$$\begin{cases} \alpha_{s,x} = \overline{\alpha_{s,x}}, \\ \alpha_{s,x} \equiv v^{\varphi(s)} p_{1,x} \mod A_{<0}, \end{cases}$$

and let

$$\beta_{s,x,y} = \begin{cases} 1 & \text{if } \ell(xy) = \ell(y) - \ell(x) \text{ and } s = \mathcal{L}(xy), \\ 0 & \text{otherwise.} \end{cases}$$

Corollary 24.3.2. *Let $k \in \{1,2,\dots,l\}$, let $s \in S_k$, let $t \in S \setminus I_{k-1}$, let $x \in W_{I_{k-1}}$ and let $w \in W$ be such that $\ell(sxtw) = \ell(s)+\ell(x)+\ell(t)+\ell(w)$. Then*

$$\mathbf{C}_s \mathbf{C}_{xtw} = \mathbf{C}_{sxtw} + \delta_{s=t}\alpha_{s,x}\mathbf{C}_{tw} + \delta_{\mathbf{k}(s)=\mathbf{k}(t)}\beta_{s,x,w}\mathbf{C}_{xw}.$$

In particular, if $\mathbf{k}(s) < \mathbf{k}(t)$, then

$$\mathbf{C}_s \mathbf{C}_{xtw} = \mathbf{C}_{sxtw}.$$

Proof. Let $C = \mathbf{C}_s\mathbf{C}_{xtw} - \mathbf{C}_{sxtw} - \alpha\mathbf{C}_{tw} - \beta\mathbf{C}_{xw}$, where

$$\alpha = \delta_{s=t}\alpha_{s,x} \quad \text{and} \quad \beta = \delta_{\mathbf{k}(s)=\mathbf{k}(t)}\beta_{s,x,w}.$$

Since $\overline{C} = C$, we only need to prove that $C \equiv 0 \mod \mathcal{H}_{<0}$. For simplification, we just write $h \equiv h'$ to mean that $h \equiv h' \mod \mathcal{H}_{<0}$. Let also $k' = \mathbf{k}(t) \geqslant k$.

By Lemma 24.3.1, we have $\mathbf{C}_{xtw} = \mathbf{C}_x\mathbf{C}_{tw}$. Therefore,

$$C \equiv \sum_{\substack{x' \leqslant x \\ w' \leqslant tw}} p_{x',x}p_{w',tw}T_{sx'}T_{w'} - T_{sxtw} - \alpha \sum_{w' \leqslant tw} p_{w',tw}T_{w'} - \beta T_{xw}.$$

First case: assume that $x = 1$. Since $stw > tw > w$ by hypothesis, this forces $s \neq t$. So $\alpha = 0$ and the above congruence may be rewritten as

$$C \equiv \sum_{w' \leqslant tw} p_{w',tw}T_sT_{w'} - T_{stw} - \beta T_w.$$

- If $w' \leqslant tw$ and $tw' < w'$, then $sw' > w'$ and so $T_sT_{w'} = T_{sw'}$. In particular, if $w' \neq tw$, then $p_{w',tw}T_sT_{w'} \in \mathcal{H}_{<0}$.
- If $w' \leqslant tw$ and $tw' > w'$, then $p_{w',tw} = v^{-a_{k'}}p_{tw',tw}$ and $T_sT_{w'} \in \mathcal{H}_{\leqslant a_k}$. In particular, $p_{w',tw}T_sT_{w'} \in \mathcal{H}_{\leqslant a_k - a_{k'} + \deg(p_{tw',tw})}$.

This shows that $C \equiv 0$ in this case, as desired.

Second case: assume from now on that $x \neq 1$. Let x' and w' be such that $x' \leqslant x$ and $w' \leqslant tw$.

- If $x' = x$ and $w' = w$, then $p_{x',x}p_{w',tw}T_{sx'}T_{w'} = v^{-a_{k'}}T_{sx}T_w$.
- If $x' = x$ and $w' = tw$, then $p_{x',x}p_{w',tw}T_{sx'}T_{w'} = T_{sxtw}$.
- If $x' = x$, $tw' < w'$ and $w' \neq tw$, then $sxt > sx$ by hypothesis, so $\mathcal{L}(w') = t \neq \mathcal{R}(sx)$ because $x \neq 1$, so $p_{x',x}p_{w',tw}T_{sx'}T_{w'} = p_{x',x}p_{w',tw}T_{sxw'} \in \mathcal{H}_{<0}$ because $p_{w',tw} \in A_{<0}$.
- If $x' \leqslant x$, $tw' > w'$ and $(x',w') \neq (x,w)$, then $p_{w',tw} = v^{-\varphi(t)}p_{tw',tw}$ and so $p_{x',x}p_{w',tw}T_{sx'}T_{w'} = v^{-a_{k'}}p_{x',x}p_{tw',tw}T_{sx'}T_{w'}$. But $T_{sx'}T_{w'} \in \mathcal{H}_{\leqslant a_k}$ by Lemma 14.1.7. So $p_{x',x}p_{w',tw}T_{sx'}T_{w'} \in \mathcal{H}_{<0}$ because $(x',w') \neq (x,w)$.

It follows from the study of all these cases that

$$C \equiv \sum_{\substack{x'<x \\ w' \leqslant tw \\ \text{such that } tw'<w'}} p_{x',x} p_{w',tw} T_{sx'} T_{w'} - \alpha \sum_{w' \leqslant tw} p_{w',tw} T_{w'} + v^{-a_{k'}} T_{sx} T_w - \beta T_{xw}.$$

Now, let $(s_m, s_{m-1}, \ldots, s_1)$ be the reduced expression of x and let $(t_1, t_2, \ldots, t_n)$ be the reduced expression of y. Note that s_1, $s_2, \ldots$, s_m, t_1, $t_2, \ldots$, t_n belong to I_{k-1}. Since $(s, s_m, s_{m-1}, \ldots, s_1)$ is a reduced expression of sx, we get from Lemma 14.1.7 that

$$v^{-a_{k'}} T_{sx} T_w \equiv \begin{cases} v^{a_k - a_{k'}} T_{xw} & \text{if } \ell(xw) = \ell(w) - \ell(x) \text{ and } s = \mathscr{L}(xw), \\ 0 & \text{otherwise.} \end{cases}$$

From the definition of β, we get

$$C \equiv \sum_{\substack{x'<x \\ w' \leqslant tw}} p_{x',x} p_{w',tw} T_{sx'} T_{w'} - \alpha \sum_{w' \leqslant tw} p_{w',tw} T_{w'}.$$

Also, if $x' \neq 1$ and $tw' < w'$, then $\mathscr{L}(w') = t \neq \mathscr{R}(sx') \in I_{k-1}$. Consequently, $p_{x',x} p_{w',tw} T_{sx'} T_{w'} = p_{x',x} p_{w',tw} T_{sx'w'} \in \mathscr{H}_{<0}$ because $p_{x',x} \in A_{<0}$. Moreover, if $w' \leqslant tw$ and $tw' > w'$, then $\alpha p_{w',tw} = v^{-a_{k'}} \alpha p_{tw',tw} \in A_{<0}$ by the definition of α and because $p_{tw',tw} \in A_{<0}$. So we get

$$C \equiv \sum_{\substack{w' \leqslant tw \\ \text{such that } tw'<w'}} p_{w',tw} (p_{1,x} T_s T_{w'} - \alpha T_{w'}).$$

Assume first that $s \neq t$. Then $\alpha = 0$ and $T_s T_{w'} = T_{sw'}$ if $tw' < w'$, so

$$C \equiv 0$$

because $p_{1,x} p_{w',tw} \in A_{<0}$ since $x \neq 1$. This proves the corollary in this case.

Assume finally that $s = t$. If $tw' < w'$, then $T_s T_{w'} \equiv T_{sw'} + v^{a_k} T_{w'}$ and so $p_{w',tw} p_{1,x} T_s T_{w'} \equiv p_{1,x} v^{a_k} T_{w'}$ because $p_{1,x} \in A_{<0}$. Therefore,

$$C \equiv \sum_{\substack{w' \leqslant tw \\ \text{such that } tw'<w'}} p_{w',tw} (v^{a_k} p_{1,x} - \alpha) T_{w'} \equiv 0,$$

by the definition of α. The proof of the corollary is complete. ■

24.4. Duflo Involutions

Since Lusztig's **a**-function is well-defined, the set

$$\mathscr{D} = \{z \in W \mid \mathbf{a}(z) = \mathbf{\Delta}(z)\}$$

is also well-defined.

Proposition 24.4.1. *If $z \in W$, then $\mathbf{a}(z) \leqslant \mathbf{\Delta}(z)$. Moreover,*

$$\mathscr{D} = \{w^{-1}sw \mid (s,w) \in \mathscr{C}ell(W,S)\}.$$

Proof. Let $z \in W$. Then there exists a unique $(s,w) \in \mathscr{C}ell(W,S)$ and $x \in X_s^{I_{\mathbf{k}(s)}}$ such that $z = xsw$. Let $k = \mathbf{k}(z) = \mathbf{k}(s)$. By Lemma 24.3.1 (and applying the involution $*$), we have $\mathbf{C}_z = \mathbf{C}_{xs}\mathbf{C}_w$ and so

$$p_{1,z} = \tau(\mathbf{C}_z) = \sum_{y \leqslant w} p_{y,w} p_{y^{-1},xs}.$$

But, if $y \leqslant w$, then $y \in W_{I_{k-1}}$ and so $y^{-1}s > y^{-1}$. So it follows from Proposition 5.6.6(a) that $p_{y^{-1},xs} = v^{-a_k} p_{y^{-1}s,xs}$. Consequently,

$$p_{1,z} = v^{-a_k} \sum_{y \leqslant w} p_{y,w} p_{y^{-1}s,xs}.$$

So $-\mathbf{\Delta}(z) = \deg(p_{1,z}) \leqslant -a_k = -\mathbf{a}(z)$, with equality if and only if $w^{-1}s = xs$, that is, if and only if $z = w^{-1}sw$. ■

Corollary 24.4.2. *If $d \in \mathscr{D}$, then $d^2 = 1$ and $n_d = 1$.*

Proof. The first result is immediate from Proposition 24.4.1 while the equality $n_d = 1$ follows from the proof of Proposition 24.4.1. ■

Since $w^{-1}sw \in C(s,w)$ for all $(s,w) \in \mathscr{C}ell(W,S)$ (because $w^{-1} \in W_{I_{\mathbf{k}(s)-1}} \subset X_s^{I_{\mathbf{k}(s)}}$), we get the following consequence:

Corollary 24.4.3. *If $(s,w) \in \mathscr{C}ell(W,S)$, then $\mathscr{D} \cap C(s,w) = \{w^{-1}sw\}$.*

24.5. Cells

We can now describe the partition of W into left and two-sided cells:

Theorem 24.5.1. *Recall that (W,S) is a free Coxeter system. Then:*

(a) *If $(s,w) \in \mathscr{C}ell(W,S)$, then $C(s,w)$ is a left cell of W. Moreover, the map*

$$\begin{array}{ccc} \mathscr{C}ell(W,S) & \longrightarrow & \mathrm{Cell}_L(W) \\ (s,w) & \longmapsto & C(s,w) \end{array}$$

is well-defined and bijective.

(b) *Let $x, y \in W$. Then $x \leqslant_{LR} y$ (resp. $x \sim_{LR} y$) if and only if $\mathbf{k}(x) \geqslant \mathbf{k}(y)$ (resp. $\mathbf{k}(x) = \mathbf{k}(y)$). So the map*

$$\begin{array}{ccc} \{0,1,\dots,l\} & \longrightarrow & \mathrm{Cell}_{LR}(W) \\ k & \longmapsto & \Gamma_k \end{array}$$

is well-defined and bijective.

Proof. Let $(s,w) \in \mathscr{C}ell(W,S)$ and let $k = \mathbf{k}(s)$. We first prove the following fact:

(♣) *The pair $(X_s \times \{sw\}, \{sw\})$ satisfies the Guilhot induction process.*

Proof (*of* (♣)). The result is trivial for $(s,w) = (1,1)$, so we may assume that $(s,w) \in \mathscr{C}ell'(W,S)$. It is clear that $(X_s \times \{sw\}, \{sw\})$ satisfies (G1) and (G3), while (G2) follows from 24.1.7.

(G4) Let $\mathscr{M} = \bigoplus_{x \in X_s} A\, T_x \mathbf{C}_{sw}$. We just need to prove that, if $x \in X_s$ and $t \in S$, then $T_t T_x \mathbf{C}_{sw} \in \mathscr{M}$. Two cases may occur:

• If $tx \in X_s$, then $T_t T_x \mathbf{C}_{sw} = (T_{tx} + \delta_{tx<x}(v^{\varphi(t)} - v^{-\varphi(t)}) T_x) \mathbf{C}_{sw} \in \mathscr{M}$.

• If $tx \notin X_s$, then it follows from Deodhar's Lemma 24.1.3 that $t = s$ and $x = 1$. So $T_t T_x \mathbf{C}_{sw} = T_s \mathbf{C}_{sw} = (v^{\varphi(s)} + v^{-\varphi(s)}) \mathbf{C}_{sw} \in \mathscr{M}$. This completes the proof of (G4).

(G5) We will prove the statement (G5′), which is equivalent to (G5) because (G2) holds. Let $x \in X_s$. We wish to prove that $T_x \mathbf{C}_{sw} \equiv T_{xsw} \mod \mathscr{H}_{<0}$. Let us write $x = x_1 x_2$ with $x_1 \in X_{I_{k-1} \cup \{s\}}$ and $x_2 \in X_{I_{k-1}}^{I_{k-1} \cup \{s\}}$. Then $T_{x_2} \mathbf{C}_{sw} \in \mathscr{H}_{I_{k-1} \cup \{s\}}$, so it is sufficient to prove that $T_{x_2} \mathbf{C}_{sw} \equiv T_{x_2 sw} \mod \mathscr{H}_{<0}$. In other words, we may, and we will, assume that $x \in W_{I_{k-1} \cup \{s\}}$. So let $w_1 < sw$. Two cases may occur:

• If $sw_1 < w_1$, then $\mathscr{L}(w_1) = \{s\} \neq \mathscr{R}(x)$, so $T_x T_{w_1} = T_{xw_1}$ by 24.1.1, so $p_{w_1,sw} T_x T_{w_1} \in \mathscr{H}_{<0}$.

• If $sw_1 > w_1$, then $p_{w_1,sw} = v^{-a_k} p_{sw_1,sw}$, so $\deg(p_{w_1,sw}) < \varphi(s)$. Also, note that $w_1 \in W_{I_{k-1}}$, so it follows from Lemma 24.2.1 that $T_x T_{w_1} \in v^{a_{k-1}} \mathscr{H}_{<0}$. But $a_k > a_{k-1}$ by definition, so $p_{w_1,sw} T_x T_{w_1} \in \mathscr{H}_{<0}$. This completes the proof of (G5′). ■

Now, let $x \in W$ and $y \in X_s \cdot sw$. It follows from (♣) and Theorem 11.1.8 that:

(◇) *If $x \leqslant_L y$, then $x \in X_s \cdot sw$.*

Assume now that $x \sim_L sw$. Since $sw \in \Gamma_k$ and Γ_k is a union of left cells by Corollary 24.2.3, we get that $x \in W_{I_k}$. So, by (◇), $x \in X_s^{I_k} \cdot sw = C(s,w)$. So we have proved that

(♡) *if $x \sim_L sw$, then $x \in C(s,w)$.*

We now prove the converse of (♡):

(♠) *If $x \in X_s^{I_k}$, then $xsw \sim_L sw$.*

First, note that if $k = 1$, then Γ_1 is the unique submaximal two-sided cell, and the result follows from Theorem 13.1.5. So we assume that $k \geqslant 2$. We argue by induction on $\ell(x)$. The result is obvious if $\ell(x) = 0$. So assume now that $\ell(x) \geqslant 1$. Since $\ell(xsw) = \ell(x) + \ell(sw)$, we have that $xsw \leqslant_L sw$ (see Example 6.1.2). We will prove by induction on $\ell_{S_k}(x)$ that $sw \leqslant_L xsw$, so that the proof of (♠) will be complete.

• If $\ell_{S_k}(x) = 0$, this means that $x \in W_{I_{k-1}}$. Now, let $t \in I_{k-1}$ and let $y = stsxsw$. By 24.1.1, then $\ell(y) = 3 + \ell(t) + \ell(x) + \ell(w)$. So we may apply Corollary 24.3.2 and obtain that $\lambda^s_{sw,y} = 1$, and so $sw \leqslant_L y$. Since $y \leqslant_L xsw$ again by Example 6.1.2, we get that $sw \leqslant_L xsw$, and so $xsw \sim_L sw$, as desired.

• Now, let us assume that $\ell_{S_k}(x) \geqslant 1$ and that the result holds for any $x' \in X_s^{I_k}$ such that $\ell_{S_k}(x') < \ell_{S_k}(x)$. We can write $x = x_1 s_1 x_2$ with $x_1 \in W_{I_{k-1}}$, $s_1 \in S_k$, $x_2 \in W_{I_k}$ and $\ell(x) = \ell(x_1) + 1 + \ell(x_2)$. Two cases may occur:

- Assume first that $x_1 = 1$. Write $x_2 = yu$ with $y \in W_{I_{k-1}}$ and $u \in (X_{I_{k-1}}^{I_k})^{-1}$, and let $t = \mathscr{L}(usw)$. Then $\lambda^t_{usw,y^{-1}s_1yusw} = 1$ by Corollary 24.3.2, so $usw \leqslant_L y^{-1}s_1yusw \leqslant_L s_1yusw = xsw$. By the induction hypothesis, $sw \leqslant_L usw$, so the result follows in this case.
- If $x_1 \neq 1$, then $\ell(x_1^{-1}s_1x) = \ell(x_1) + 1 + \ell(x)$ and it follows from Corollary 24.3.2 that $\lambda^{s_1}_{s_1x_2sw,x_1^{-1}s_1xsw} \neq 0$, so $s_1x_2sw \leqslant_L x_1^{-1}s_1xsw \leqslant_L xsw$, and we are back to the previous case.

We can now conclude the proof of the theorem.

(a) By (♡) and (♠), $C(s,w)$ is a left cell for all $(s,w) \in \mathscr{C}ell(W,S)$. So (a) follows from 24.1.8 and 24.1.5.

(b) We first show that Γ_k is a two-sided cell. So let $x, x' \in \Gamma_k$. By 24.1.8, there exist (s,w) and (s',w') in $\mathscr{C}ell_k(W,S)$ such that $x \in C(s,w)$ and $x' \in C(s',w')$. But $w^{-1} \in X_s^{I_k}$, so $w^{-1}s \sim_L s$ by (b). In other words, $sw \sim_R s$. Similarly, $w's' \sim_R s'$. Now, $s' \in X_s^{I_k}$ and $s \in X_{s'}^{I_k}$ because $\varphi(s) = \varphi(s') = a_k$. So $ss' \sim_L s'$ and $s's \sim_L s$ by (a). Therefore,

$$x \sim_L sw \sim_R s \sim_R ss' \sim_L s' \sim_R s'w \sim_L x',$$

so $x \sim_{LR} x'$. This shows that Γ_k is contained in a single two-sided cell. Hence, Γ_k is a two-sided cell by Corollary 24.2.3.

Now, if $s \in S_k$ and $t \in S_{k+1}$, then t and ts belong to Γ_k, so $t \sim_{LR} ts$ by the above discussion. But $ts \leqslant_L s$ by Example 6.1.2, so $\Gamma_{k+1} \leqslant_L \Gamma_k$. This completes the proof of (b). ■

Corollary 24.5.2. *The number of left cells is infinite if and only if one of the following two conditions is satisfied:*

(a) $l = 2$ *and* $|S_1| \geqslant 2$.
(b) $l \geqslant 3$.

24.6. Proof of the Conjectures

Theorem 24.6.1. *Recall that* (W,S) *is a free Coxeter system. Then:*

(a) *Conjecture 6.5.2 holds (statements* (L1), (L2),..., (L8)*).*
(b) *Conjecture 6.5.3 holds.*
(c) *The Semicontinuity Conjecture 7.2.1 holds.*
(d) *Conjectures 14.1.5 and 14.2.1 hold.*
(e) *Lusztig's Conjecture 14.2.15 holds (statements* (P1), (P2),..., (P15)*).*

Proof. (L1) follows from Corollaries 24.4.2 and 24.4.3.

(L2) follows from (L1).

(L3) follows from the proof of Theorem 24.5.1(a).

(L4) follows from Theorem 24.5.1.

(L5) follows from the fact that, for all $s \in S$, $X_s^{I_{\mathbf{k}(s)}}$ is left-connected (by Deodhar's Lemma 3.6.12).

(L6) Let $k \in \{0,1,\dots,l\}$, let $(s,w) \in \mathscr{C}ell_k(W,S)$, let $x \in X_s^{I_k}$ and let $t \in S$ be such that $txsw \in \Gamma_k$. We need to prove that $txsw \in C(s,w)$. First, $t \in I_k$ so, by Deodhar's Lemma 24.1.3, two cases may occur:

• If $tx \in X_s^{I_k}$, then $txsw \in C(s,w)$ by construction.

• If $tx \notin X_s^{I_k}$, then $t = s$ and $x = 1$, so that $txsw = w \notin \Gamma_k$: this case cannot happen.

(L7) is obvious, as $S_k \subset \Gamma_k$.

(L8) is obvious too. This completes the proof of (a).

(b) Let $\mathscr{C}$ be a conjugacy class of involutions which meets Γ_k (we may assume that $k \geqslant 1$, for otherwise the result is obvious). Then there exists a $t \in I_k$ such that $\mathscr{C}$ is the conjugacy class of t (see Proposition 24.1.2). Now, let $(s,w) \in \mathscr{C}ell_k(W,S)$. It $t = s$, then $w^{-1}sw \in C(s,w) \cap \mathscr{C}$ (see Corollary 24.4.3). If $t \neq s$, then $t \in X_s^{I_k}$ and $w^{-1}st > w^{-1}s$, so $\mathscr{R}(w^{-1}st) = \{t\}$, which proves that $w^{-1}st \in X_s^{I_k}$. So $w^{-1}stsw \in C(s,w) \cap \mathscr{C}$.

(c) is left as an exercise for the reader (see Exercise 24.1).

(d) follows from Corollary 24.2.2 and Proposition 24.2.4.

(P1) follows from Proposition 24.4.1.

(P2) and (P3) Let $x,y \in W$. Since (P1) holds, it follows from Lemma 14.3.2(a) that

$$\sum_{d \in \mathscr{D}} \gamma_{x,y,d}\, n_d = \delta_{xy=1}.$$

But, if $d \in \mathscr{D}$, then $n_d = 1$ by Proposition 24.4.1 and $\gamma_{x,y,d} \in \{0,1\}$ by Corollary 24.2.5. So

(24.6.2)
$$|\{d \in \mathscr{D} \mid \gamma_{x,y,d} \neq 0\}| = \delta_{xy=1}.$$

This proves (P2) and (P3).

(P4) follows from Theorem 24.5.1.

(P5) follows from (P1), (P3) and Proposition 14.3.3(b).

(P6) follows from Corollary 24.4.2.

(P7) follows from (P2), (P3), (P4), (P5) and Proposition 14.3.3(d) (because **a** takes only finitely many values by Proposition 24.2.4).

(P8) follows from (P7) and Proposition 14.3.3(e).

(P9), (P10) and (P11) follow from the description of cells and of the preorder $\leqslant_{LR}$ in Theorem 24.5.1.

(P12) follows from (P3), (P4) and (P8) for W and all its standard parabolic subgroups and from Proposition 14.3.3(i).

(P13) The first assertion follows from Corollary 24.4.3. The second one follows from 24.6.2, (P7) and (P8).

(P14) follows from (P6), (P13) and Proposition 14.3.3(k).

(P15) By Corollary 14.3.6 and its proof, it is sufficient to check that, if s, $t \in S$ and $w \in W$ are such that $sw > w$ and $wt > w$, then

(♯)
$$\mathbf{C}_s(\hat{\mathbf{c}}_w \mathbf{C}_t) = (\mathbf{C}_s \hat{\mathbf{c}}_w)\mathbf{C}_t$$

in $\mathscr{H}[W(a_k)]$, where $k = \mathbf{k}(w)$. If $\mathbf{k}(s) > k$, then $\mathbf{C}_s$ acts by 0 on $\mathscr{H}^{LR}[\Gamma_k]$ so (♯) holds. This means that we may assume that $\mathbf{k}(s) = j \leqslant k$. Similarly, we may assume that $\mathbf{k}(s') = j' \leqslant k$. Also, we may assume that $k \geqslant 1$.

Let us write $w = uyu'$, with $u \in W_{Ij-1}$, $u' \in W_{I_{j'-1}}$ and $y \in X_{I_{j-1}, I_{j'-1}}$, so that $\ell(w) = \ell(u) + \ell(y) + \ell(u')$ (see Corollary 24.1.4). Since $\mathbf{k}(w) = k$, this forces $\mathbf{k}(y) = k$ and in particular $x \neq 1$. Two cases may occur:

- If $\ell(y) \geqslant 2$, let us write $y = txt'$, with $t = \mathscr{L}(y) \in I_k \setminus I_{j-1}$, $t' = \mathscr{R}(y) \in I_k \setminus I_{j'-1}$, $x \in W_{I_k}$, so that $\ell(y) = \ell(x) + 2$. It follows from Corollary 24.3.2 that

$$\begin{aligned}
(\mathbf{C}_s \hat{\mathbf{c}}_w)\mathbf{C}_t &= (1 \otimes_{\mathbb{Z}} 1)\hat{\mathbf{c}}_{sutxt'u's'} + (1 \otimes_{\mathbb{Z}} \delta_{s'=t'}\alpha_{s',u'})\hat{\mathbf{c}}_{sutxt'} \\
&\quad + (1 \otimes_{\mathbb{Z}} \delta_{\mathbf{k}(s')=\mathbf{k}(t')}\beta_{s',u'^{-1},x^{-1}})\hat{\mathbf{c}}_{sutxu'} + (\delta_{s=t}\alpha_{s,u} \otimes_{\mathbb{Z}} 1)\hat{\mathbf{c}}_{txt'u's'} \\
&\quad + (\delta_{s=t}\alpha_{s,u} \otimes_{\mathbb{Z}} \delta_{s'=t'}\alpha_{s',u'})\hat{\mathbf{c}}_{txt'} + (\delta_{s=t}\alpha_{s,u} \otimes_{\mathbb{Z}} \delta_{\mathbf{k}(s')=\mathbf{k}(t')}\beta_{s',u'^{-1},x^{-1}})\hat{\mathbf{c}}_{txu'} \\
&\quad + (\delta_{\mathbf{k}(s)=\mathbf{k}(t)}\beta_{s,u,x} \otimes_{\mathbb{Z}} 1)\hat{\mathbf{c}}_{uxt'u's'} + (\delta_{\mathbf{k}(s)=\mathbf{k}(t)}\beta_{s,u,x} \otimes_{\mathbb{Z}} \delta_{s'=t'}\alpha_{s',u'})\mathbf{c}_{uxt'} \\
&\quad + (\delta_{\mathbf{k}(s)=\mathbf{k}(t)}\beta_{s,u,x} \otimes_{\mathbb{Z}} \delta_{\mathbf{k}(s')=\mathbf{k}(t')}\beta_{s',u'^{-1},(ux)^{-1}})\hat{\mathbf{c}}_{uxu'}.
\end{aligned}$$

By symmetry, a similar computation holds for $\mathbf{C}_s(\hat{\mathbf{c}}_w\mathbf{C}_{s'})$ and it follows that

$$(\mathbf{C}_s\hat{\mathbf{c}}_w)\mathbf{C}_t - \mathbf{C}_s(\hat{\mathbf{c}}_w\mathbf{C}_t) \\ = \delta_{\mathbf{k}(s)=\mathbf{k}(t)}\delta_{\mathbf{k}(s')=\mathbf{k}(t')}(\beta_{s,u,x}\beta_{s',u'^{-1},(ux)^{-1}} - \beta_{s,u,x'u'}\beta_{s',u'^{-1},x^{-1}})(1\otimes_{\mathbb{Z}} 1)\hat{\mathbf{c}}_{uxu'}.$$

It is now easy to check from the definition of the $\beta_{s,z,z'}$ that

$$\beta_{s,u,x}\beta_{s',u'^{-1},(ux)^{-1}} - \beta_{s,u,x'u'}\beta_{s',u'^{-1},x^{-1}} = 0,$$

but this can also be derived from the fact that $(\mathbf{C}_s\mathbf{c}_w)\mathbf{C}_t = \mathbf{C}_s(\mathbf{c}_w\mathbf{C}_t)$ in $\mathscr{H}^{LR}[\Gamma_k]$.

• If $\ell(y)=1$, then the result follows from the same computation as above, by setting $t=t'=y$ and removing all the β-terms. ■

24.7. The Infinite Dihedral Group

Hypothesis. *In addition to the general hypothesis of this chapter, we assume in this section that* $|S|=2$ *and we write* $S=\{s,t\}$, $a=\varphi(s)$ *and* $b=\varphi(t)$. *We also assume that* $0<a\leqslant b$ *(the case* $a>b$ *can easily be deduced by using the automorphism of* W *which exchanges* s *and* t*).*

As all of Lusztig's Conjectures (Pk) hold, the asymptotic algebra $\mathscr{J}$ is well-defined. We aim to describe it in terms of the representation rings of the algebraic groups $\mathbf{SL}_2(\mathbb{C})$ and $\mathbf{PGL}_2(\mathbb{C})$.

24.7.A. *Preliminary: Representations of* $\mathbf{SL}_2(\mathbb{C})$ *and* $\mathbf{PGL}_2(\mathbb{C})$

If $\mathbf{G} = \mathbf{SL}_2(\mathbb{C})$ or $\mathbf{PGL}_2(\mathbb{C})$, we denote by $\mathscr{K}_0(\mathbf{G})$ the Grothendieck ring of the category of rational representations of the algebraic group $\mathbf{G}$. Note that $\mathscr{K}_0(\mathbf{G})$ is commutative. If $d\geqslant 0$, we denote by $\mathbf{v}_d$ the class of $\mathbf{S}^d(\mathbb{C}^2)$ in $\mathscr{K}_0(\mathbf{SL}_2(\mathbb{C}))$, where $\mathbb{C}^2$ denotes the natural representation of dimension 2 of $\mathbf{SL}_2(\mathbb{C})$, and $\mathbf{S}^d(M)$ denotes the d-th symmetric power of M. Then $\mathbf{S}^d(\mathbb{C}^2)$ is simple and it factors through a representation of $\mathbf{PGL}_2(\mathbb{C})$ if and only if d is even. It is well-known that this describes all the irreducible representations of $\mathbf{SL}_2(\mathbb{C})$, hence

(24.7.1) $$\mathscr{K}_0(\mathbf{SL}_2(\mathbb{C})) = \bigoplus_{d\geqslant 0}\mathbb{Z}\mathbf{v}_d \quad\text{and}\quad \mathscr{K}_0(\mathbf{PGL}_2(\mathbb{C})) = \bigoplus_{m\geqslant 0}\mathbb{Z}\mathbf{v}_{2m}.$$

The multiplication in both rings (induced by the tensor product of representations) is completely determined by the following rules: first, $\mathbf{v}_0$ is the unit element,

$$\text{(24.7.2)} \qquad \mathbf{v}_1\mathbf{v}_d = \mathbf{v}_{d+1} + \mathbf{v}_{d-1} \qquad \text{and} \qquad \mathbf{v}_2\mathbf{v}_{2m} = \mathbf{v}_{2m+2} + \mathbf{v}_{2m} + \mathbf{v}_{2m-2}$$

for all $d, m \geqslant 1$. In particular, $\mathscr{K}_0(\mathbf{SL}_2(\mathbb{C}))$ is generated by $\mathbf{v}_1$ while $\mathscr{K}_0(\mathbf{PGL}_2(\mathbb{C}))$ is generated by $\mathbf{v}_2$.

24.7.B. The Equal Parameter Case

Hypothesis. *Assume in this subsection, and only in this subsection, that $a = b$.*

Then it follows from Theorem 24.5.1 that W contains two two-sided cells (namely $\{1\}$ and Γ^{uni}), and three left cells (namely $\{1\}$, Γ_s^{uni} and Γ_t^{uni}). Moreover, by Proposition 24.4.1,

$$\text{(24.7.3)} \qquad \mathscr{D} = \{1, s, t\}.$$

We now determine the structure of the ring $\mathscr{J}[\Gamma_s^{\text{uni}} \cap (\Gamma_s^{\text{uni}})^{-1}]$. First, note that

$$\Gamma_s^{\text{uni}} \cap (\Gamma_s^{\text{uni}})^{-1} = \{s_{2m+1} \mid m \geqslant 0\}.$$

Immediate computations yield

$$\text{(24.7.4)} \qquad \mathbf{C}_{s_3} = \mathbf{C}_s\mathbf{C}_t\mathbf{C}_s - \mathbf{C}_s \qquad \text{and} \qquad \mathbf{C}_t\mathbf{C}_{s_3} = \mathbf{C}_{s_4} + \mathbf{C}_{s_2}.$$

Moreover, if $z \in W$ is such that $sz > z$ (resp. $tz > z$), then

$$\text{(24.7.5)} \qquad \mathbf{C}_s\mathbf{C}_{s_2z} = \mathbf{C}_{s_3z} + \mathbf{C}_{sz}$$

(resp.

$$\text{(24.7.6)} \qquad \mathbf{C}_t\mathbf{C}_{t_2z} = \mathbf{C}_{t_3z} + \mathbf{C}_{tz} \qquad).$$

Indeed, this follows from Corollary 24.3.2 applied in the case where $x = y = 1$ and $u = s$ (resp. $u = t$). As a consequence, we get

$$\text{(24.7.7)} \quad \mathbf{C}_{s_3}\mathbf{C}_{s_{2m+1}} = \mathbf{C}_{s_{2m+1}}\mathbf{C}_{s_3} = \begin{cases} (v^a + v^{-a})\mathbf{C}_{s_3} & \text{if } m = 0, \\ (v^a + v^{-a})(\mathbf{C}_{s_{2m+3}} + \mathbf{C}_{s_{2m+1}} + \mathbf{C}_{s_{2m-1}}) & \text{if } m \geqslant 1. \end{cases}$$

This yields

$$\text{(24.7.8)} \qquad \mathbf{t}_{s_3}\mathbf{t}_{s_{2m+1}} = \mathbf{t}_{s_{2m+1}}\mathbf{t}_{s_3} = \begin{cases} \mathbf{t}_{s_3} & \text{if } m = 0, \\ \mathbf{t}_{s_{2m+3}} + \mathbf{t}_{s_{2m+1}} + \mathbf{t}_{s_{2m-1}} & \text{if } m \geqslant 1. \end{cases}$$

By 24.7.1 and 24.7.2, we have proved the following result:

Proposition 24.7.9. *The $\mathbb{Z}$-linear map*

$$\begin{aligned} \mathcal{J}[\Gamma_s^{\mathrm{uni}} \cap (\Gamma_s^{\mathrm{uni}})^{-1}] &\longrightarrow \mathcal{K}_0(\mathbf{PGL}_2(\mathbb{C})) \\ \mathbf{t}_{s_{2m+1}} &\longmapsto \mathbf{v}_{2m} \end{aligned}$$

is an isomorphism of rings.

We now describe the complete ring $\mathcal{J}[\Gamma^{\mathrm{uni}}]$. For this, let **spin** denote the unique involutive automorphism of the ring $\mathcal{K}_0(\mathbf{SL}_2(\mathbb{C}))$ such that $\mathbf{spin}(\mathbf{v}_1) = -\mathbf{v}_1$. In fact,

$$\mathbf{spin}(\mathbf{v}_d) = (-1)^d \mathbf{v}_d \tag{24.7.10}$$

for all $d \geqslant 0$. Therefore,

$$\mathcal{K}_0(\mathbf{SL}_2(\mathbb{C}))^{\mathbf{spin}} = \mathcal{K}_0(\mathbf{PGL}_2(\mathbb{C})). \tag{24.7.11}$$

We denote by $\mathbf{spin}_2$ the automorphism of $\mathrm{Mat}_2(\mathcal{K}_0(\mathbf{SL}_2(\mathbb{C})))$ induced by **spin** and we denote by $\mathbf{sp}_2$ the automorphism of $\mathrm{Mat}_2(\mathcal{K}_0(\mathbf{SL}_2(\mathbb{C})))$ obtained by composing $\mathbf{spin}_2$ with the inner automorphism induced by the diagonal matrix $\mathrm{diag}(1,-1)$.

Proposition 24.7.12. *The $\mathbb{Z}$-linear map*

$$\begin{aligned} \mathcal{J}[\Gamma^{\mathrm{uni}}] &\longrightarrow \mathrm{Mat}_2(\mathcal{K}_0(\mathbf{SL}_2(\mathbb{C})))^{\mathbf{sp}_2} \\ \mathbf{t}_w &\longmapsto \begin{cases} \begin{pmatrix} \mathbf{v}_{2m} & 0 \\ 0 & 0 \end{pmatrix} & \text{if } w = s(ts)^m, \\ \begin{pmatrix} 0 & \mathbf{v}_{2m+1} \\ 0 & 0 \end{pmatrix} & \text{if } w = (st)^{m+1}, \\ \begin{pmatrix} 0 & 0 \\ \mathbf{v}_{2m+1} & 0 \end{pmatrix} & \text{if } w = (ts)^{m+1}, \\ \begin{pmatrix} 0 & 0 \\ 0 & \mathbf{v}_{2m} \end{pmatrix} & \text{if } w = t(st)^m, \end{cases} \end{aligned}$$

is an isomorphism of algebras.

Proof. Let ψ denote the $\mathbb{Z}$-linear map defined in this proposition. It is easily checked that it is an isomorphism of $\mathbb{Z}$-modules, so we only need to prove that it is a morphism of rings. Note that $\psi(e_{\Gamma^{\mathrm{uni}}}) = \psi(\mathbf{t}_s + \mathbf{t}_t) = 1$. So it remains to show that $\psi(jk) = \psi(j)\psi(k)$ for all $j, k \in \mathcal{J}[\Gamma^{\mathrm{uni}}]$. But, an immediate computation yields

$$\mathbf{t}_{st}^2 = \mathbf{t}_{ts}^2 = 0, \quad \mathbf{t}_{st}\mathbf{t}_{ts} = \mathbf{t}_{sts} + \mathbf{t}_s \quad \text{and} \quad \mathbf{t}_{ts}\mathbf{t}_{st} = \mathbf{t}_{tst} + \mathbf{t}_t.$$

This shows in particular that $\mathbf{t}_s$, $\mathbf{t}_t$, $\mathbf{t}_{st}$ and $\mathbf{t}_{ts}$ generate the algebra $\mathscr{J}[\Gamma^{\mathrm{uni}}]$. So we may assume that $j \in \{\mathbf{t}_s, \mathbf{t}_{st}, \mathbf{t}_{ts}, \mathbf{t}_t\}$. Then the results follow from formulas 24.7.5 and 24.7.6. ■

24.7.C. The Unequal Parameter Case

Hypothesis. *Assume in this subsection, and only in this subsection, that $a < b$.*

Then it follows from Theorem 24.5.1 that W contains three two-sided cells (namely $\{1\}$, $\{s\}$ and $\Gamma^{\mathrm{uni}} \setminus \{s\}$), and four left cells (namely $\{1\}$, $\{s\}$, $\Gamma_s^{\mathrm{uni}} \setminus \{s\}$ and Γ_t^{uni}). Moreover, by Proposition 24.4.1,

(24.7.13) $$\mathscr{D} = \{1, s, t, sts\}.$$

An immediate computation yields

(24.7.14) $$\begin{cases} \mathbf{C}_t \mathbf{C}_{st} = \mathbf{C}_{tst} + (v^{b-a} + v^{a-b})\mathbf{C}_t, \\ \mathbf{C}_t \mathbf{C}_{sts} = \mathbf{C}_{tsts} + (v^{b-a} + v^{a-b})\mathbf{C}_{ts}. \end{cases}$$

If $z \in W$ is such that $tz > z$, then

(24.7.15) $$\begin{cases} \mathbf{C}_s \mathbf{C}_{tz} = \mathbf{C}_{stz}, \\ \mathbf{C}_t \mathbf{C}_{ststz} = \mathbf{C}_{tststz} + (v^{b-a} + v^{a-b})\mathbf{C}_{tstz} + \mathbf{C}_{tz}. \end{cases}$$

Indeed, the first equality is a special case of Lemma 24.3.1 while the second one follows from Corollary 24.3.2.

Proposition 24.7.16. *The $\mathbb{Z}$-linear map*

$$\begin{array}{ccc} \mathscr{J}[\Gamma^{\mathrm{uni}} \setminus \{s\}] & \longrightarrow & \mathrm{Mat}_2(\mathscr{K}_0(\mathbf{SL}_2(\mathbb{C}))) \\ \mathbf{t}_w & \longmapsto & \begin{cases} \begin{pmatrix} \mathbf{v}_d & 0 \\ 0 & 0 \end{pmatrix} & \text{if } w = t(st)^d, \\ \begin{pmatrix} 0 & \mathbf{v}_d \\ 0 & 0 \end{pmatrix} & \text{if } w = (ts)^{d+1}, \\ \begin{pmatrix} 0 & 0 \\ \mathbf{v}_d & 0 \end{pmatrix} & \text{if } w = (st)^{d+1}, \\ \begin{pmatrix} 0 & 0 \\ 0 & \mathbf{v}_d \end{pmatrix} & \text{if } w = s(ts)^{d+1} \end{cases} \end{array}$$

is an isomorphism of rings.

Proof. The proof is similar as the one of Proposition 24.7.12. ∎

We conclude with the following remark:

Proposition 24.7.17. *If $a < b$, then the map $\sigma : W \to W$ such that $\sigma(w) = w$ if $w \in \{1, s\}$ and $\sigma(w) = ws$ if $w \notin \{1, s\}$ is strongly left cellular.*

Proof. Using the anti-automorphism *, it follows immediately from 24.7.15 that the right multiplication by $\mathbf{C}_s$ on $\mathscr{H}^{LR}[\Gamma^{\mathrm{uni}} \setminus \{s\}]$ induces the left cellular map σ described in the proposition. The strongness is also obvious. ∎

24.8. Rank 3

We assume here, and only here, that $|S| = 3$. We write $S = \{s, t, u\}$ and

$$\begin{cases} \varphi(s) = a, \\ \varphi(t) = b, \\ \varphi(u) = c. \end{cases}$$

By symmetry, we may (and we will) assume that $0 < a \leqslant b \leqslant c$. The partition of W into left and two-sided cells is then given by Figures 24.a and 24.b, following the convention described in Sections 3.14 and 6.6.

Notes

The Kazhdan–Lusztig theory for free Coxeter groups (unequal parameters, partition into cells, Lusztig's Conjectures) is due to Shi and Yang [ShYa1]: this chapter is mostly taken from their paper, except that we have made use of the Guilhot induction process in order to simplify some of their arguments.

The particular case of infinite dihedral groups (see Section 24.7), and particularly the description of the asymptotic algebras in terms of representation rings of $\mathbf{SL}_2(\mathbb{C})$ and $\mathbf{PGL}_2(\mathbb{C})$ is due to Lusztig [Lus20, §7, §17, §18]. In the equal parameter case, this is a special case of the description of the asymptotic algebra of the lowest two-sided cell of an affine Weyl group thanks to the representation ring of the corresponding adjoint group over $\mathbb{C}$ (see [Lus6, Corollary 8.7], [Xi2] or [Lus20, §18.20][‡]).

[‡] We refer here to the updated version of [Lus20] which is available on `arXiv`: version 2, June 2014.

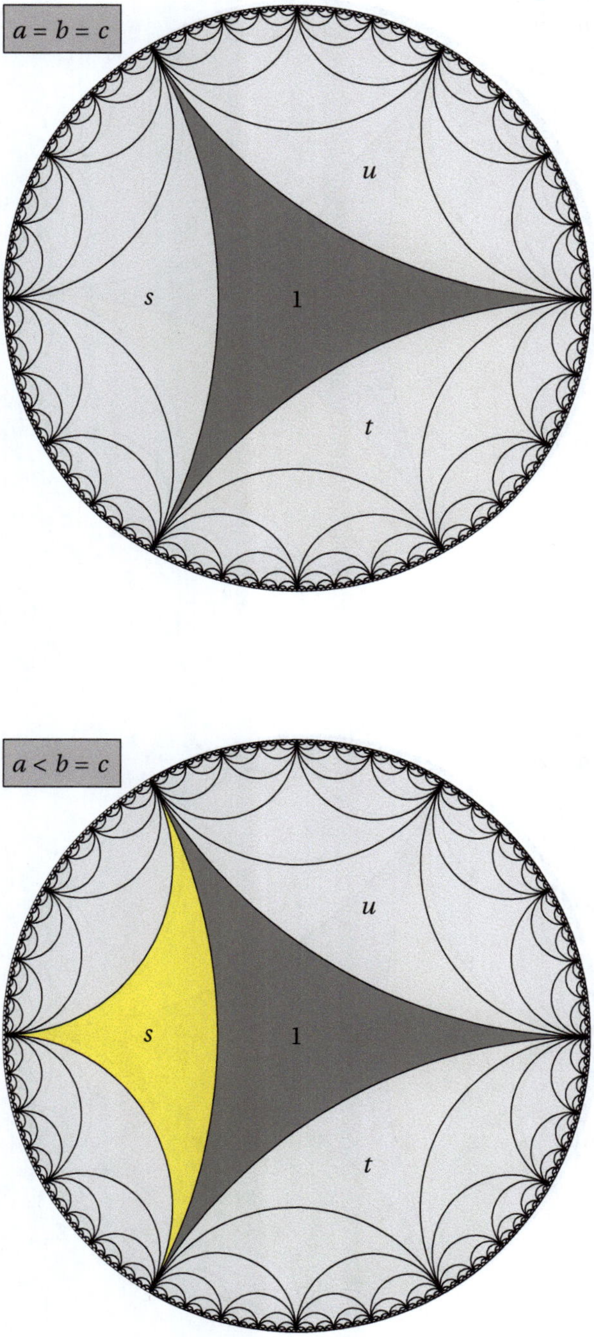

Fig. 24.a. Cells in free Coxeter group of rank 3

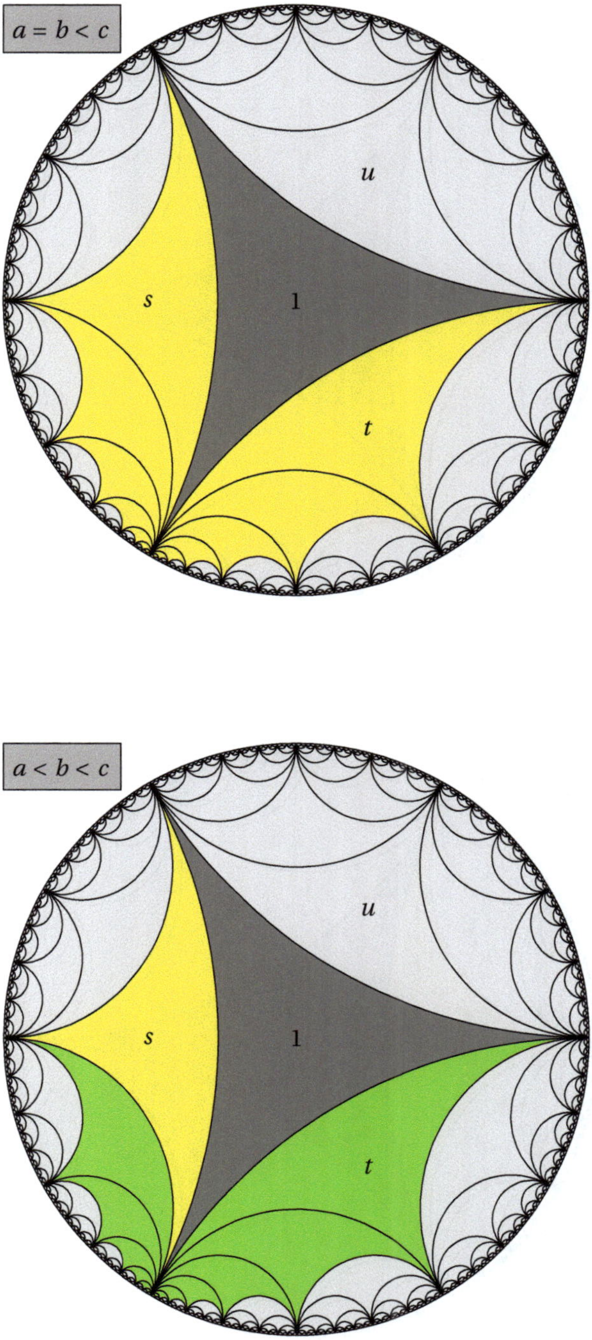

Fig. 24.b. Cells in free Coxeter group of rank 3 (continued)

Exercises

Exercise 24.1. Prove Theorem 24.6.1(c).

Exercise 24.2. Assume here that $|S| = 2$ and that φ is constant. Prove that there is no non-trivial left cellular map.

Exercise 24.3. Assume here that $|S| = 2$ and that φ is not constant. Prove that there is only one non-trivial bijective left cellular map (the one given by Proposition 24.7.17).

Chapter 25
Rank 3

Hypothesis and notation. *In this chapter, and only in this chapter, we assume that* (W,S) *is irreducible and that* $|S| = 3$ *and we write* $S = \{s,t,u\}$ *in such a way that* $m_{su} \leqslant m_{tu} \leqslant m_{st}$. *We set* $(p,q,r) = (m_{su}, m_{tu}, m_{st})$.

Under these hypotheses, we give the partitions into left, right and two-sided cells for all possible values of φ whenever W is finite of type B_3 or H_3 and whenever W is an affine Weyl group of type $\tilde{G}_2$. We also explore some of the conjectures stated in this book (particularly the Semicontinuity Conjecture) and sometimes properties of the asymptotic algebra. We also give this partition for some hyperbolic Coxeter groups (some of them are conjectural). The case of finite Weyl groups can be obtained by computer calculations. In type $\tilde{G}_2$, we do not provide a proof (this would go far beyond the aims of this book): note that the equal parameter case is due to Lusztig [Lus12] while the general case was settled by Guilhot [Gui4]. In the hyperbolic case, we treat only a few examples.

We set

$$\xi = \frac{1}{p} + \frac{1}{q} + \frac{1}{r}.$$

Since $|S| = 3$, three cases may occur (see Example 3.12.13):

(S) If $\xi > 1$, then (W,S) is spherical.
(A) If $\xi = 1$, then (W,S) is an affine Weyl group.
(H) If $\xi < 1$, then (W,S) is hyperbolic.

In the pictures, we follow the convention of Section 6.6.

C. Bonnafé, *Kazhdan–Lusztig Cells with Unequal Parameters*,
Algebra and Applications 24, https://doi.org/10.1007/978-3-319-70736-5_25

25.1. Spherical Groups

The type A_3 has been treated in Chapter 22 (see Figure 22.a). So we will only consider here the types B_3 and H_3.

25.1.A. Type B_3

Hypothesis and notation. *We assume in this subsection, and only in this subsection, that* (W,S) *is of type* B_3 *and we set* $b = \varphi(s)$, $a = \varphi(t)$, *so that the Coxeter graph of* (W,S), *together with the values of* φ, *are given as follows:*

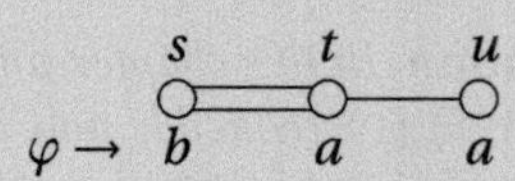

We also assume that $a, b \geqslant 0$.

The essential hyperplanes are given by Figure 25.a.

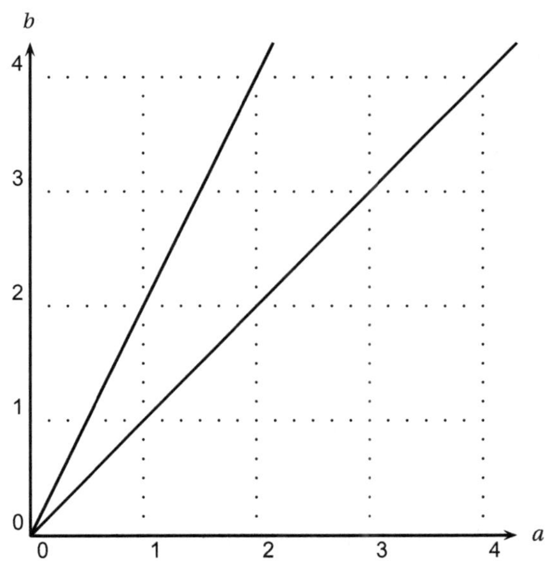

Fig. 25.a. Essential hyperplanes in type B_3

The partition into cells of W is given, according to the values of a and b, in Figure 25.d (at the end of this chapter). Here we just give the a figure describing the triangles corresponding to elements of s (as usual, the grey triangle corresponds to the neutral element).

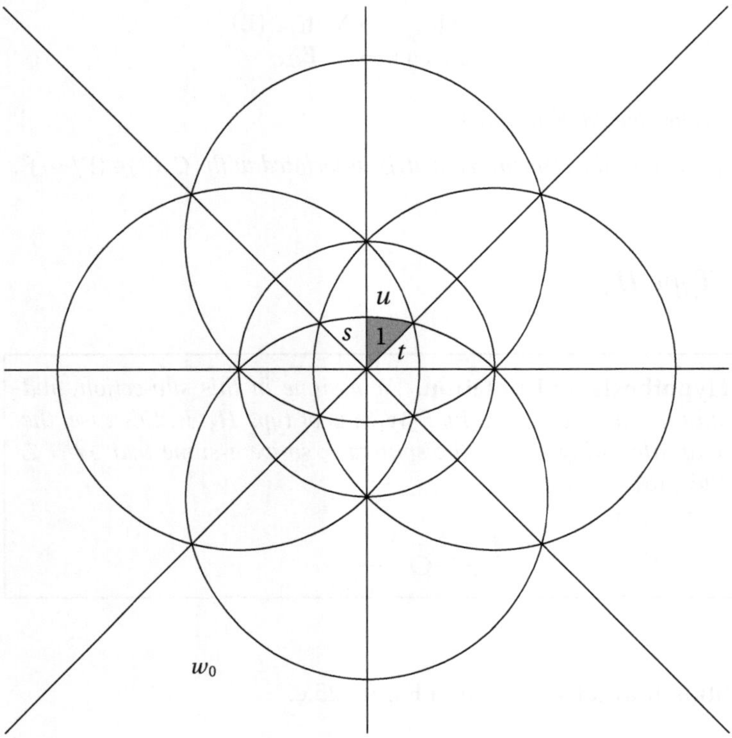

Note the following facts, that can be obtained by brute force computations:

Proposition 25.1.1. *Assume that (W,S) is of type B_3. Then Conjectures 6.5.2, 6.5.3 and 14.2.15 hold. Also, the Semicontinuity Conjecture 7.2.1 holds.*

Proposition 25.1.2. *Assume that (W,S) is of type B_3, that $a \neq 0$ and that $b \notin \{0, a, 2a\}$. Let Γ be a two-sided cell. Then:*

(a) *All the involutions of W are Duflo involutions.*

(b) *If C and C' are two left cells contained in Γ, then $C' \cap C^{-1}$ contains exactly one element (which is denoted by $w(C,C')$).*

(c) *Let $\Gamma/\sim_L$ denote the set of left cells contained in Γ and let $\mathrm{Mat}_{\Gamma/\sim_L}(\mathbb{Z})$ denote the algebra of matrices whose rows and columns are indexed by $\Gamma/\sim_L$. Then the $\mathbb{Z}$-linear map*

$$\begin{array}{rcl} \mathcal{J}[\Gamma] & \longrightarrow & \mathrm{Mat}_{\Gamma/\sim_L}(\mathbb{Z}) \\ \mathbf{t}_{w(C,C')} & \longmapsto & \mathbf{E}_{C,C'} \end{array}$$

is an isomorphism of algebras.

Here, $\mathbf{E}_{C,C'}$ denotes the elementary matrix associated with $(C,C') \in (\Gamma/\sim_L)^2$.

25.1.B. *Type H_3*

Hypothesis and notation. *We assume in this subsection, and only in this subsection, that (W,S) is of type H_3. In this case, the only interesting case is the split case, so we assume that $\mathcal{A} = \mathbb{Z}$ and $\varphi(S) = \{1\}$.*

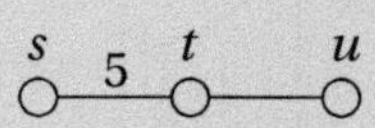

The partition into cells is given in Figure 25.e.

25.2. Affine Weyl Groups

All the results which are stated in this section are due to Guilhot [Gui1], [Gui2], [Gui3] and [Gui4], except for the split case which was already done by Lusztig [Lus12]. As a consequence of these results, one gets:

Theorem 25.2.1 (Guilhot). *Conjecture 6.5.2 as well as the Semicontinuity Conjecture 7.2.1 hold if (W,S) is of type $\tilde{B}_2$ or $\tilde{G}_2$.*

25.2.A. Type $\tilde{B}_2$

Hypothesis and notation. *We assume in this subsection, and only in this subsection, that (W,S) is of type $\tilde{B}_2$ and that the Coxeter graph of (W,S) and the values of the weight function φ are given by*

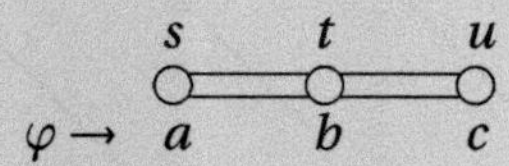

where a, b and c are non-negative real numbers. We also assume that $a \geqslant c$.

We will not provide the partition of W into left and two-sided cells for all values of (a,b,c), as there are too many cases to be considered. The reader who would be interested in seeing all the beautiful pictures corresponding to all the cases will find them in [Gui5]. We will only provide the essential hyperplane arrangement (see Figure 25.b).

The case where $b=0$. We assume here, and only here, that $b=0$. We then use the notation of §6.2C with $I=\{t\}$ and $J=\{s,u\}$. Then

$$\tilde{J}=\{s,tst,u,tut\}$$

and the Coxeter graph of $(\widetilde{W},\tilde{J})$, together with the values of $\tilde{\varphi}$, are given by

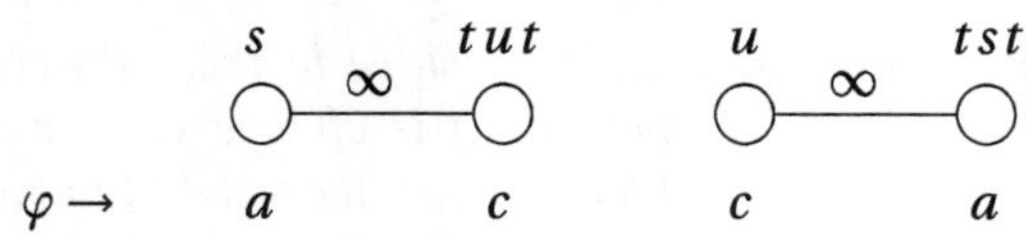

Then Proposition 6.2.8 together with the description of the partition into cells of infinite dihedral groups given in Theorem 24.5.1 (because it is a particular case of free Coxeter groups) gives a description of the partition into cells in this case.

The case where $b=1$. If $b \neq 0$, then a rescaling using the results of §6.2.A shows that we may assume that $b=1$. Then it has been shown by Guilhot [Gui4] that the traces of the essential hyperplanes on the (affine) hyperplane defined by the equation $b=1$ are given by Figure 25.b (where we restrict to the north-east quarter defined by a, $c \geqslant 0$).

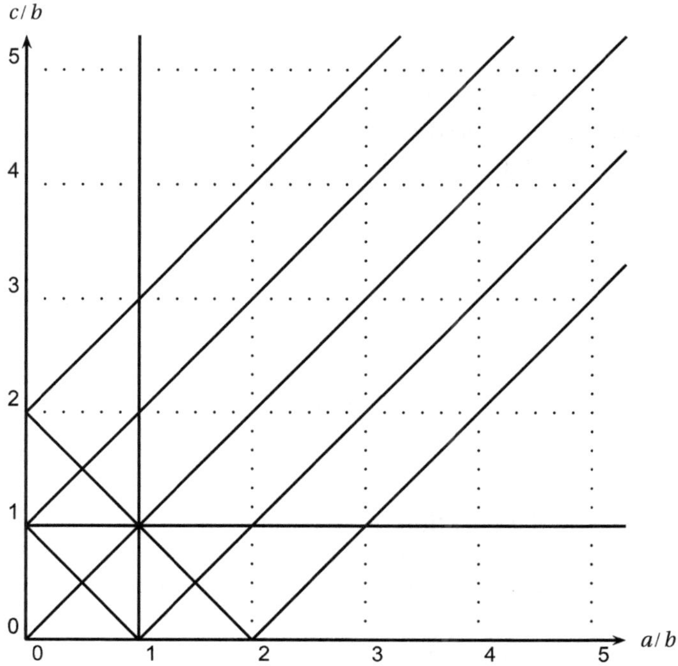

Fig. 25.b. Essential hyperplanes in type $\tilde{B}_2$

25.2.B. Type $\tilde{G}_2$

Hypothesis and notation. *We assume in this subsection, and only in this subsection, that* (W,S) *is of type* $\tilde{G}_2$ *and that the Coxeter graph of* (W,S) *and the values of the weight function* φ *are given by*

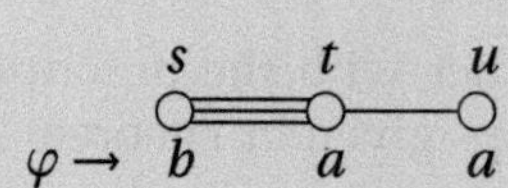

where a *and* b *are non-negative real numbers.*

We aim to describe the partition of W into left, right and two-sided cells according to the values of (a,b). According to Guilhot [Gui4], the essential hyperplanes (in the north-east quarter defined by $a,\ b \geqslant 0$) are given by Figure 25.c.

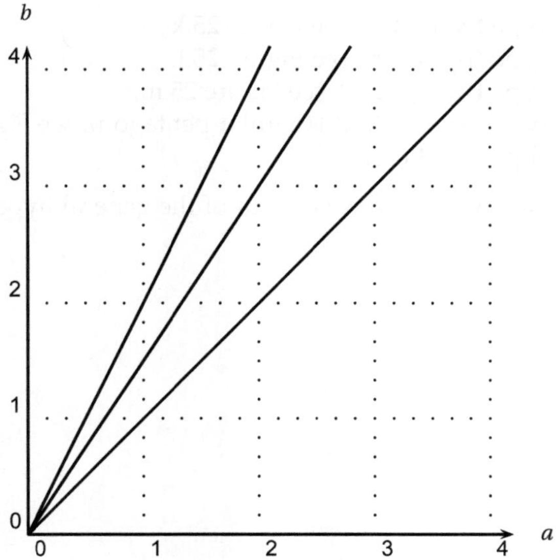

Fig. 25.c. Essential hyperplanes in type $\tilde{G}_2$

The partition of W into cells, according to the possible values of a, b, is given by Figures 25.g and 25.h (see [Gui4]).

25.3. Hyperbolic Coxeter Systems

> **Hypothesis.** *In this section,* ***it we assume that we are in the split case****, i.e. that* $\varphi(s)=1$ *for all* $s \in S$.

For simplification, we will say that (W,S) is of type $\mathbf{Hyp}(p_1,p_2,\dots,p_n)$ if $|S|=n$ and (W,S) is the reflection group along the sides of a hyperbolic n-gon with angles π/p_1, $\pi/p_2,\dots$, π/p_n. If $n=3$ and $S=\{s,t,u\}$, this corresponds to the case where $m_{st}=p_1$, $m_{tu}=p_2$, $m_{su}=p_3$. We will give the partition into cells in the following cases (in the second, third, fourth and fifth case, these are just ***conjectural*** partitions, supported by extensive computer calculations made by Gunnells*):

- If (W,S) is of type $\mathbf{Hyp}(3,3,4)$, see Figure 25.i: historically, this is the first non-finite, non-affine Coxeter group for which the partition was computed (by Bédard [Béd]).
- If (W,S) is of type $\mathbf{Hyp}(2,3,\infty)$, see Figure 25.j (in this case, $W \simeq \mathbf{GL}_2(\mathbb{Z})$ as an abstract group, as it is shown in Exercise 3.8).

* see his webpage http://people.math.umass.edu/~gunnells/cells/cells.html, where the reader can find beautiful pictures of hyperbolic Coxeter groups, even in dimension 3.

- If (W,S) is of type $\mathbf{Hyp}(2,3,7)$, see Figure 25.k.
- If (W,S) is of type $\mathbf{Hyp}(4,4,4)$, see Figure 25.l.
- If (W,S) is of type $\mathbf{Hyp}(2,2,2,3)$, see Figure 25.m.
- If (W,S) is of type $\mathbf{Hyp}(2,2,2,2,2)$ (regular pentagon), see Figure 25.n: this is due to Belolipetsky [Belo].

Note that, in the last two cases, we are out of the general hypothesis of this chapter, as $|S| > 3$.

25.4. Pictures

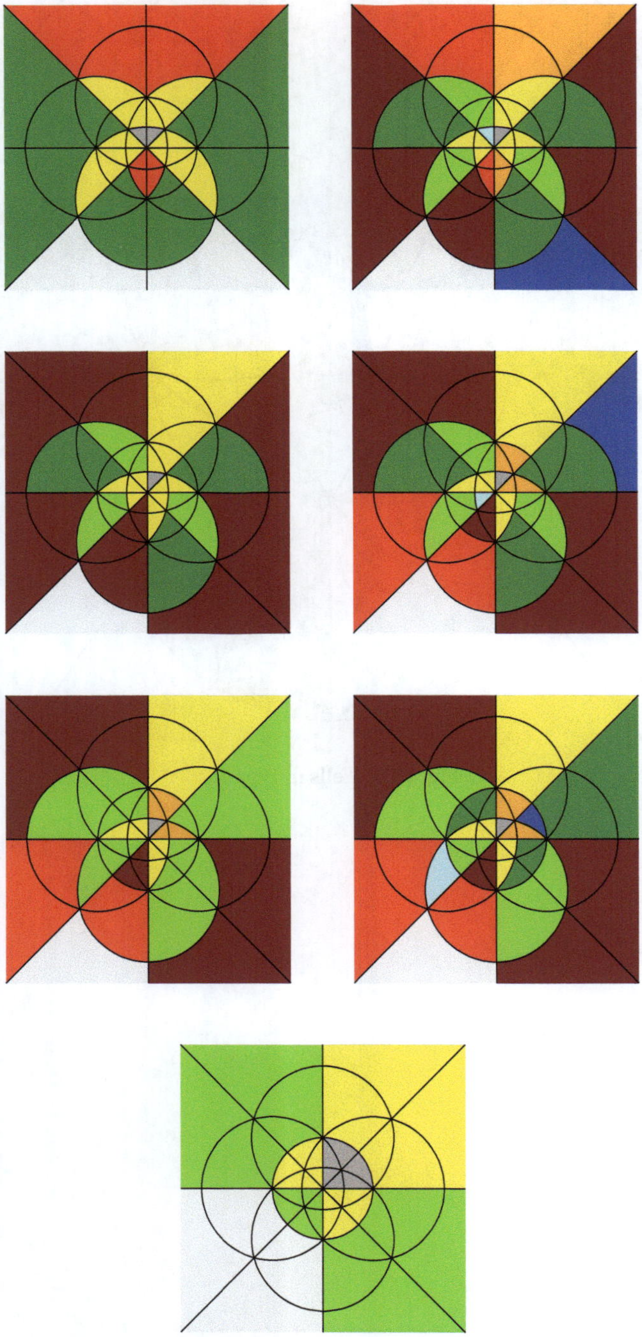

Fig. 25.d. Cells in type B_3

Fig. 25.e. Cells in type H_3

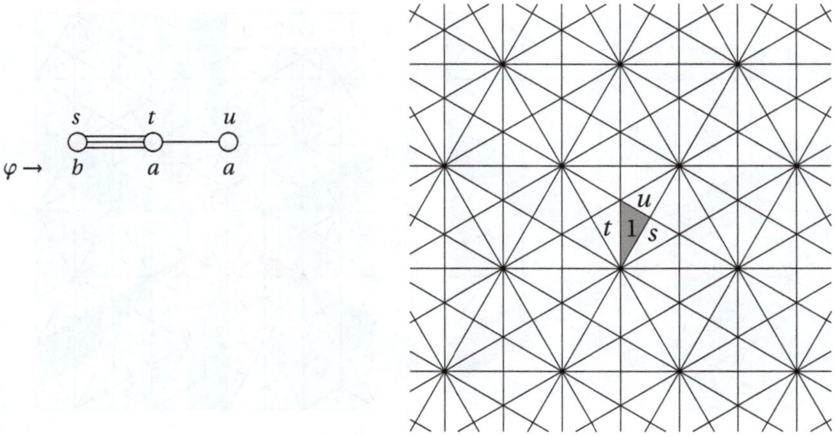

Fig. 25.f. Simple reflections in type $\tilde{G}_2$

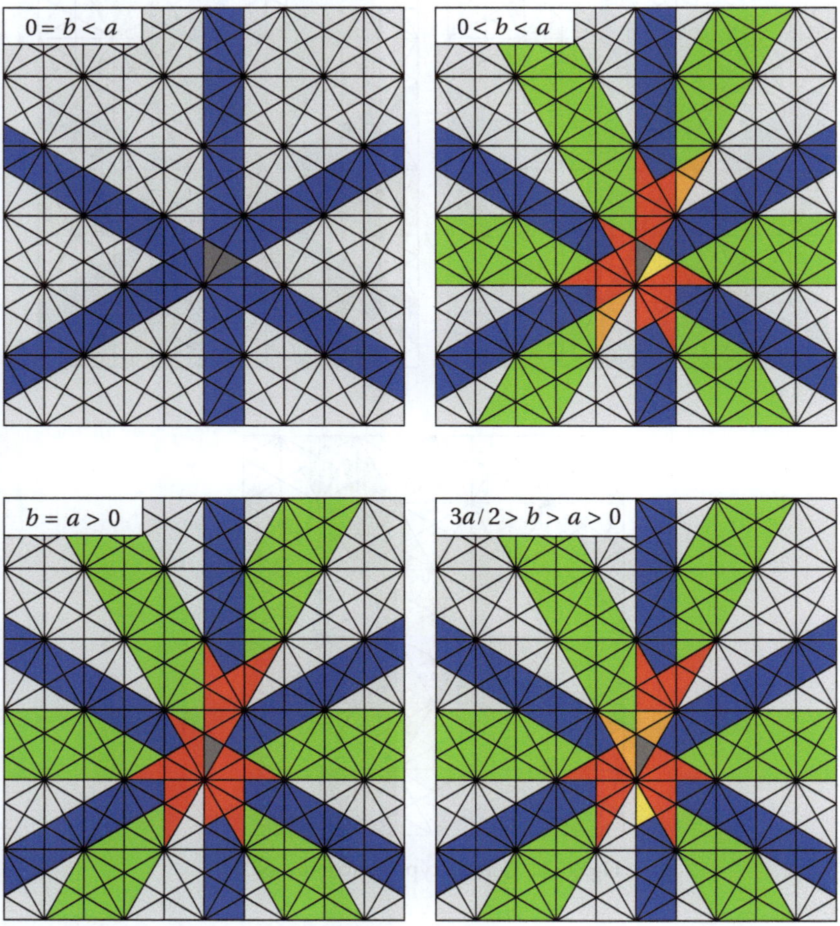

Fig. 25.g. Cells in type $\tilde{G}_2$

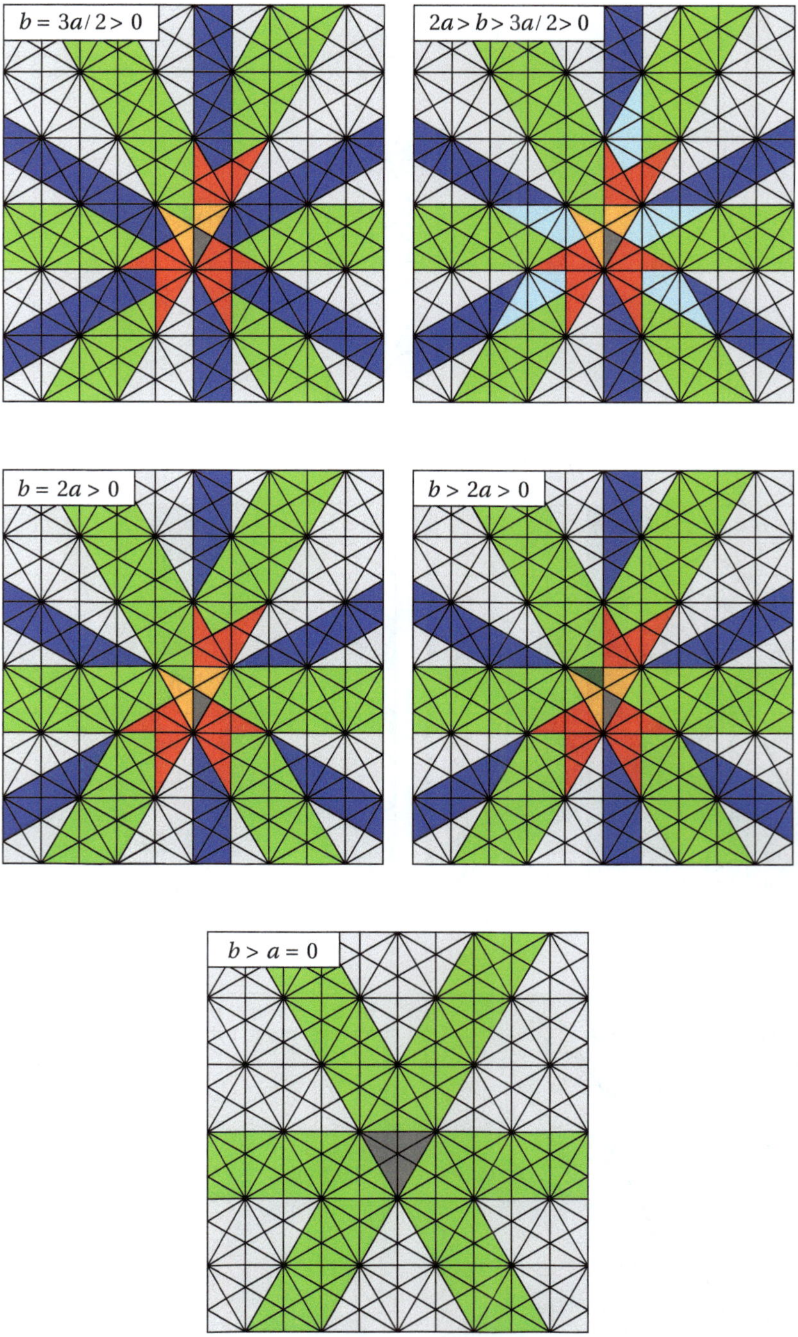

Fig. 25.h. Cells in type $\tilde{G}_2$ (continued)

Fig. 25.i. Cells in type **Hyp**(3, 3, 4)

Fig. 25.j. Conjectural cells in type **Hyp**$(2, 3, \infty)$

Fig. 25.k. Conjectural cells in type **Hyp**$(2,3,7)$

Fig. 25.l. Conjectural cells in type **Hyp**$(4,4,4)$

Fig. 25.m. Conjectural cells in type **Hyp**(2, 2, 2, 3)

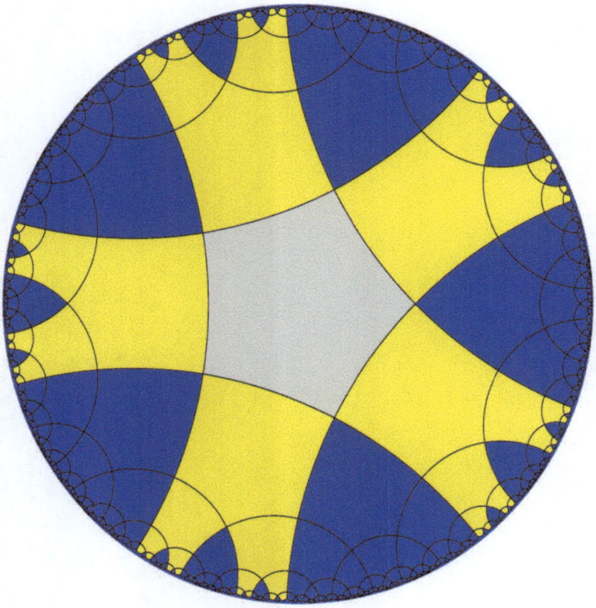

Fig. 25.n. Cells in type **Hyp**(2, 2, 2, 2, 2)

Chapter 26
Some Bibliographical Comments

This book is essentially centered on general methods for computing the partition of a Coxeter group into left, right, or two-sided cells. It also provides some examples (dihedral groups, free Coxeter groups, some groups of rank 3,...). However, it cannot give a complete account, together with complete proofs, of all the cases in which the partition into cells is known. In this chapter, we provide a (non-exhaustive) list of such cases, together with references. We wish to thank warmly Jian-Yi Shi, Misha Belolipetsky and Paul Gunnells for their very helpful comments.

26.1. Finite Groups

26.1.A. Split Case

The partition into cells of finite Coxeter groups is known in all cases whenever φ is constant:

- The type A was solved in Kazhdan–Lusztig's original paper [KaLu1, Theorem 1.4 and §5].
- Types B and D were obtained by Barbasch–Vogan [BaVo] and Garfinkle [Gar1], [Gar2], [Gar3]. See also Garfinkle–Vogan [GaVo].
- Finite dihedral groups are easy and can be handled by direct computations (see Chapter 21).
- The case of exceptional finite groups relies heavily on computer computations: see Alvis [Alv] (types H_3 and H_4), Takahashi [Tak] (type F_4), Tong [Ton] (type E_6), Chen–Shi [ChSh] (type E_7) and Chen [CheY3] (type E_8). See also the highly efficient programs of du Cloux [duCl1], [duCl3], and Geck–Halls [GeHa].

C. Bonnafé, *Kazhdan–Lusztig Cells with Unequal Parameters*,
Algebra and Applications 24, https://doi.org/10.1007/978-3-319-70736-5_26

26.1.B. *Unequal Parameter Case*

The finite irreducible Coxeter systems with at least (and so, at most) two conjugacy classes of reflections are the ones of type $I_2(2m)$ (with $m \geqslant 2$), F_4 or B_n (with $n \geqslant 2$). In this case, the partition into cells is not completely known:

- Again, finite dihedral groups are easy and can be handled by direct computations (see Chapter 21).
- Type F_4 was solved by Geck [Gec4], relying on computer computations.
- In type B_n, Geck, Lam, Iancu and the author proposed conjectures describing the partition into cells for all possible parameters in terms of domino insertion algorithms [BGIL]: if we write $S = \{t, s_1, s_2, \dots, s_{n-1}\}$, $b = \varphi(t)$ and $a = \varphi(s_i)$ as in Example 7.3.3, then these conjectures are known to hold in the following cases:
    - if $n \leqslant 6$ (by computer computations);
    - if $b = a$ (the equal parameter case, see above);
    - if $a = 2b$ or if $3a = 2b$ (see Lusztig [Lus9, Theorem 11]): these cases may be viewed as a quasi-split case (by embedding in a Coxeter group of type A), and follow (not directly) from Proposition 15.3.3);
    - if $b > (n-1)a$ (see Iancu and the author [BoIa], [Bon1]).
    - if $b = (n-1)a$, the left cells have been obtained by Howse [Hows].

26.2. Affine Weyl Groups

26.2.A. *Split Case*

As usual, the equal parameter case is much more understood. The complete description of left, right and two-sided cells is known in the following cases:

- The infinite dihedral group (i.e. the Coxeter group of type $\tilde{A}_1$) is easy and can be handled by direct computations (see Chapter 24).
- If $|S| = 3$ (i.e. if W is of type $\tilde{A}_2$, $\tilde{B}_2$ or $\tilde{G}_2$), this was solved by Lusztig in [Lus12].
- Type $\tilde{A}$ was obtained by Shi [Shi1].
- Type $\tilde{B}_3$ is due to Du [Du1]; type $\tilde{B}_4$ is due to Zhang [Zha].
- Type $\tilde{C}_3$ is due to Bédard [Béd]; type $\tilde{C}_4$ is due to Shi [Shi4].
- Two-sided cells in type $\tilde{D}_4$ were obtained by Du [Du2] while left cells were obtained by Chen [CheCD1] and Shi [Shi2], independently; type $\tilde{D}_5$ is due to Chen [CheCD2].
- Type $\tilde{F}_4$ is due to Shi [Shi3].

26.2.B. *Unequal Parameter Case*

The complete description of left, right and two-sided cells is known in the following cases:

- The case where $|S|=3$ (i.e. if W is of type $\tilde{A}_2$, $\tilde{B}_2$ or $\tilde{G}_2$) has been solved by Guilhot, for all possible parameters [Gui1], [Gui3], [Gui4].
- The case where (W,S,φ) is quasi-split of type $\tilde{C}_3$, viewed as a fixed point subgroup of an affine Weyl group of type $\tilde{A}_7$, is due to Huang–Shi [HuSh].
- The case where (W,S,φ) is quasi-split of type $\tilde{C}_3$, viewed as a fixed point subgroup of an affine Weyl group of type $\tilde{A}_6$, is due to Yue [Yue].

26.3. Other Groups

Again, we will say that (W,S) is of type $\mathbf{Hyp}(p,q,r)$ if $|S| = 3$, $S = \{s,t,u\}$, $m_{st}=p$, $m_{tu}=q$, $m_{su}=r$, with $2 \leqslant p \leqslant q \leqslant r \leqslant \infty$.

26.3.A. *Split Case*

The partition into cells is completely known in the following cases:

- Type $\mathbf{Hyp}(3,3,4)$ is due to Bédard [Béd]: historically, this was the first example of a non-finite and non-affine Coxeter group in which cells were computed (and the first example of an infinite number of left cells).
- In [BeGu], Belolipetsky and Gunnells propose a conjecture for describing explicitly the set of Duflo involutions as well as the partition into left, right and two-sided cells. This conjecture is known to hold in the following cases:
  - If W is generated by reflections in the hyperbolic plane about the sides of right-angled polygons, this is due to Belolipetsky [Belo].
  - If W is *crystallographic* (i.e. stabilizes a $\mathbb{Z}$-lattice in the reflection representation E) and $\mathscr{C}ox(W,S)$ is a complete graph, this is due to Xi [Xi3].

26.3.B. *Unequal Parameter Case*

Partition into cells are known for all parameters if (W,S) is a universal Coxeter system [ShYa1] (this case is treated in Chapter 24).

Appendix A
Symmetric Algebras

One of the fundamental properties of Hecke algebras is the existence of a *symmetrizing trace* (in the sense of Definition A.1.2 of this Appendix). In this chapter, we recall the definition of *symmetric algebra* and a few basic properties.

> *We fix in this chapter a commutative ring A as well as an A-algebra $\mathcal{H}$, that we assume to be free of finite rank over A. We set $\mathcal{H}^* = \mathrm{Hom}_A(\mathcal{H}, A)$.*

A.1. Symmetrizing Form

The A-module $\mathcal{H}^*$ is naturally endowed with a structure of $(\mathcal{H},\mathcal{H})$-bimodule as follows: if $f \in \mathcal{H}^*$ and $h, h' \in \mathcal{H}$, then we set

$$(hf)(h') = f(h'h) \quad \text{and} \quad (fh)(h') = f(hh').$$

If $\tau \in \mathcal{H}^*$, then τ induces two A-linear maps

$$\begin{array}{rccl} \tau_L^\vee \colon & \mathcal{H} & \longrightarrow & \mathcal{H}^* \\ & h & \longmapsto & \big(h' \mapsto \tau(hh')\big) \end{array}$$

and

$$\begin{array}{rccl} \tau_R^\vee \colon & \mathcal{H} & \longrightarrow & \mathcal{H}^* \\ & h & \longmapsto & \big(h' \mapsto \tau(h'h)\big). \end{array}$$

It is then easy to check that $\tau_L^\vee$ (resp. $\tau_R^\vee$) is a morphism of right (resp. left) $\mathcal{H}$-modules.

C. Bonnafé, *Kazhdan–Lusztig Cells with Unequal Parameters*,
Algebra and Applications 24, https://doi.org/10.1007/978-3-319-70736-5

Definition A.1.1. *The linear form $\tau : \mathcal{H} \to A$ is said to be* ***central*** *if $\tau(hh') = \tau(h'h)$ for all h and h' in $\mathcal{H}$ (i.e. if $\tau_L^\vee = \tau_R^\vee$).*

If τ is central, we denote by $\tau^\vee : \mathcal{H} \to \mathcal{H}^*$ the map $\tau_L^\vee = \tau_R^\vee$: it is a morphism of $(\mathcal{H}, \mathcal{H})$-bimodules.

Definition A.1.2. *If $\tau \in \mathcal{H}^*$, we say that τ is* ***symmetrizing*** *if it is central and if $\tau^\vee$ is an isomorphism.*

The following two propositions follow immediately from the definition.

Proposition A.1.3. *Let $(b_i)_{i \in I}$ be an A-basis of $\mathcal{H}$. Then τ is symmetrizing if and only if there exists a basis $(b_i^\vee)_{i \in I}$ such that*

$$\tau(b_i b_j^\vee) = \tau(b_i^\vee b_j) = \delta_{i=j}$$

for all $i, j \in I$.

Proposition A.1.4. *Let τ be a symmetrizing form on $\mathcal{H}$ and let B be a commutative A-algebra. Then the scalar extension τ_B of τ to $B\mathcal{H} = B \otimes_A \mathcal{H}$ is also symmetrizing.*

Remark A.1.5. Let τ be a symmetrizing form on $\mathcal{H}$ and let $(b_i)_{i \in I}$ be an A-basis of $\mathcal{H}$ (recall that I is then necessarily finite). Then there exists a unique basis $(b_i^\vee)_{i \in I}$ of $\mathcal{H}$ such that $\tau(b_i b_j^\vee) = \delta_{i=j}$ for all $i, j \in I$. Then $(b_i^\vee)_{i \in I}$ is called the *dual basis* of $(b_i)_{i \in I}$ (with respect to τ). □

The following two examples are very standard.

Example A.1.6. Let n be a non-zero natural number. Then the trace form $\mathrm{Tr} : \mathrm{Mat}_n(A) \to A$ is symmetrizing. Indeed, if E_{ij} denotes the elementary matrix all of whose coefficients are zero, except the (i, j)-one which is 1, then $(E_{ij})_{1 \leqslant i,j \leqslant n}$ and $(E_{ji})_{1 \leqslant i,j \leqslant n}$ are two A-bases of $\mathrm{Mat}_n(A)$ satisfying the condition of Proposition A.1.3 with respect to $\tau = \mathrm{Tr}$. □

Example A.1.7. Let G be a finite group and assume that $\mathcal{H} = AG$. Let $\tau : AG \to A$ be the unique A-linear map such that $\tau(g) = \delta_{g=1}$. If g and h are two elements of G, then $gh = 1$ if and only if $hg = 1$ (i.e. if and only if $h = g^{-1}$) and so τ is central. Moreover, τ is symmetrizing: indeed, the two A-bases $(g)_{g \in G}$ and $(g^{-1})_{g \in G}$ satisfy the condition of Proposition A.1.3 with respect to τ. □

Definition A.1.8. *The algebra $\mathcal{H}$ is said to be* ***symmetric*** *if it admits a symmetrizing form.*

Example A.1.9. If $n \geqslant 1$, then $\mathrm{Mat}_n(A)$ is symmetric. If G is a finite group, then AG is symmetric. □

A.2. Dual Bases

From now on, and until the end of this chapter, we fix a central form τ on $\mathscr{H}$ and two A-bases $(b_i)_{i\in I}$ and $(b_i^\vee)_{i\in I}$ of $\mathscr{H}$, and we assume that $\tau(b_i b_j^\vee)=\delta_{i=j}$ for all $i, j \in I$.

In particular, τ is a symmetrizing form on $\mathscr{H}$. Also,

(A.2.1)
$$h=\sum_{i\in I}\tau(hb_i^\vee)b_i=\sum_{i\in I}\tau(hb_i)b_i^\vee$$

for all $h\in\mathscr{H}$.

Let us fix now another A-basis $(c_j)_{j\in J}$ of $\mathscr{H}$. Write

$$c_j=\sum_{i\in I}\gamma_{ji}b_i \quad \text{and} \quad b_i=\sum_{j\in J}\beta_{ij}c_j.$$

Then

(A.2.2)
$$\sum_{i\in I}\gamma_{ji}\beta_{ij'}=\delta_{j=j'} \quad \text{and} \quad \sum_{j\in J}\beta_{ij}\gamma_{ji'}=\delta_{i=i'}$$

for all $i, i'\in I$ and $j, j'\in J$.

If $j\in J$, we denote by $c_j^\vee$ the unique element of $\mathscr{H}$ such that $\tau(c_j^\vee c_{j'})=\delta_{j=j'}$ for all $j'\in J$ (see Remark A.1.5). We can check easily that

(A.2.3)
$$c_j^\vee=\sum_{i\in I}\beta_{ij}b_i^\vee.$$

Exercise

Exercise A.1. Let $\mathscr{H}$ and $\mathscr{H}'$ be two symmetric A-algebras. Show that $\mathscr{H}\oplus\mathscr{H}'$ and $\mathscr{H}\otimes_A\mathscr{H}'$ are symmetric.

Exercise A.2. Assume that the A-algebra $\mathscr{H}$ is symmetric (with symmetrizing form τ) and let $(b_i)_{i\in I}$ and $(b_i^\vee)_{i\in I}$ be two A-bases of $\mathscr{H}$ such that $\tau(b_i b_j^\vee)=\delta_{i=j}$ for all i, j.

Let $h\in\mathscr{H}$ and write $hb_i=\sum_{j\in I}h_{ij}b_j$. Show that $b_i^\vee h=\sum_{j\in I}h_{ji}b_j^\vee$.

Appendix B
Reflection Subgroups of Coxeter Groups

The aim of this appendix is, for the sake of completeness, to give a complete proof of the decomposition of a Coxeter group as a semidirect product of two Coxeter groups. This result is due independently to Gal [Gal, Proposition 2.1] and to Dyer and the author [BoDy, Theorem 1.1]: we present not only its proof but also some of its consequences, with an emphasis on properties that have some impact on the theory of Hecke algebras and Kazhdan–Lusztig theory.

We rely a lot on the description of reflection subgroups of Coxeter groups as Coxeter groups, an important fact due independently to Deodhar [Deo2], Tits [Tit] and Dyer [Dye]: we also provide a complete proof of this result, following Dyer's point of view (*reflection systems*) in a simplified context (groups generated by elements of order 2).

If not explicitly written, all the results contained in Sections B.1 and B.2 are due to Dyer [Dye], while all the results of Section B.3 are due to Dyer and the author [BoDy] or Gal [Gal].

B.1. Reflection Systems

Hypothesis and notation. *In this section, and only in this section, we fix a group $\mathcal{W}$ and a subset $\mathcal{S}$ of $\mathcal{W}$ consisting of elements of order 2. We assume that $\mathcal{S}$ generates $\mathcal{W}$ and we set*

$$\mathcal{T} = \bigcup_{w \in \mathcal{W}} w \mathcal{S} w^{-1}.$$

Finally, we endow $\mathcal{P}(\mathcal{T})$ with the group structure given by the symmetric difference $\dotplus$ and we denote by $\ell_{\mathcal{S}} : \mathcal{W} \to \mathbb{N}$ the length function with respect to the generating set $\mathcal{S}$.

C. Bonnafé, *Kazhdan–Lusztig Cells with Unequal Parameters*,
Algebra and Applications 24, https://doi.org/10.1007/978-3-319-70736-5

The next definition is taken from [Dye, Definition 2.1], but our context is simplified because the elements of $\mathscr{S}$ have order 2.

Definition B.1.1. *The pair $(\mathscr{W},\mathscr{S})$ is called a* **reflection system** *if there exists a map $\mathscr{N}:\mathscr{W}\to\mathscr{P}(\mathscr{T})$ such that*

(RS1) $\mathscr{N}(s)=\{s\}$ *for all* $s\in\mathscr{S}$.

(RS2) $\mathscr{N}(xy)=\mathscr{N}(x)\dotplus {}^{x}\mathscr{N}(y)$ *for all* $x,\ y\in\mathscr{W}$.

In this case, $\mathscr{N}$ is called the **reflection cocycle** *of $(\mathscr{W},\mathscr{S})$.*

As $\mathscr{W}$ is generated by $\mathscr{S}$, note that there is at most one map satisfying (RS1) and (RS2).

Example B.1.2. If $(\mathscr{W},\mathscr{S})$ is a Coxeter system, then $(\mathscr{W},\mathscr{S})$ is a reflection system, with $\mathscr{N}:W\to\mathscr{P}(\mathscr{T})$ being given by $\mathscr{N}(w)=\{t\in\mathscr{T}\mid \ell_{\mathscr{S}}(tw)<\ell(w)\}$: this follows from (3.3.1) and Corollary 3.3.5. We will show in this section that the converse holds. □

From now on, and until the end of this section, we assume that $(\mathscr{W},\mathscr{S})$ is a reflection system, with associated reflection cocycle $\mathscr{N}$.

Note that (RS2) implies that

(B.1.3)
$$\mathscr{N}(1)=\varnothing$$

and

(B.1.4)
$$\mathscr{N}(w^{-1})={}^{w^{-1}}\mathscr{N}(w)$$

for all $w\in\mathscr{W}$. It then follows that

(B.1.5)
$$\mathscr{N}(wxw^{-1})=\mathscr{N}(w)\dotplus {}^{wxw^{-1}}\mathscr{N}(w)\dotplus {}^{w}\mathscr{N}(x).$$

Lemma B.1.6. *Assume that $(\mathscr{W},\mathscr{S})$ is a reflection system, with corresponding reflection cocycle $\mathscr{N}$. Let $w,\ w'\in\mathscr{W}$ and $t\in\mathscr{T}$. Then:*

(a) $t\in\mathscr{N}(t)$.
(b) $\ell_{\mathscr{S}}(w)=|\mathscr{N}(w)|$.
(c) $\ell_{\mathscr{S}}(ww')\equiv\ell_{\mathscr{S}}(w)+\ell_{\mathscr{S}}(w')\ \mathrm{mod}\ 2$.
(d) $\ell_{\mathscr{S}}(t)$ is odd.
(e) $\ell_{\mathscr{S}}(tw)\neq\ell_{\mathscr{S}}(w)$.
(f) $\ell_{\mathscr{S}}(tw)>\ell_{\mathscr{S}}(w)$ if and only if $t\notin\mathscr{N}(w)$.

Proof. (a) Let $s\in\mathscr{S}$ and $x\in\mathscr{W}$ be such that $t=xsx^{-1}$. It then follows from B.1.5 and (RS1) that

$$\mathscr{N}(t)=\mathscr{N}(x)\dotplus {}^{t}\mathscr{N}(x)\dotplus\{t\}.$$

But $t \in \mathcal{N}(x)$ if and only if $t \in {}^t\mathcal{N}(x)$. So $t \in \mathcal{N}(t)$.

(b) Write $w = s_1 s_2 \cdots s_r$ with $s_i \in \mathcal{S}$ and $r = \ell_{\mathcal{S}}(w)$ and set

$$t_j = (s_1 s_2 \cdots s_{j-1}) s_j (s_1 s_2 \cdots s_{j-1})^{-1}.$$

Then, by (RS1) and (RS2),

$$\mathcal{N}(w) = \{t_1\} \dotplus \{t_2\} \dotplus \cdots \dotplus \{t_r\}.$$

So $\ell_{\mathcal{S}}(w) \geqslant |\mathcal{N}(w)|$. Assume now that $\ell_{\mathcal{S}}(w) > |\mathcal{N}(w)|$. Then there exist i, j such that $1 \leqslant i < j \leqslant r$ and $t_i = t_j$. In other words, $s_i = (s_i s_{i+1} \cdots s_{j-1}) s_j (s_i s_{i+1} \cdots s_{j-1})^{-1}$, that is, $s_i s_{i+1} \cdots s_j = s_{i+1} s_{i+2} \cdots s_{j-1}$, which shows that $\ell_{\mathcal{S}}(w) \leqslant r - 2$, contrary to our hypothesis. So $\ell_{\mathcal{S}}(w) = r = |\mathcal{N}(w)|$.

(c) It follows from (RS2) that $|\mathcal{N}(ww')| \equiv |\mathcal{N}(w)| + |\mathcal{N}(w')| \bmod 2$, so (c) follows from (b).

(d) follows immediately from (c).

(e) follows immediately from (d).

(f) Here we continue to use the notation of the proof of (b). Then $\mathcal{N}(w) = \{t_1, t_2, \ldots, t_r\}$. So, if $t \in \mathcal{N}(w)$, then $t = t_j$ for some j and then $tw = s_1 s_2 \cdots s_{j-1} s_{j+1} s_{j+2} \cdots s_r$ and so $\ell_{\mathcal{S}}(tw) \leqslant r - 1 < r = \ell_{\mathcal{S}}(w)$.

Conversely, if $t \notin \mathcal{N}(w)$, then $t \in \mathcal{N}(tw)$ by (a) and (RS2), so the previous argument shows that $\ell_{\mathcal{S}}(ttw) < \ell_{\mathcal{S}}(tw)$, as desired. ■

Corollary B.1.7. *Assume that $(\mathcal{W}, \mathcal{S})$ is a reflection system, with corresponding reflection cocycle $\mathcal{N}$. Then $\mathcal{N}(w) = \{t \in \mathcal{T} \mid \ell_{\mathcal{S}}(tw) < \ell_{\mathcal{S}}(w)\}$.*

Exchange Condition B.1.8. *Assume that $(\mathcal{W}, \mathcal{S})$ is a reflection system. Let $w = s_1 s_2 \cdots s_r$ with $s_i \in \mathcal{S}$ and $r = \ell_{\mathcal{S}}(w)$ and let $s \in \mathcal{S}$ be such that $\ell_{\mathcal{S}}(sw) \leqslant \ell_{\mathcal{S}}(w)$. Then there exists a $j \in \{1, 2, \ldots, r\}$ such that $s s_1 s_2 \cdots s_j = s_1 s_2 \cdots s_{j-1}$.*

Proof. Let $\mathcal{N}$ be the corresponding reflection cocycle. As usual, let $t_j = (s_1 s_2 \cdots s_{j-1}) s_j (s_1 s_2 \cdots s_{j-1})^{-1}$. Then it follows from (RS1) and (RS2) that

$$\mathcal{N}(w) = \{t_1\} \dotplus \{t_2\} \dotplus \cdots \dotplus \{t_r\}.$$

By Lemma B.1.6(b), $|\mathcal{N}(w)| = r$ and $|\mathcal{N}(sw)| \leqslant r$. So $t_i \neq t_j$ if $1 \leqslant i < j \leqslant r$ and $\mathcal{N}(w) = \{t_1, t_2, \ldots, t_r\}$. Also, again by (RS1) and (RS2), we get

$$\mathcal{N}(sw) = \{s\} \dotplus {}^s\mathcal{N}(w)$$

and, by Lemma B.1.6(b), we have $|\mathcal{N}(sw)| \leqslant |\mathcal{N}(w)|$. Therefore, $s \in s\mathcal{N}(w)s$ that is, $s \in \mathcal{N}(w)$. So there exists a j such that $s = t_j$, which exactly means that $s s_1 s_2 \cdots s_j = s_1 s_2 \cdots s_{j-1}$. ■

Theorem B.1.9 (Dyer). *$(\mathcal{W}, \mathcal{S})$ is a Coxeter system if and only if $(\mathcal{W}, \mathcal{S})$ is a reflection system.*

Proof. If $(\mathscr{W},\mathscr{S})$ is a Coxeter system, then it is a reflection system by Example B.1.2. Conversely, if $(\mathscr{W},\mathscr{S})$ is a reflection system, then it satisfies the Exchange Condition B.1.8, so it is a Coxeter system (see Theorem 3.1.6). ■

B.2. Reflection Subgroups of Coxeter Groups

From now on, and until the end of this chapter, we fix a Coxeter system, that is, a reflection system, (W,S) and we keep all the notation introduced in Chapter 3 (ℓ, T, N, W_I, X_I,...).

Following [Dye, Definition 3.1], we set

$$\mathrm{can}(\Gamma) = \{t \in T \mid N(t) \cap \Gamma = \{t\}\}$$

for any subgroup Γ of W. Note that

(B.2.1) $$\mathrm{can}(\Gamma) \subset \Gamma \qquad \text{and} \qquad \Gamma \cap S = \mathrm{can}(\Gamma) \cap S.$$

Definition B.2.2. *A subgroup $\mathscr{W}$ of W is called a* **reflection subgroup** *if it is generated by elements of T (i.e. if $\mathscr{W} = \langle \mathscr{W} \cap T\rangle$).*

Theorem B.2.3 (Deodhar, Dyer, Tits). *Let $\mathscr{W}$ be a reflection subgroup of W, let $\mathscr{T} = \mathscr{W} \cap T$ and let*

$$\mathscr{S} = \mathrm{can}(\mathscr{W}) = \{t \in T \mid N(t) \cap \mathscr{W} = \{t\}\}.$$

Then:

(a) $\mathscr{T} = \bigcup_{w \in \mathscr{W}} w\mathscr{S}w^{-1}$.
(b) *$(\mathscr{W},\mathscr{S})$ is a reflection system: the corresponding reflection cocyle $\mathscr{N} : \mathscr{W} \to \mathscr{P}(\mathscr{T})$ is given by $w \mapsto N(w) \cap \mathscr{T} = N(w) \cap \mathscr{W}$.*
(c) *$(\mathscr{W},\mathscr{S})$ is a Coxeter system.*

Proof. We need the following intermediate results:

Lemma B.2.4. *Let Γ be any subgroup of W. Then:*

(a) *If $s \in S$, then*

$$\mathrm{can}({}^s\Gamma) = \begin{cases} {}^s\mathrm{can}(\Gamma) & \text{if } s \notin \Gamma, \\ \mathrm{can}(\Gamma) & \text{if } s \in \Gamma. \end{cases}$$

(b) *If* $t \in \Gamma \cap T$, *then there exist* $t_1, t_2, \ldots, t_{r+1}$ *in* $\operatorname{can}(\Gamma)$ *such that*

$$t = (t_1 t_2 \cdots t_r) t_{r+1} (t_1 t_2 \cdots t_r)^{-1}.$$

(c) *Let* $N_\Gamma : W \to \mathscr{P}(T \cap \Gamma)$, $\gamma \mapsto N(\gamma) \cap \Gamma$ *and let* $w \in W$ *and* $\gamma \in \Gamma$. *Then* $N_\Gamma(\gamma w) = N_\Gamma(\gamma) \dotplus {}^{\gamma} N_\Gamma(w)$.

Proof. (a) If $s \in \Gamma$, then the result is obvious. So let us assume that $s \notin \Gamma$. It is sufficient to show that ${}^s\operatorname{can}(\Gamma) \subset \operatorname{can}({}^s\Gamma)$ (indeed, by applying the same result to ${}^s\Gamma$, one would then get ${}^s\operatorname{can}({}^s\Gamma) \subset \operatorname{can}(\Gamma)$, which is equivalent to the reverse inclusion).

Let $t \in \operatorname{can}(\Gamma)$. Then, by B.1.5,

$$N(sts) = \{s\} \dotplus \{ststs\} \dotplus {}^s N(t).$$

But, as $t \in \Gamma$, $tst \notin \Gamma$ and so $ststs \notin {}^s\Gamma$. Therefore,

$$N(sts) \cap {}^s\Gamma = {}^s N(t) \cap {}^s\Gamma = {}^s\big(N(t) \cap \Gamma\big) = \{sts\}.$$

(b) We prove it by induction on $\ell(t)$ (recall that $\ell(t)$ is odd). If $\ell(t) = 1$, then $t \in \Gamma \cap S \subset \operatorname{can}(\Gamma)$ (see B.2.1), so the result is obvious. If $\ell(t) \geqslant 3$, then there exists an $s \in S$ such that $\ell(sts) = \ell(t) - 2$ (see Proposition 3.8.1). By the induction hypothesis, there exists $t_1, t_2, \ldots, t_{r+1}$ in $\operatorname{can}({}^s\Gamma)$ such that $sts = (t_1 t_2 \cdots t_r) t_{r+1} (t_1 t_2 \cdots t_r)^{-1}$. Two cases may occur:

• If $s \in \Gamma$, then $s \in \operatorname{can}(\Gamma) = \operatorname{can}({}^s\Gamma)$ by B.2.1 and then $t = (st_1 t_2 \cdots t_r) t_{r+1} (st_1 t_2 \cdots t_r)^{-1}$, as desired.

• If $s \notin \Gamma$, then $t_i' = st_i s \in \operatorname{can}(\Gamma)$ by (a), and so $t = (t_1' t_2' \cdots t_r') t_{r+1}' (t_1' t_2' \cdots t_r')^{-1}$, as desired.

(c) is obvious. □

Let us now come back to the proof of the theorem. Let $\mathscr{W}' = \langle \mathscr{S} \rangle$ and $\mathscr{T}' = \bigcup_{w' \in \mathscr{W}'} w' \mathscr{S} w'^{-1}$. Then $\mathscr{W}' \subset \mathscr{W}$ and $\mathscr{T}' \subset \mathscr{T}$. But

$$\mathscr{S} \subset \mathscr{W} \cap T \subset \mathscr{T}' \subset \mathscr{W}' = \langle \mathscr{S} \rangle \subset \mathscr{W} = \langle \mathscr{W} \cap T \rangle$$

(the second inclusion follows from Lemma B.2.4(b)). So $\mathscr{W} = \mathscr{W}'$ an $\mathscr{T} = \mathscr{T}'$. This proves statement (a) of the theorem, and this shows that $\mathscr{S}$ generates $\mathscr{W}$. Now (b) follows from Lemma B.2.4(c) and (c) follows from Theorem B.1.9. ■

We keep the notation of the previous section ($\mathscr{N}, \ell_{\mathscr{S}}, \ldots$).

Corollary B.2.5 (Dyer). *Let* $\mathscr{W}$ *be a reflection subgroup of* W, *let* $\mathscr{T} = \mathscr{W} \cap T$ *and let*

$$\mathscr{S} = \operatorname{can}(\mathscr{W}) = \{t \in T \mid N(t) \cap \mathscr{W} = \{t\}\}.$$

(a) *If* $t \in \mathscr{T}$ *and* $w \in \mathscr{W}$, *then* $\ell(tw) < \ell(w)$ *if and only if* $\ell_{\mathscr{S}}(tw) < \ell_{\mathscr{S}}(w)$.

(b) *Let $x \in W$. The following are equivalent:*

(1) *x has minimal length (with respect to ℓ) in the coset $\mathcal{W}x$.*
(2) *$N(x) \cap \mathcal{W} = \varnothing$ (that is, if and only if $\ell(tx) > \ell(x)$ for all $t \in \mathcal{T}$).*
(3) *For all $w \in \mathcal{W}$, $N(w) \cap \mathcal{W} = N(wx) \cap \mathcal{W}$.*

(c) *There is a unique element of minimal length in $\mathcal{W}x$.*

Proof. (a) follows immediately from Theorem B.2.3(b) and Lemma B.1.6(f).

(b) It is clear that (1) implies (2). The fact that (2) implies (1) follows from the cocyle condition (RS2). Now assume that (3) holds and write $x = wx'$, with $w \in \mathcal{W}$ and x' of minimal length in $\mathcal{W}x = \mathcal{W}x'$. Then $N(x') \cap \mathcal{W} = \varnothing$ (since (1) implies (2) for x') and so, by (3), $N(w^{-1}) \cap \mathcal{W} = \varnothing$, which is possible only if $w = 1$, that is, if $x = x'$.

It remains to show that (2) implies (3). So assume that $N(x) \cap \mathcal{W} = \varnothing$. Then $N(wx) = N(w) \dotplus {}^{w}N(x)$, so

$$N(wx) \cap \mathcal{W} = (N(w) \cap \mathcal{W}) \dotplus {}^{w}(N(x) \cap \mathcal{W}) = N(w) \cap \mathcal{W},$$

as expected.

(c) follows from (b) and its proof. ■

Definition B.2.6. *If $\mathcal{W}$ is a reflection subgroup of W, then* $\mathrm{can}(\mathcal{W})$ *is called the* ***canonical generating set*** *of $\mathcal{W}$.*

B.3. Semidirect Product Decomposition

We fix in this section a partition $S = I \dot\cup J$ where I and J are W-closed subsets of S and we set

$$\tilde{J} = \{wtw^{-1} \mid w \in W_I \text{ and } t \in J\}.$$

We denote by $\widetilde{W}$ the subgroup of W generated by $\tilde{J}$.

The aim of this section is to prove the following theorem (see [BoDy, Theorem 1.1]; statements (a) and (c) were proved independently by S. Gal [Gal]):

Theorem B.3.1. *Under the above hypothesis,*

(a) $W = W_I \ltimes \widetilde{W}$.
(b) $\tilde{J} = \mathrm{can}(\widetilde{W})$.
(c) *$(\widetilde{W}, \tilde{J})$ is a Coxeter system.*
(d) *If $w \in W_I$, then w is the unique element of minimal length in $\widetilde{W}w = w\widetilde{W}$.*

Proof. First, note the following easy property, which follows from the fact that I and J are both W-closed and from Lemma 3.2.7:

(♣) *if $s \in I$ and $t \in J$, then $m_{s,t}$ is even or infinite.*

Now, let

$$\begin{aligned} \varphi : W &\longrightarrow W_I \\ s &\longmapsto \begin{cases} s & \text{if } s \in I, \\ 1 & \text{if } s \in J. \end{cases} \end{aligned}$$

Then, by (♣), $(\varphi(s)\varphi(t))^{m_{st}} = 1$ for all $s, t \in S$, so φ extends to a morphism of groups $W \to W_I$, still denoted by φ.

(a) The restriction of φ to W_I is the identity morphism and $J \subset \operatorname{Ker}\varphi$, so

$$W = W_I \ltimes (\operatorname{Ker}\varphi) \quad \text{and} \quad \widetilde{W} \subset \operatorname{Ker}\varphi.$$

To prove (a), it is now enough to prove that $W = \widetilde{W} \cdot W_I$. Let $w \in W$: we will prove by induction on $\ell(w)$ that $w \in W_I\widetilde{W}$. The result is obvious if $w \in W_I$. So assume that $w \in W \setminus W_I$ and write $w = s_1 s_2 \cdots s_r$ with $s_i \in S$ and $r = \ell(w)$. Let j be minimal such that $s_j \notin I$ (such a j exists) and let $t = (s_1 s_2 \cdots s_{j-1}) s_j (s_1 s_2 \cdots s_{j-1})^{-1}$. Then $t \in \tilde{J}$ and $\ell(tw) < \ell(w)$ so, by the induction hypothesis, $tw \in \widetilde{W} \cdot W_I$ and so $w \in \widetilde{W} \cdot W_I$, as desired. This proves (a) and also that

(♢) $$\widetilde{W} = \operatorname{Ker}\varphi.$$

(b) Let $T(I) = \bigcup_{w \in W} wIw^{-1}$ and $T(J) = \bigcup_{w \in W} wJw^{-1}$. Then $T = T(I) \dot\cup T(J)$ is a partition of T (because I and J are W-closed) and

$$T(J) = \bigcup_{w \in \widetilde{W}} w\tilde{J}w^{-1}.$$

So, by Theorem B.2.3(a),

(♡) $$T(J) = \widetilde{W} \cap T.$$

Let $w \in W_I$ and $s \in J$. Then

$$N(wsw^{-1}) = N(w) \dotplus {}^{wsw^{-1}}N(w) \dotplus {}^{w}N(s).$$

But $N(w)$ is contained in $T(I)$, while $N(s) = \{s\}$ is contained in $T(J)$. So it follows from (♡) that $N(wsw^{-1}) \cap \widetilde{W} = \{wsw^{-1}\}$, so $\tilde{J} \subset \operatorname{can}(\widetilde{W})$. As $(\widetilde{W}, \operatorname{can}(\widetilde{W}))$ is a Coxeter system by Theorem B.2.3(c), the set $\operatorname{can}(\widetilde{W})$ is a minimal (for inclusion) set of generators of $\widetilde{W}$ (see [Bou, Chapter IV, §1, Corollary 3]), so we get (b).

(c) follows from (b) and from Theorem B.2.3(c).

(d) Let $w \in W_I$ and let $t \in T(J) = \widetilde{W} \cap T$. Since $N(w) \subset T(I)$, this shows that $\ell(tw) > \ell(w)$, so that w has minimal length in $w\widetilde{W} = \widetilde{W}w$ by Corollary B.2.5(b). ■

Let us denote by $\tilde{\ell} : \widetilde{W} \to \mathbb{N}$ the length function with respect to $\tilde{J}$. Recall that the function $\ell_J : W \to \mathbb{N}$ has been defined in Section 3.4.

Corollary B.3.2. *If $a \in W_I$ and $w \in \widetilde{W}$, then $\ell_J(aw) = \ell_J(wa) = \ell_J(w) = \tilde{\ell}(w)$ and $\tilde{\ell}(awa^{-1}) = \tilde{\ell}(w)$.*

Proof. The equalities $\ell_J(aw) = \ell_J(wa) = \ell_J(w)$ are obvious, as well as the equality $\tilde{\ell}(awa^{-1}) = \tilde{\ell}(w)$. It remains to show that $\ell_J(w) = \tilde{\ell}(w)$. We argue by induction on $\tilde{\ell}(w)$, the result being obvious if $\tilde{\ell}(w) = 0$. So assume that $\tilde{\ell}(w) \geqslant 1$. Then there exists a $\tilde{s} \in \tilde{J}$ such that $\tilde{\ell}(\tilde{s}w) < \tilde{\ell}(w)$. Write $\tilde{s} = asa^{-1}$, with $a \in W_I$ and $s \in J$. Then $\tilde{\ell}(\tilde{s}w) = \tilde{\ell}(sa^{-1}wa) < \tilde{\ell}(w) = \tilde{\ell}(a^{-1}wa)$. By the induction hypothesis, one gets $\ell_J(sa^{-1}wa) = \tilde{\ell}(sa^{-1}wa)$. But $\ell_J(sa^{-1}wa) = \ell_J(a^{-1}wa) - 1 = \ell_J(w) - 1$ and $\tilde{\ell}(\tilde{s}w) = \tilde{\ell}(w) - 1$. So $\ell_J(w) = \tilde{\ell}(w)$, as desired. ■

We now examine some properties of the set $\tilde{J}$. First, note that

(B.3.3) *if $s, s' \in J$ are W_I-conjugate, then $s = s'$.*

Indeed, if $s' = wsw^{-1}$ with $w \in W_I$, then $\{s'\} = \mathrm{supp}(s') = \mathrm{supp}(wsw^{-1}) \subset I \cup \{s\}$ (by the Cancellation Law 3.2.1). Therefore, there is a well-defined map

$$\nu : \tilde{J} \longrightarrow J$$

such that $\nu(\tilde{s})$ is the unique element of J such that $\tilde{s}$ and $\nu(\tilde{s})$ are W_I-conjugate. It follows from this definition that:

Corollary B.3.4. *Let $\tilde{s}$ and $\tilde{s}'$ be two elements of $\tilde{J}$. If $\tilde{s}$ and $\tilde{s}'$ are $\widetilde{W}$-conjugate, then $\nu(\tilde{s})$ and $\nu(\tilde{s}')$ are W-conjugate.*

Recall from Proposition 3.8.2 that

$$C_{W_I}(s) = W_{I \cap s^{\perp}}$$

for all $s \in J$, where $s^{\perp} = \{t \in S \mid st = ts\}$. Therefore, if $\mathcal{J}$ denotes the set of pairs (x, s) where $s \in J$ and $x \in X^I_{I \cap s^{\perp}}$, then the map

(B.3.5)
$$\begin{array}{ccc} \mathcal{J} & \longrightarrow & \tilde{J} \\ (x,s) & \longmapsto & xsx^{-1} \end{array}$$

is bijective.

Example B.3.6. Even if S is finite, it might happen that $\tilde{J}$ is infinite. For instance, if $|S| = 3$, $S = \{s, t, u\}$ and $m_{st} = m_{su} = m_{tu} = \infty$, and if we take $J = \{s\}$ and $I = \{t, u\}$, then W_I is infinite, $I \cap s^{\perp} = \varnothing$ and the bijection B.3.5 shows that the map $W_I \to \tilde{J}$, $w \mapsto wsw^{-1}$ is bijective. □

References

[A'Ca] N. A'Campo, Sur les valeurs propres de la transformation de Coxeter. Invent. Math. **33**, 61–67 (1976)

[Alv] D. Alvis, The left cells of the Coxeter group of type H_4. J. Algebra **107**, 160–168 (1987)

[Ari] S. Ariki, Robinson-Schensted Correspondence and Left Cells, Combinatorial Methods in Representation Theory, (Kyoto, 1998). Advanced Studies in Pure Mathematics, vol. 28 (Kinokuniya, Tokyo 2000), pp. 1–20

[Bah] P. Bahls, The Isomorphism Problem in Coxeter Groups (Imperial College Press, London, 2005), xiv+176 pp

[BaVo] D. Barbasch, D. Vogan, Primitive ideals and orbital integrals of complex classical groups. Math. Ann. **259**, 153–199 (1982)

[Béd] R. Bédard, Cells for two coxeter groups. Commun. Algebr. **14**, 1253–1286 (1986)

[BeBe] A. Beilinson, J. Bernstein, Localisation de g-modules. C.R. Acad. Sci. Paris Sér. I, Math. 292, 15–18 (1981)

[Belo] M. Belolipetsky, Cells and representations of right-angled coxeter groups. Sel. Math. **10**, 325–339 (2004)

[BeGu] M. Belolipetsky, P.E. Gunnells, Cells in coxeter groups I. J. Algebr. **385**, 134–144 (2013)

[BeGuSc] M. Belolipetsky, P.E. Gunnells, R.A. Scott, Kazhdan-Lusztig cells in planar hyperbolic Coxeter groups and automata. Intern. J. Algebr. Comput. **24**, 757–772 (2014)

[Bell] G. Bellamy, Cuspidal representations of rational Cherednik algebras at $t = 0$. Math. Z. **269**, 609–627 (2011)

[BeTh] G. Bellamy, U. Thiel, Cuspidal Calogero-Moser and Lusztig families for Coxeter groups. J. Algebr. **462**, 197–252 (2016)

[BjBr] A. Björner, F. Brenti, Graduate Texts in Mathematics, in Combinatorics of Coxeter groups, vol. 231 (Springer, New York, 2005). xiv+363 pp

[Bon1] C. Bonnafé, Two-sided cells in type B (asymptotic case). J. Algebr. **304**, 216–236 (2006)

[Bon2] C. Bonnafé, Semicontinuity properties of Kazhdan-Lusztig cells. N. Z. J. Math. **39**, 171–192 (2009)

[Bon3] C. Bonnafé, Automorphisms of Coxeter groups and Lusztig's conjectures for Hecke algebras with unequal parameters. Nagoya J. Math. **195**, 153–164 (2009)

C. Bonnafé, *Kazhdan–Lusztig Cells with Unequal Parameters*,
Algebra and Applications 24, https://doi.org/10.1007/978-3-319-70736-5

[Bon4] C. Bonnafé, On Kazhdan-Lusztig, cells in type B, J. Algebr. Combin. 31, 53–82 (2010); Erratum to: On Kazhdan-Lusztig cells in type B. J. Algebr. Combin. 35 515–517 (2012)

[Bon5] C. Bonnafé, Cells and cacti. Int. Math. Res. Not. **19**, 5775–5800 (2016)

[BoDy] C. Bonnafé, M. Dyer, Semidirect product decomposition of Coxeter groups. Commun. Algebr. **38**, 1549–1574 (2010)

[BoGe] C. Bonnafé, M. Geck, Conjugacy classes of involutions and Kazhdan-Lusztig cells. Represent. Theory **18**, 155–182 (2014)

[BoGe2] C. Bonnafé, M. Geck, Hecke algebras with unequal parameters and Vogan's left cell invariants, to appear in *The Proceedings of the Conference, Representations of reductive groups* (Celebrating David Vogan's sixtieth birthday)

[BGIL] C. Bonnafé, M. Geck, L. Iancu, T. Lam, On domino insertion and Kazhdan-Lusztig cells in type B_n, in Representation Theory Of Algebraic Groups And Quantum Groups. Progress in Mathematics, vol. 284 (Birkhäuser/Springer, New York, 2010), pp. 33–54

[BoIa] C. Bonnafé, L. Iancu, Left cells in type B_n with unequal parameters. Represent. Theory **7**, 587–609 (2003)

[BoRo1] C. Bonnafé, R. Rouquier, Calogero-Moser versus Kazhdan-Lusztig cells. Pacific J. Math. **261**, 45–51 (2013)

[BoRo2] C. Bonnafé, R. Rouquier, Cellules de Calogero-Moser, preprint (2013), arXiv:1302.2720

[BoRo3] C. Bonnafé, R. Rouquier, Cherednik algebras and Calogero-Moser cells, preprint (2017). arXiv:1708.09764

[Bou] N. Bourbaki, Groupes et algèbres de Lie, chap. 4, 5 et 6, (Hermann, Paris, 1968)

[Bre1] K. Bremke, Kazhdan-Lusztig polynomials and cells for affine Weyl groups and unequal parameters, Ph.D. Thesis, Massachusetts Institute of Technology, 1996

[Bre2] K. Bremke, On generalized cells in affine Weyl groups. J. Algebra **191**, 149–173 (1997)

[BrKa2] J.-L. Brylinski, M. Kashiwara, Kazhdan-Lusztig conjecture and holonomic systems. Invent. Math. **64**, 387–410 (1981)

[BrKa1] J.-L. Brylinski, M. Kashiwara, Démonstration de la conjecture de Kazhdan-Lusztig sur les modules de Verma. C.R. Acad. Sci. Paris Sér. A-B 291, A373–A376 (1980)

[Br] M. Broué, Higman criterion revisited. Michigan Math. J. **58**, 125–179 (2009)

[CheCD1] C.D. Chen, The decomposition into left cells of the affine Weyl group of type $\tilde{D}_4$. J. Algebra **163**, 692–728 (1994)

[CheCD2] C.D. Chen, The left cells of the affine Weyl group of type $\widetilde{D}_5$. Commun. Algebra **29**, 11–30 (2001)

[CheY3] Y. Chen, Left cells in the Weyl group of type E_8. J. Algebra **231**, 805–830 (2000)

[ChSh] Y. Chen, J.Y. Shi, Left cells in the Weyl group of type E_7. Commun. in Algebra **26**, 3837–3852 (1998)

[CuIwKij] C.W. Curtis, N. Iwahori & R. Kilmoyer, Hecke algebras and characters of parabolic type of finite groups with (B, N)-pairs. Publ. Math. I.H.E.S. 81–116 (1972)

[Dav] M. Davis, The Geometry and Topology of Coxeter Groups. London Mathematical Society Monographs Series, vol. 32 (Princeton University Press, Princeton, NJ, 2008), pp. xvi+584

[DDPW] B. Deng, J. du, B. Parwill, J. Wang, Finite Dimensional Algebras and Quantum Groups. Mathematical Surveys and Monographs, vol. 150 (American Mathematical Society, Providence, RI, 2008), pp. xxvi+759

[Deo1] V. Deodhar, On the Kazhdan-Lusztig conjectures. Nederl. Akad. Wetensch. Indag. Math. **44**, 1–17 (1982)

[Deo2] V. Deodhar, A note on subgroups generated by reflections in Coxeter groups. Arch. Math. **53**, 543–546 (1989)

[DeGaKa] V. Deodhar, O. Gabber, V. Kac, Structure of some categories of representations of infinite-dimensional Lie algebras. Adv. Math. **45**, 92–116 (1982)

[Dou] J.M. Douglass, Cells and the reflection representation of Weyl groups and Hecke algebras. Trans. Am. Math. Soc. **318**, 373–399 (1990)

[Du1] J. Du, The decomposition into cells of the affine Weyl group of type $\tilde{B}_3$. Commun. Algebra **16**, 1383–1409 (1988)

[Du2] J. Du, Cells in the affine Weyl group of type $\tilde{D}_4$. J. Algebra **128**, 384–404 (1990)

[duCl1] F. du Cloux, The state of the art in the computation of Kazhdan-Lusztig polynomials. Appl. Algebra Engrg. Commun. Comput. 7, 211–219 (1996); (Computational methods in Lie theory (Essen, 1994))

[duCl2] F. du Cloux, Computing Kazhdan-Lusztig polynomials for arbitrary Coxeter groups. Experiment. Math. **11**, 371–381 (2002)

[duCl3] F. du Cloux, Positivity results for the Hecke algebras of noncrystallographic finite Coxeter groups. J. Algebra **303**, 731–741 (2006)

[Dye] M. Dyer, Reflection subgroups of Coxeter groups. J. Algebra **135**, 57–73 (1990)

[Eli] B. Elias, Folding with Soergel bimodules, in *Categorification and Higher Representation Theory*. Contemp. Math., vol. 683 (Amer. Math. Soc., Providence, RI, 2017), pp. 287–332

[ElWi] B. Elias, G. Williamson, The Hodge theory of Soergel bimodules. Ann. Math. **180**, 1089–1136 (2014)

[Ful] W. Fulton, Young Tableaux, London Mathematical Society Student Texts, vol. 35 (Cambridge University Press, Cambridge, 1997), pp. ix + 260

[Gal] S. Gal, On normal subgroups of Coxeter groups generated by standard parabolic subgroups. Geometriae Dedicata **115**, 65–78 (2005)

[Gar1] D. Garfinkle, On the classification of primitive ideals for complex classical Lie algebras I. Compo. Math. **75**, 135–169 (1990)

[Gar2] D. Garfinkle, On the classification of primitive ideals for complex classical Lie algebras II. Compo. Math. **81**, 307–336 (1992)

[Gar3] D. Garfinkle, On the classification of primitive ideals for complex classical Lie algebras III. Compo. Math. **88**, 187–234 (1993)

[GaVo] D. Garfinkle, D. Vogan, On the structure of Kazhdan-Lusztig cells for branched Dynkin diagrams. J. Algebra **153**, 99–120 (1992)

[Gec1] M. Geck, *Beiträge zur Darstellungstheorie von Iwahori-Hecke-Algebren* (RWTH Aachen, Habilitationsschrift, 1993)

[Gec2] M. Geck, Constructible characters, leading coefficients and left cells for finite Coxeter groups with unequal parameters. Represent. Theory **6**, 1–30 (2002)

[Gec3] M. Geck, On the induction of Kazhdan-Lusztig cells. Bull. London Math. Soc. **35**, 608–614 (2003)

[Gec4] M. Geck, Computing Kazhdan-Lusztig cells for unequal parameters. J. Algebra **281**, 342–365 (2004)

[Gec5] M. Geck, Left cells and constructible representations. Represent. Theory **9**, 385–416 (2005)

[Gec6] M. Geck, Relative Kazhdan-Lusztig cells. Represent. Theory **10**, 481–524 (2006)

[Gec7] M. Geck, Hecke algebras of finite type are cellular. Invent. Math. **169**, 501–517 (2007)

[Gec8] M. Geck, On Iwahori-Hecke algebras with unequal parameters and Lusztig's isomorphism theorem. Pure Appl. Math. Quart. **7**, 587–620 (2011)

[Gec9] M. Geck, PyCox: computing with (finite) Coxeter groups and Iwahori-Hecke algebras. LMS J. Comput. Math. **15**, 231–256 (2012)

[Gec10] M. Geck, Kazhdan-Lusztig cells and the Frobenius-Schur indicator. J. Algebra **398**, 329–342 (2014)

[GeHa] M. Geck, A. Halls, On the Kazhdan-Lusztig cells in type E_8. Math. Comput. **84**, 3029–3049 (2015)

[GeIa1] M. Geck, L. Iancu, Lusztig's **a**-function in type B_n in the asymptotic case. Nagoya Math. J. **182**, 199–240 (2006)

[GeIa2] M. Geck, L. Iancu, Coxeter groups and automorphisms, preprint (2015), arXiv:1412.5428

[GePf] M. Geck, G. Pfeiffer, Characters of Finite Coxeter Groups and Iwahori-Hecke Algebras. London Mathematical Society Monographs, New Series, vol. 21 (The Clarendon Press, Oxford University Press, New York, 2000), pp. xvi+446

[GeRo] M. Geck, R. Rouquier, Centers and simple modules for Iwahori-Hecke algebras, in Finite Reductive Groups: Related Structures and Representations, ed. by M. Cabanes. Progress in Mathematics, vol. 141 (Birkhäuser, 1997), pp. 251–272

[Gor] I. Gordon, Baby Verma modules for rational Cherednik algebras. Bull. London Math. Soc. **35**, 321–336 (2003)

[GoMa] I. Gordon, M. Martino, Calogero-Moser space, restricted rational Cherednik algebras and two-sided cells. Math. Res. Lett. **16**, 255–262 (2009)

[GrLe] J.J. Graham, G.I. Lehrer, Cellular algebras. Invent. Math. **123**, 1–34 (1996)

[Gui1] J. Guilhot, On the determination of Kazhdan-Lusztig cells for affine Weyl groups with unequal parameters. J. Algebra **318**, 893–917 (2007)

[Gui2] J. Guilhot, On the lowest two-sided cell in affine Weyl groups. Represent. Theory **12**, 327–345 (2008)

[Gui3] J. Guilhot, Generalized induction of Kazhdan-Lusztig cells. Ann. Inst. Fourier **59**, 1385–1412 (2009)

[Gui4] J. Guilhot, Kazhdan-Lusztig cells in affine Weyl groups of rank 2. Int. Math. Res. Not. **17**, 3422–3462 (2010)

[Gui5] J. Guilhot, Some computations about Kazhdan-Lusztig cells in affine Weyl groups of rank 2, preprint (2009), arXiv:0810.5165, version 2

[Gun1] P.E. Gunnells, Cells in Coxeter groups. Notices Am. Math. Soc. **53**, 528–535 (2006)

[Gun2] P.E. Gunnells, Automata and cells in affine Weyl groups. Represent. Theory **14**, 627–644 (2010)

[GuPa] J. Guilhot, J. Parkinson, A proof of Lusztig's conjectures for affine type G_2 with arbitrary parameters, preprint (2017), arXiv:1711.06551

[Hée] J.-Y. Hée, Systèmes de racines sur un anneau commutatif totalement ordonné. Geometriae Dedicata **37**, 65–102 (1991)

[Hil] H. Hiller, Geometry of Coxeter Groups. Research Notes in Mathematics, vol. 54. (Pitman (Advanced Publishing Program), Boston, Mass.-London, 1982), pp. iv+213

[Hoh] C. Hohlweg, Minimal and maximal elements in two-sided cells of S_n and Robinson-Schensted correspondence. Discret. Math. **304**, 79–87 (2005)

[Hot] R. Hotta, On Springer's representations. J. Fac. Sci. Univ. Tokyo Sect. IA Math. 28(1981), 863–876 (1982)

[Howl] R.B. Howlett, Coxeter groups and M-matrices. Bull. Lond. Math. Soc. **14**, 137–141 (1982)

[Hows] E. Howse, Vogan classes in type B_n, preprint (2017), arXiv:1709.02061

[HuSh] Q. Huang, J.Y. Shi, Some cells in the weighted Coxeter group $(\tilde{C}_n, \tilde{\ell}_{2n+1})$. J. Algebra **395**, 63–81 (2013)

[Hum] J. Humphreys, Reflection Groups and Coxeter Groups. Cambridge Studies in Advanced Mathematics, vol. 29 (Cambridge University Press, Cambridge, 1990), pp. xii+204

[Jan] J.C. Jantzen, Moduln mit einem höchsten Gewicht. Lecture Notes in Mathematics, vol. 750 (Springer, Berlin, 1979)

[Jos1] A. Joseph, A characteristic variety for the primitive spectrum of a semisimple Lie algebra, in Non-commutative Harmonic Analysis (Actes Colloq., Marseille-Luminy, 1976). Lecture Notes in Mathematics, vol. 587 (Springer, Berlin, 1977), pp. 102–118

[Jos2] A. Joseph, W-module structure in the primitive spectrum of the enveloping algebra of a semisimple Lie algebra, in Noncommutative Harmonic Analysis (Proc. Third Colloq., Marseille-Luminy, 1978). Lecture Notes in Mathematics, vol. 728 (Springer, Berlin, 1979), pp. 116–135

[Jos3] A. Joseph, Goldie rank in the enveloping algebra of a semisimple Lie algebra III. J. Algebra **73**, 295–326 (1981)

[KaLu1] D. Kazhdan, G. Lusztig, Representations of Coxeter groups and Hecke algebras. Invent. Math. **53**, 165–184 (1979)

[KaLu2] D. Kazhdan & G. Lusztig, Schubert varieties and Poincaré duality, geometry of the laplace operator, in Proceedings of Symposia in Pure Mathematics, Univ. Hawaii, Honolulu, Hawaii, 1979, vol. 36 (American Mathematical Society, Providence, R.I., 1980), pp. 185–203

[Kil] R. Kilmoyer, On some irreducible representations of a finite group with a (B, N)-pair, Ph.D. Thesis, Massachusetts Institute of Technology, 1970

[LeMi] B. Leclerc, H. Miyachi, Constructible characters and canonical bases. J. Algebra **277**, 298–317 (2004)

[Los] I. Losev, Cacti and cells, preprint (2015), arxiv:1506.04400. To appear in J. Eur. Math. Soc

[Lus1] G. Lusztig, A class of irreducible representations of a Weyl group. Nederl. Akad. Wetensch. Indag. Math. **41**, 323–335 (1979)

[Lus2] G. Lusztig, Unipotent representations of a finite Chevalley group of type E_8. Q. J. Math. **30**, 315–338 (1979)

[Lus3] G. Lusztig, Unipotent characters of the symplectic and odd orthogonal groups over a finite field. Invent. Math. **64**, 263–296 (1981)

[Lus4] G. Lusztig, Green polynomials and singularities of unipotent classes. Adv. Math. **42**, 169–178 (1981)

[Lus5] G. Lusztig, On a theorem of Benson and Curtis. J. Algebra **71**, 490–498 (1981)

[Lus6] G. Lusztig, Singularities, character formulas and a q-analog of weight multiplicities, in Analysis and topology on singular spaces, II, III (Luminy, 1981). Astérisque **101–102**, 208–229 (1983)

[Lus7] G. Lusztig, A class of irreducible representations of a Weyl group II. Nederl. Akad. Wetensch. Indag. Math. **44**, 219–226 (1982)

[Lus8] G. Lusztig, Some examples of square integrable representations of semisimple p-adic groups. Trans. Amer. Math. Soc. **277**, 623–653 (1983)

[Lus9] G. Lusztig, Left cells in Weyl groups, Lie group representations, I (College Park, Md., 1982/1983), Lecture Notes in Mathmatics, vol. 1024, (Springer, Berlin, 1983), pp. 99–111

[Lus10] G. Lusztig, Open problems in algebraic groups, in Problems from the conference on algebraic groups and representations held at Katata ed. by Ryoshi Hotta, August 29-September 3, 1983, p. 14

[Lus11] G. Lusztig, Characters of reductive groups over a finite field, Annals of Mathematics Studies, vol. 107, (Princeton University Press, Princeton, NJ, 1984). xxi+384 pp

[Lus12] G. Lusztig, *Cells in affine Weyl groups, Algebraic groups and related topics, Advanced in Studies Pure Mathematics*, vol. 6 (Tokyo and Amsterdam, North-Holland and Kinokuniya, 1985), pp. 255–287

[Lus13] G. Lusztig, Leading coefficients of character values of Hecke algebras, The Arcata conference on representations of finite groups (Arcata, California, 1986), pp. 235–262, Proceedings of Symposia in Pure Mathematics, vol. 47, Part 2, (American Mathematical Society, Providence, RI, 1987)

[Lus14] G. Lusztig, Cells in affine Weyl groups II. J. Algebr. **109**, 536–548 (1987)
[Lus15] G. Lusztig, Cells in affine Weyl groups III. J. Fac. Sci. Univ. Tokyo **34**, 223–243 (1987)
[Lus16] G. Lusztig, Cells in affine Weyl groups IV. J. Fac. Sci. Univ. Tokyo **36**, 297–328 (1989)
[Lus17] G. Lusztig, Canonical bases arising from quantized enveloping algebras. J. Amer. Math. Soc. **3**, 447–498 (1990)
[Lus18] G. Lusztig, On quantum groups. J. Algebr. **131**, 466–475 (1990)
[Lus19] G. Lusztig, Introduction to quantum groups, Progress in Mathematics, vol. 110, (Birkhäuser Boston, 1993)
[Lus20] G. Lusztig, Hecke algebras with unequal parameters (We will sometime refer to the updated version posted on arXiv in June 2014: arXiv:math/0208154v2). CRM Monograph Series, vol. 18 (American Mathematical Society, Providence, RI, 2003), p. 136
[Lus21] G. Lusztig, Action of the longest element on a Hecke algebra cell module. Pacific J. Math. **279**, 383–396 (2015)
[Lus22] G. Lusztig, Comments to my papers, preprint (2017), arXiv:1707.09368
[MaRo] G. Malle, R. Rouquier, Familles de caractères de groupes de réflexions complexes. Represent. Theory **7**, 610–640 (2003)
[Mar] M. Martino, Blocks of restricted rational Cherednik algebras for $G(m,d,n)$. J. Algebr. **397**, 209–224 (2014)
[Mat] A. Mathas, On the left cell representations of Iwahori-Hecke algebras of finite Coxeter groups. J. London Math. Soc. **54**, 475–488 (1996)
[Ric] R.W. Richardson, Conjugacy classes of involutions in Coxeter groups. Bull. Austr. Math. Soc. **26**, 1–15 (1982)
[Sag] B.E. Sagan, The symmetric group: representations, combinatorial algorithms, and symmetric functions, Graduate Texts in Mathematics, vol. 203, 2nd edn. (Springer, Berlin, 2001). xv + 241 pages
[Sch] M.-P. Schützenberger, La correspondance de Robinson, Combinatoire et représentation du groupe symétrique (Actes Table Ronde CNRS), University Louis-Pasteur Strasbourg, Strasbourg, vol. 579, Lecture Notes in Mathematics (Springer, Berlin, 1976), pp. 59–113
[Shi1] J.Y. Shi, in The Kazhdan-Lusztig cells in certain affine Weyl groups. Lecture Notes in Mathematics, vol. 1179, (1986)
[Shi2] J.Y. Shi, Left cells in the affine Weyl group $W_a(\tilde{D}_4)$. Osaka J. Math **31**, 27–50 (1994)
[Shi3] J.Y. Shi, Left cells in the affine Weyl group of type $\tilde{F}_4$. J. Algebr. **200**, 173–206 (1998)
[Shi4] J.Y. Shi, Left cells in the affine Weyl group of type $\tilde{C}_4$. J. Algebr. **202**, 745–776 (1998)
[Shi5] J.Y. Shi, The Laurent polynomials $M^s_{y,w}$ in the Hecke algebra with unequal parameters. J. Algebr. **357**, 1–19 (2012)
[ShYa1] J.Y. Shi, G. Yang, The weighted universal Coxeter group and some related conjectures of Lusztig. J. Algebr. **441**, 678–694 (2015)
[ShYa2] J.Y. Shi, G. Yang, The boundedness of the weighted Coxeter group with complete graph. Proc. Amer. Math. Soc. **144**, 4573–4581 (2016)
[Soe1] W. Soergel, Kategorie $\mathcal{O}$, perverse Garben und Moduln über den Koinvarianten zur Weylgruppe. J. Am. Math. Soc. **3**, 421–445 (1990)
[Soe2] W. Soergel, Kazhdan-Lusztig Polynome und unzerlegbare Bimoduln über Polynomringen. J. Inst. Math. Jussieu **6**, 501–525 (2007)
[Spe] D.E. Speyer, Powers of Coxeter elements in infinite groups are reduced. Proc. Am. Math. Soc. **137**, 1295–1302 (2009)

[Spr1] T.A. Springer, Quelques applications de la cohomologie d'intersection, Bourbaki Seminar, Vol. 1981/1982, Astérisque 92–93, 249–273, (Soc. Math. France, Paris, 1982)

[Spr2] T.A. Springer, Some remarks on involutions in Coxeter groups. Commun. Algebr. **10**, 631–636 (1982)

[Ste] R. Steinberg, Lectures on Chevalley groups, (Yale University, 1967)

[Tak] K. Takahashi, The left cells and their W-graphs of Weyl group of type F_4. Tokyo J. Math. **13**, 327–340 (1990)

[Tit] J. Tits, Sur le groupe des automorphismes de certains groupes de Coxeter. J. Algebr. **113**, 346–357 (1988)

[Thé] J. Thévenaz, *G-Algebras and Modular Representation Theory* (Clarendon Press, Oxford, 1995)

[Ton] C.Q. Tong, Left cells in Weyl group of type E_6. Commun. Algebra **23**, 5031–5047 (1995)

[Vog1] D.A. Vogan, A generalized τ-invariant for the primitive spectrum of a semisimple Lie algebra. Math. Ann. **242**, 209–224 (1979)

[Vog2] D.A. Vogan, Irreducible characters of semisimple Lie groups II. The Kazhdan-Lusztig conjectures. Duke Math. J. **46**, 805–859 (1979)

[Vog3] D.A. Vogan, Ordering of the primitive spectrum of a semisimple Lie algebra. Math. Ann. **248**, 195–203 (1980)

[Vog4] D.A. Vogan, Irreducible characters of semisimple Lie groups IV. Character-multiplicity duality. Duke Math. J. **49**, 943–1073 (1982)

[Xi1] N.H. Xi, Induced cells. Proc. Am. Math. Soc. **108**, 25–29 (1990)

[Xi2] N.H. Xi, The based ring of the lowest two-sided cell of an affine Weyl group. J. Algebr. **134**, 356–368 (1990)

[Xi2] N.H. Xi, *Representations of affine Hecke algebras*, vol. 1587 (Lecture Notes in Mathematics (Springer-Verlag, Berlin, 1994)

[Xi3] N. Xi, Lusztig's **a**-function for Coxeter groups with complete graphs. Bull. Inst. Math. Acad. Sin. **7**, 71–90 (2012)

[Yue] M.S. Yue, Cells of the weighted Coxeter group $(\tilde{C}_3, \tilde{\ell}_6)$, J. East China Norm. Univ. Natur. Sci. Ed. 27–38 (2016)

[Zha] X.F. Zhang, Cells decomposition in the affine Weyl group $W_a(\tilde{B}_4)$. Comm. Algebr. **22**, 1955–1974 (1994)

[Zho] P. Zhou, Lusztig's **a**-function for Coxeter groups of rank 3. J. Algebr. **384**, 169–193 (2013)

Index

$<_L, <_R, <_{LR}$, **5**
$\approx_n^{\nabla,\gamma}$, **136**
$\bowtie, \bowtie_a$, **169**
$\leqslant, <$, **27**
$\leqslant_L, \leqslant_R, \leqslant_{LR}$, **4, 93**
$\leqslant_L^G, \leqslant_R^G, \leqslant_{LR}^G$, **235**
$\leqslant_L^I, \leqslant_R^I, \leqslant_{LR}^I$, **119**
$\leqslant_L^\sigma, \leqslant_R^\sigma, \leqslant_{LR}^\sigma$, **190**
$\leqslant_L^\varphi, \leqslant_R^\varphi, \leqslant_{LR}^\varphi$, **93**
$\leqslant_L^{\mathscr{F}^\#}, \leqslant_L^{\mathscr{F}}$, **148**
$\dot{+}$, **xix**
$\dot{\cup}$, **xix**
$\xleftarrow{L}, \xleftarrow{R}$, **94**
$\overset{L}{- - \rightarrow}, \overset{R}{- - \rightarrow}, \overset{LR}{- - \rightarrow}$, **250**
$\sim_L, \sim_R, \sim_{LR}$, **4, 93**
$\sim_L^G, \sim_R^G, \sim_{LR}^G$, **235**
$\sim_L^I, \sim_R^I, \sim_{LR}^I$, **119**
$\sim_L^\sigma, \sim_R^\sigma, \sim_{LR}^\sigma$, **190**
$\sim_L^\varphi, \sim_R^\varphi, \sim_{LR}^\varphi$, **93**
$\sim_L^{\mathscr{F}^\#}, \sim_L^{\mathscr{F}}$, **148**
$\smile_L^\varphi, \smile_R^\varphi$, **140**
$\sqsubseteq^{\mathscr{F}^\#}, \sqsubset^{\mathscr{F}^\#}$, **147**
$\sqsubseteq_L^I, \sqsubset_L^I$, **120**
$\dot{\smile}_L^\varphi, \dot{\smile}_R^\varphi$, **134**
$\trianglelefteq$, **266**
✂$_H$, **112**
✂$_{\mathfrak{A}}$, **112**
∇_2, **136**
$\mathbf{1}_W$, **48, 245**
A, **13, 56**
A-mod, **xix**
A^+, **xix**
$A^\times$, **xix**
A_n, **40**
$A_{\leqslant 0}, A_{<0}$, **13**
A_{anti}, **13**
$\tilde{A}_n$, **42**
aug, aug$_*$, aug$_*^\varphi$, **68**
$\overline{a}$, **13**
a, **243, 294**
a_0, a_k, a_l, **281**
$a_{x,y,z}$, **3**
$\mathbf{a}(z)$, **177, 183**
$\mathbf{a}_I(z)$, **177, 207**
$\mathbf{a}_\sigma(z)$, **197**
(A, B)-bimod, **xix**
$\mathscr{A}$, **13**
$\mathscr{A}_{>0}, \mathscr{A}_{\geqslant 0}, \mathscr{A}_{<0}, \mathscr{A}_{\leqslant 0}$, **13**
$\boldsymbol{\alpha}$, **53**
$\boldsymbol{\alpha}_I$, **55**
BB', **191**
$B[d]$, **191**
B_M, **37**
B_W, **38**
B_n, **40**
Braid(W, M), **24**
$\tilde{B}_n$, **42**
$\mathbf{B}_s, \mathbf{B}_w$, **192**
B$^+(W, S)$, **25**
b, **243, 294**
$\mathscr{B}r$, **27**
$\mathscr{B}^{\mathscr{F}^\#}$, **147**
$\boldsymbol{\mathscr{B}im}$, **191**
$\boldsymbol{\mathscr{B}im}^{(\sigma)}$, **193**
$C(P)$, **270**
$C(s), C(t)$, **106**
C_s, **171**
$\mathrm{Cell_L(W)}, \mathrm{Cell_R(W)}, \mathrm{Cell_{LR}(W)}$, **94**
$\mathrm{Cell_L(X)}, \mathrm{Cell_R(X)}, \mathrm{Cell_{LR}(X)}$, **5**
$\mathrm{Cell_L^\varphi(W)}, \mathrm{Cell_R^\varphi(W)}, \mathrm{Cell_{LR}^\varphi(W)}$, **94**
$\tilde{C}_n$, **42**

C. Bonnafé, *Kazhdan–Lusztig Cells with Unequal Parameters*,
Algebra and Applications 24, https://doi.org/10.1007/978-3-319-70736-5

$\tilde{\mathbf{C}}_w$, **77**
$\mathbf{C}^G_w$, **235**
$\mathbf{C}^\sigma_w$, **190**
$\mathbf{C}^<_w, \mathbf{C}^>_w$, **247**
$\mathbf{C}_y$, **74**
$\mathbf{C}^\varphi_y$, **76**
$\mathbf{C}^=_y$, **246**
Cact_W, **219**
$\mathrm{C}_\Gamma(\mathscr{E})$, **xix**
$\mathbb{C}_0$, **191**
can, **60**, **330**
can_G, **237**
$c^\vee_x$, **8**
$\boldsymbol{c}(z)$, **209**
$\boldsymbol{c}_I$, **219**
$\hat{\mathbf{c}}_w$, **186**
$\mathbf{c}_w, \mathbf{c}^?_w$, **95**
char^σ, **192**
$\mathrm{coef}_a(\beta)$, **73**
$\mathscr{C}$, **3**
$\mathscr{C}ell(W,S), \mathscr{C}ell'(W,S)$, **283**
$\mathscr{C}ell_k(W,S)$, **283**
$\mathscr{C}ox(W,S)$, **36**
$\mathfrak{C}_1, \overline{\mathfrak{C}}_1$, **43**
$\mathfrak{C}^\Sigma_1, \overline{\mathfrak{C}}^\Sigma_1$, **43**
Γ^{uni}, **155**
Γ^{uni}_s, **155**
$\Gamma^{\mathrm{uni}}_\Omega$, **170**
Γ_k, **283**
$\gamma_{x,y,z,w}$, **182**
$\gamma_{x,y,z}$, **177**
$\gamma^I_{x,y,z}$, **207**
$[\gamma,\gamma']$, **xix**
D_n, **40**
$\tilde{D}_n$, **42**
$\mathbf{D}_y$, **140**
$\mathrm{D}(\Gamma)$, **xix**
deg, **73**
$\mathbf{d}_y$, **140**
$\mathscr{D}$, **178**
$\mathscr{D}(P)$, **265**
$\mathscr{D}_I$, **198**
$\boldsymbol{\Delta}(z)$, **177**
$\boldsymbol{\Delta}_G(z)$, **235**
$\boldsymbol{\Delta}_\sigma(z)$, **190**
δ^L, δ^R, **119**
$\delta_{\mathscr{P}}$, **xix**
$\delta_{x=y}$, **xix**
$\mathscr{D}_\sigma$, **197**
$\partial^L_I, \partial^R_I$, **207**
∂^L, ∂^R, **208**
E, **38**
$E(s)$, **166**
E^φ, **67**
E_6, E_7, E_8, **40**
E_X, **39**
$\tilde{E}$, **190**
$\tilde{E}_6, \tilde{E}_7, \tilde{E}_8$, **42**
e_Γ, **226**
$[\mathscr{E}/\sim]$, **xix**
$|\mathscr{E}|$, **xix**
${}^\gamma\mathscr{E}$, **xix**
$\mathscr{E}^{-1}$, **xix**
$\varepsilon, \varepsilon_W$, **21**
ε_I, **48**
$\varepsilon_s, \varepsilon_t$, **246**
η, **22**, **212**, **214**
η^I, **219**
F_4, **40**
$\tilde{F}_4$, **42**
$\mathscr{F}, \mathscr{F}^\#$, **147**
G, **34**, **325**
G_2, **40**
$\tilde{G}_2$, **42**
$\mathbf{GL}_R(M)$, **xx**
$\mathbf{GL}_n(R)$, **xx**
g_*, **59**
grdim, grTr, **191**
$\mathscr{G}^{\mathrm{uni}}_s$, **162**
H_3, H_4, **40**
H_s, **111**
$\mathrm{Hom}_{\mathscr{B}im^{(\sigma)}}((B,\phi),(B',\phi'))$, **193**
$\mathbf{Hom}^{\mathbf{sh}}(B,B')$, **191**
H^d, **192**
$h\bullet j$, **228**
h^*, **56**
$h^\dagger$, **58**
$h_{x,y,z,w}$, **182**
$h_{x,y,z}$, **78**
$h^G_{x,y,z}$, **235**
$h^\sigma_{x,y,z}$, **190**
$\overline{h}$, **58**
$\mathscr{H}_R, \mathscr{H}_R(W), \mathscr{H}_R(W,S)$, **60**
$\hat{\mathscr{H}}^{LR}[\Gamma]$, **186**
$\tilde{\mathscr{H}}$, **62**, **77**, **98**
$\mathscr{H}(\mathfrak{I})$, **5**
$\mathscr{H}^*$, **323**
$\mathscr{H}^?[C]$, **6**
$\mathscr{H}^L[C](I,l)$, **125**
$\mathscr{H}^{LR}_\#[\Gamma]$, **186**
$\mathscr{H}^{\leqslant_? C}, \mathscr{H}^{<_? C}$, **6**
$\mathscr{H}^{\leqslant_? C}, \mathscr{H}^{<_? C}, \mathscr{H}^?[C]$, **95**
$\mathscr{H}^{\vee,?}[C]$, **8**
$\mathscr{H}^{\vee,\geqslant_? C}, \mathscr{H}_{\vee,>_? C}$, **8**
$\mathscr{H}_A(W,S,\boldsymbol{\alpha})$, **53**
$\mathscr{H}_G$, **235**
$\mathscr{H}_I$, **55**

$\mathscr{H}_\sigma$, **190**
$\mathfrak{h}^?[C]$, 8
$I \dot\cup J$, **61, 77, 97, 332**
$I(\mathscr{F})$, **112**
$I_2(m)$, **39**
I_λ, **266**
I_k, **281**
I_{-1}, **281**
I_n, **xx**
ind, ind_φ, ind_W, $\mathrm{ind}_{W,\varphi}$, **57**
i_L^Γ, i_R^Γ, **186**
$\tilde{J}$, **61, 77, 97, 332**
$\mathscr{J}$, **225**
$\mathscr{J}[X]$, **225**
$\mathscr{J}_G$, **240**
$\mathbf{k}(w)$, **283**
$\mathscr{K}(n), \mathscr{K}^L(n), \mathscr{K}^R(n)$, **268**
$\mathscr{K}_0(\mathbf{G})$, **294**
$\mathbf{LC}(W,S,\varphi)$, **135**
$\ell, \ell_W, \ell_{W,S}$, **19**
ℓ^+, **25**
ℓ_I, **26**
$\ell_{\mathscr{S}}$, **327**
$\tilde{\ell}$, **61, 334**
l, **243, 281**
$\mathbf{l}_w^I$, **123**
$\mathscr{L}(w)$, **31**
$\mathscr{L}^\varphi(w)$, **134**
λ^*, **264**
λ, **212, 214**
λ_I, **219**
$\lambda_{x,y}^s$, **78**
$\lambda_{x,y}^{(n)}$, **209**
$M[d]$, **191**
M_*, **95**
M_i, **191**
M_x, **15**
$\mathrm{Mat}_n(R)$, **xx**
${}^tM$, **xx**
Mon(S), **25**
m, **243**
m_{st}, **19**
$\mathscr{M}$, **147**
$\mathscr{M}^{\mathscr{X}}, \mathscr{M}^{\mathscr{X}}_{<0}$, **13**
$\mathscr{M}_{<0}$, **13**
$\mathscr{M}_{\mathrm{anti}}$, **13**
$\mathscr{M}_{<0}$, **148**
μ_L, μ_R, **119**
$\boldsymbol{\mu}_2$, **119**
$N(w)$, **22**
$\mathrm{N}_\Gamma(\mathscr{E})$, **xix**
$\mathbb{N}, \mathbb{N}^*$, **xix**
n, **189**
n_z, **177**
n_z^σ, **190**
$\mathscr{N}$, **328**
ν, **61**
$o(\gamma)$, **xix**
ω_0, **139**
ω_I, **219**
$\omega_{s,t}$, **163**
$P \leftarrow i$, **264**
P_λ, **266**
$P_{x,y}$, **75**
$P_{x,y}^\varphi$, **76**
Prod$(a,b;m)$, **24**
$\mathbf{P}(w)$, **265**
Part, **264**
Part(n), **264**
p, q, r, **303**
$p_{x,y}$, **75**
$p_{x,y}^G$, **235**
$p_{x,y}^I$, **121**
$p_{x,y}^\sigma$, **190**
$p_{x,y}^\varphi$, **76**
$\mathrm{pr}_L^I, \mathrm{pr}_R^I$, **119**
$(\mathscr{P}), (\mathscr{P}_{\mathbf{sph}})$, **208**
$\mathscr{P}(W)$, **27**
$\mathscr{P}(\mathscr{E})$, **xix**
$\mathscr{P}_{\mathrm{ir,sph}}(S)$, **219**
$\mathscr{P}_{\mathrm{sph}}(S)$, **31**
$\mathscr{P}_{\mathrm{sph}}^\flat(S)$, **32**
$\pi_C^?$, **9**
π, **157**
$\bar{\pi}_C^?$, **9**
Φ, **22**
ϕ, ϕ^{op}, **227**
φ, **56**
$\varphi > 0$, **76**
φ^{lin}, **60**
φ_-, **61, 76, 97**
φ_G, **235**
φ_I, **86, 119**
φ_σ, **189**
$\varphi_{x,y,z}^{(d)}$, **194**
$\tilde{\varphi}$, **61, 77, 98**
$\mathbf{Q}(w)$, **265**
$q_{x,y}^I$, **124**
$R'M$, **xx**
$R\mathscr{M}$, **xx**
RSK, **265**
$r_{x,y}, r_{x,y}^\varphi$, **62**
$\mathbf{r}_{,x,y}$, **62**
$\mathrm{rep}_L^I, \mathrm{rep}_R^I$, **119**
$\mathscr{R}(w)$, **31**
$\mathscr{R}^\varphi(w)$, **134**
ρ, **212, 214**

ρ_I, **219**
ρ_s^φ, **67**
$\rho_{x,y}^s$, **78**
$(S/G)_{\mathrm{sph}}$, **34**, **235**
$(S/\sigma)_{\mathrm{sph}}$, **189**
$(S/\bowtie)^\blacktriangledown$, **169**
S, **19**, **20**
$S[a]$, **169**
S^φ, **134**
S_+, S_-, **61**, **76**, **97**
S_G, **34**, **235**
S_σ, **189**
S_k, **281**
$S_{\neq}$, **99**
$\bar{S}, \bar{s}$, **60**
$\mathbf{S}$, **190**
$\mathbf{S} \rtimes W$, **191**
$\mathbf{S}^{>0}$, **190**
$\mathbf{SL}_n(R)$, **xx**
$\mathbf{ST}_2(n)$, **264**
sgn, sgn_φ, sgn_W, $\mathrm{sgn}_{W,\varphi}$, **57**
sgn_I, sgn_I^φ, **68**
supp, **26**
$s^\perp$, **32**
s_r, **243**
$\mathbf{s}_x$, **166**
$\mathbf{sp}_2$, **spin**, $\mathbf{spin}_2$, **296**
sh(P), **264**
sh(w), **265**
$\mathscr{S}$, **327**
$\mathscr{S}_s$, **166**
$\mathscr{S}\!\boldsymbol{oe}$, $\mathscr{S}\!\boldsymbol{oe}_\heartsuit$, **192**
$\mathscr{S}\!\boldsymbol{oe}^{(\sigma)}$, $\mathscr{S}\!\boldsymbol{oe}_\heartsuit^{(\sigma)}$, **193**
$\mathfrak{S}_\lambda$, **266**
$\mathfrak{S}_W$, **223**
$\mathfrak{S}_n$, **xix**
$\mathfrak{S}_{\mathscr{E}}$, **xix**
$\Sigma(W,S)$, **43**
σ, **189**
$\sigma^*(M)$, **193**
σ^{op}, **131**
σ_I, **219**
$\sigma_=, \sigma_<, \sigma_>$, **127**, **250**
σ_w, **193**
$\sigma_\varphi^L, \sigma_\varphi^R$, **223**
$\sigma_{s,t}$, **163**
$\boldsymbol{\sigma}_{\phi,\phi'}, \boldsymbol{\sigma}_{\phi,\phi',i}$, **194**
T, **22**
T_s, **53**
T_w, **54**
T_w^G, **235**
T_w^σ, **190**
T_w^φ, **56**
$T_w^{\boldsymbol{a}}$, **54**
Tr, Tr_H^G, **237**
$\widetilde{T}_w$, **62**, **77**
t_r, **243**
$t_{x,y,z}$, **175**
$\mathbf{t}_w$, **225**
$\mathbf{t}_w^G$, **240**
$trans_L^G$, $trans_R^G$, $trans_{LR}^G$, **112**
$\mathscr{T}$, **327**
$\mathfrak{T}_s$, **109**
Θ, **76**
Θ_{S_-}, **61**
τ, **56**
$\tau_w^{\nabla,\gamma}$, **135**
$\tau^\vee$, **324**
$\tau_L^\vee, \tau_R^\vee$, **323**
τ_s, **111**
θ_Γ, **207**
$U(W,S)$, **43**
U, U_s, U_w, **22**
$\mathscr{U}nbr(W,S)$, **157**
V, **111**
val, **73**
v^a, **13**
$\boldsymbol{v}$, **105**, **106**
$\boldsymbol{v}, \boldsymbol{V}$, **106**
$\boldsymbol{v}_{\bar{s}}, \boldsymbol{v}_s$, **60**
$\mathscr{V}_\nabla$, **135**
W, **19**, **20**
$W(I)$, **30**
$W(\geqslant a), W(> a), W(a)$, **181**
W^+, **245**
W^G, **34**
W^σ, **189**
$W^{\leqslant_? C}, W^{<_? C}$, **94**
W_I, **27**
$W_{I_{-1}}$, **281**
$W_{\mathscr{F}}$, **112**
$W_{s_1,s_2,\ldots,s_n}$, **28**
Weight$(S,\mathscr{E})$, Weight$_W(S,\mathscr{E})$, **26**
$\widetilde{W}$, **61**, **77**, **97**, **332**
w^∞, **19**
w_0, **139**
w_I, w_0, **31**
$\mathscr{W}$, **327**
$\mathscr{W}_s$, **166**
$X^{\geqslant_? C}, X^{>_? C}$, **8**
$X^{\leqslant_? C}, X^{<_? C}$, **6**
$X_1, X_1^{I_0}$, **283**
$X_I^{\leqslant l}, X_I^l$, **125**
X_λ, **266**
$X_{IJ}, X_{IJ}^K, X_I, X_I^K$, **28**
ξ, **189**
$Y(\lambda)$, **264**
$Z(A)$, **xix**

Z(Γ), **xix**
$\mathbb{Z}^L_\gamma, \mathbb{Z}^R_\gamma, \mathbb{Z}^{LR}_{\gamma\boxtimes\gamma'}$, **97**
char, **192**
w_λ, **266**

A
Algebra
  asymptotic, **226**
  symmetric, **324**

B
Braided map, **24**
Brauer
  morphism, **237**
  quotient, **237**
Bruhat order, **27**

C
Cactus group, **219**
Cancellation law, **21**
Carpet, **135**
φ-cell
  left, **94**
  right, **94**
  two-sided, **94**
Cell
  left, **5**, **94**
  module, **95**
  right, **5**, **94**
  two-sided, **5**, **94**
Cell module, **7**
𝔄-chambers, **112**
Coplactic
  class, **135**
  equivalence, **135**
Complete, **112**
Connected
  left, **101**
  right, **101**
Convex
  left, **5**, **95**
  right, **5**, **95**
  two-sided, **5**, **95**
Coxeter
  complex, **43**
  element, **50**
  graph, **36**
  matrix, **36**
  system, **20**

D
Degree, **73**
Deodhar's Lemma, **29**, **282**
Duflo involutions, **180**

E
Enhanced descent set, **134**
Essential hyperplane
  arrangement, **113**
Exchange condition, **20**

F
𝔄-faces, **112**
Forest, **40**
Form
  central, **324**
  symmetrizing, **324**

G
Guilhot induction process, **147**

H
Hecke algebra, **53**
  generic, **60**
Hyperbolic, **42**

I
Ideal
  left, **5**, **94**
  right, **5**, **94**
  two-sided, **5**, **94**
(I, J)-reduced, **28**
Irreducible, **38**

K
Kazhdan–Lusztig
  basis, **74**
  cell, **94**
  cell module, **95**
  polynomials, **75**
Kazhdan–Lusztig–Geck basis, **120**
Knuth relation
  left, **135**
  right, **135**

L
Left cellular, **126**
  strongly, **126**
Left-connected component, **101**
Left extension, **119**
Length, **19**

M
Matsumoto's Theorem, **25**, **26**

P
Parabolic
  subalgebra, **55**
  subgroup, **28**

Partitions, **264**
  size, **264**
Plactic
  class, **135**
  equivalence, **135**

Q
Quasi-split, **190**

R
Rational
  hyperplane, **111**
Rational subspace, **111**
Reduced expression, **20**
Reflection
  cocycle, **328**
  representation, **38**, **67**
  subgroup, **330**
  system, **328**
Regular monoid, **26**
Right cellular
  strongly, **126**
Right extension, **119**
Robinson–Schensted–Knuth correspondence, **265**
Row insertion algorithm, **264**

S
Support, **26**
Semicontinuity Conjecture, **112**
Semicontinuous
  lower, **115**
  upper, **115**
Semilinear, **13**
Shift, **191**
Sign character, **21**
Soergel bimodules
  perverse, **192**
Spherical, **31**
Split case, **86**
Standard
  basis, **54**
  parabolic subgroup, **28**
Submaximal, **155**

T
Tableau, **264**
  semistandard, **264**
  standard, **264**
Tame, **41**
Tits cone, **43**
Tits' Deformation Theorem, **66**
Tree, **40**
Two-sided cellular, **126**

U
Unbraided path, **157**

V
Valuation, **73**
Vogan
  classes, **136**
  invariant, **136**

W
W-closed, **26**
W-graph, **109**
Weight function, **26**

Y
Young
  diagram, **264**
  subgroup, **266**

MIX
Papier aus verantwortungsvollen Quellen
Paper from responsible sources
FSC® C105338

If you have any concerns about our products,
you can contact us on
ProductSafety@springernature.com

In case Publisher is established outside the EU,
the EU authorized representative is:
Springer Nature Customer Service Center GmbH
Europaplatz 3, 69115 Heidelberg, Germany

Printed by Libri Plureos GmbH
in Hamburg, Germany